DESIGNING GEODATABASES

FOR TRANSPORTATION

J. Allison Butler

ESRI PRESS
REDLANDS, CALIFORNIA

ESRI Press, 380 New York Street, Redlands, California 92373-8100

Copyright © 2008 ESRI

All rights reserved. First edition 2008
10 09 08   1 2 3 4 5 6 7 8 9 10

Printed in the United States of America

Library of Congress Cataloging-in-Publication Data
Butler, J. Allison, 1953–
  Designing geodatabases for transportation / J. Allison Butler—1st ed.
    p. cm.
  Includes index.
  ISBN 978-1-58948-164-0 (pbk. : alk. paper)
  1. Transportation engineering—Data processing. 2. Transportation—Planning—Data processing. 3. Geodatabases
4. Geographic information systems. I. Title.
  TA1165.B88 2008
  629.040285'57—dc22   2008018404

The information contained in this document is the exclusive property of ESRI. This work is protected under United States copyright law and the copyright laws of the given countries of origin and applicable international laws, treaties, and/or conventions. No part of this work may be reproduced or transmitted in any form or by any means, electronic or mechanical, including photocopying or recording, or by any information storage or retrieval system, except as expressly permitted in writing by ESRI. All requests should be sent to Attention: Contracts and Legal Services Manager, ESRI, 380 New York Street, Redlands, California 92373-8100, USA.

The information contained in this document is subject to change without notice.

U.S. Government Restricted/Limited Rights: Any software, documentation, and/or data delivered hereunder is subject to the terms of the License Agreement. In no event shall the U.S. Government acquire greater than restricted/limited rights. At a minimum, use, duplication, or disclosure by the U.S. Government is subject to restrictions as set forth in FAR §52.227-14 Alternates I, II, and III (JUN 1987); FAR §52.227-19 (JUN 1987) and/or FAR §12.211/12.212 (Commercial Technical Data/Computer Software); and DFARS §252.227-7015 (NOV 1995) (Technical Data) and/or DFARS §227.7202 (Computer Software), as applicable. Contractor/Manufacturer is ESRI, 380 New York Street, Redlands, California 92373-8100, USA.

ESRI, the ESRI Press logo, www.esri.com, ArcGIS, ArcSDE, ArcObjects, ArcMap, ArcIMS, and ArcInfo are trademarks, registered trademarks, or service marks of ESRI in the United States, the European Community, or certain other jurisdictions. Other companies and products mentioned herein are trademarks or registered trademarks of their respective trademark owners.

Ask for ESRI Press titles at your local bookstore or order by calling 1-800-447-9778. You can also shop online at www.esri.com/esripress. Outside the United States, contact your local ESRI distributor.

ESRI Press titles are distributed to the trade by the following:

In North America:
Ingram Publisher Services
Toll-free telephone: 1-800-648-3104
Toll-free fax: 1-800-838-1149
E-mail: customerservice@ingrampublisherservices.com

In the United Kingdom, Europe, and the Middle East:
Transatlantic Publishers Group Ltd.
Telephone: 44 20 7373 2515
Fax: 44 20 7244 1018
E-mail: richard@tpgltd.co.uk

Cover design by	Antoinette Beltran
Interior design by	Donna Celso
Copyediting by	Julia Nelson
Print production by	Cliff Crabbe

Contents

Foreword   vii
Preface   ix
Acknowledgments   xiii
About the author   xvii

Chapter 1   Introduction   1

Part 1 *Basic geodatabase design concepts*

Chapter 2   Data modeling   11

Chapter 3   Geodatabases   37

Chapter 4   Best practices in transportation database design   71

Chapter 5   Geometric networks   97

Part 2 *Understanding transportation geodatabase design issues*

Chapter 6   Data editing   115

Chapter 7   Linear referencing methods   143

Chapter 8   Advanced dynamic segmentation functions   161

Chapter 9   Traffic monitoring systems   183

Chapter 10   Classic transportation data models   205

Chapter 11   The original UNETRANS data model   227

Contents

Part 3 Enterprise-level solutions and modal data models

Chapter 12   Improving the UNETRANS data model   237

Chapter 13   The revised UNETRANS network data model   265

Chapter 14   State DOT highway inventory: Editing   287

Chapter 15   State DOT highway inventory: Publishing   331

Chapter 16   A multipurpose transit geodatabase   365

Chapter 17   Navigable waterways   383

Chapter 18   Railroads   411

Final thoughts   431
Index   433

Foreword

Designing Geodatabases for Transportation is an outstanding contribution to the ESRI thematic geodatabase design series of "best practices." This transportation domain data model achieves the two primary goals of the series:

- Provide best-practice templates for implementing geodatabases for transportation applications.
- Communicate a practical database-design process for geographic information systems for transportation (GIS-T).

Al Butler's contribution of the transportation application domain data model provides guidance on "what really works." He shares his knowledge of transportation, GIS, and database management with the GIS-T user and developer communities.

In the early 1990s, Al Butler and I were critiquing each other's early efforts to create GIS-T data models. Then we decided to collaborate to achieve our common vision of a GIS-T data model that could accommodate:

- Multiple networks, some of which might include forest roads, alleys, or private roads.
- Multiple cartographic representations necessitated by differences in scale and source.
- Multimodal transportation systems.
- Multiple segmentations for various applications.
- Multiple linear referencing systems for infrastructure management and dynamic routing applications.
- A common data model to support multiple users, including facility managers and fleet managers, and multiple applications, ranging from long-range planning to real-time operations.
- A geodatabase approach for transaction-based updating and maintenance and a reduced reliance on versioning.

Our collaborative effort resulted in an enterprise GIS-T data model using E-R diagrams. However, this effort was perceived narrowly as a state department of transportation and highway relational representation. Meanwhile, higher-profile efforts, such as the Federal Geographic Data Committee (FGDC), National Cooperative Highway Research Program (NCHRP) 20-27, Unified Network for Transportation (UNETRANS), and Geospatial One-Stop utilized high-level object-oriented data-modeling approaches. These approaches were not adopted because they required a single authority to dictate a single network, single geometry, or single feature ID schema. Although Al and I participated in these efforts and tried to influence their design, none fulfilled our vision to incorporate all of the above-listed objectives. In addition, the high-level, object-oriented approach allowed ambiguities that created problems at the implementation stage.

While I have faded into retirement, Al has persisted and has shown the way to achieve the vision. His resulting book should be viewed as a manual of best practices that will provide guidance, not serve as a prescription. Most significantly, he calls for separating the editing geodatabase environment from the published application-specific datasets. This separation serves to reduce redundant updates that tend to cause inconsistent representations. Al provides detailed instructions for how to structure and process data so that it can be used to support multiple applications without having to separately and duplicatively maintain the data.

Al Butler shows how to address the many problems unique to transportation data and related business processes. Transportation geodatabases have been difficult to construct in the past, due in large part to a lack of basic guidance. This book fills that void, explaining how to construct transportation geodatabases in a manner that fulfills our vision for flexibility to meet multiple needs. I am impressed by both its breadth and depth. *Designing Geodatabases for Transportation* will prove to be an important and powerful template in the series.

I greatly appreciate the vision of ESRI, particularly Jack Dangermond, for supporting development of this transportation template addition to the geodatabase design series. I have known Jack for many years and continue to be impressed by his deep commitment to the development of GIS.

Kenneth J. Dueker, AICP
Emeritus professor of urban studies and planning
Portland State University
Portland, Oregon

Preface

Designing Geodatabases for Transportation is the first book published on transportation spatial database design, a fact you may find quite surprising. To be sure, much work has been done to develop transportation data models, but the result of these efforts has remained fairly theoretical and conceptual. Geographic information systems (GIS) have included spatial databases for transportation since the vector data structure was first used. Indeed, one of the oldest and most famous vector data structures is the TIGER/Line file, which uses transportation features as a principal organizing element. As with many transportation datasets that are used as a reference layer, the TIGER/Line file was designed to convey something else: census data. But it has become the de facto source of transportation data for most local governments and several commercial vendors. Yet, as old and common as transportation datasets are, there has been virtually no organized presentation of how they might be broadly constructed. This book is intended to provide such a presentation by offering a set of geodatabase design problems and alternative solutions that cover the range of travel modes and GIS users.

Although the context is transportation and its many modes of travel, this book is about data maintenance, emphasizing a "measure once, cut twice" design philosophy that offers guidance for all types of spatial databases. *Designing Geodatabases for Transportation* is a useful guide for all data themes, not just transportation. The book's editing geodatabase design approach, which supports record-level metadata and the ability to recover the state of the database at any point in time, has widespread applicability. Part 1 also includes fundamental guidance on the principles of agile design, database normalization, key geodatabase classes, and ArcGIS functionality. It takes you "under the hood" so you can understand better why the ideas offered in parts 2 and 3 can work for you.

The foundation established in part 1 shifts to a purely transportation orientation in part 2, concentrating on the fundamental needs of the GIS for transportation (GIS-T) field. The original industry data models developed in NCHRP (National Cooperative Highway Research Program) 20-27, UNETRANS (Unified Network for Transportation), and other efforts are presented and critiqued. Part 2 illustrates a variety of linear facility segmentation schemes, with design approaches offered for each. Most importantly, part 2 shows how to integrate multiple segmentation schemes in a single geodatabase.

In part 3, useful ideas from the historical references and various design approaches of part 2 are incorporated into a revised UNETRANS enterprise GIS-T data model and in a number of design approaches directed at specific functional requirements. The revised UNETRANS data model drops the original's geometric network foundation for the new street network capabilities of ArcGIS. There is no need for the user to adopt the entire data model, which is offered only as a general guide for how a large organization might assemble all the pieces.

The multimodal UNETRANS model leads to mode-specific chapters. Two chapters provide guidance for building a highway inventory geodatabase at a state DOT. Chapter 14 covers the editing environment, showing how one model can be implemented in different ways depending on the agency's business rules. Chapter 15 deals with the publishing process, showing how to use ArcGIS functions or to structure the SQL statements that extract data from the normalized editing geodatabase and publish it in denormalized tables and fully attributed feature classes.

The transit chapter expands on the earlier discussion of transportation networks and the capabilities of a street network foundation for the new UNETRANS model. The chapter shows real-world examples for readers interested in the transit mode. The closing chapters on navigable waterways and railroads are intended to serve as primers for the generalist to understand transportation features and their attributes in these modes of travel.

Throughout the book, design concepts presented in a problem-solution approach enable you to quickly identify the parts you need. Each discussion begins with the simplest approach and grows increasingly complex by adding functional requirements to the initial problem. In general, the simpler solutions are most appropriate for smaller agencies with limited transportation feature attributes, while the more complex solutions will likely be applicable at larger organizations that must meet a variety of application requirements.

For all the recent technological advances in the GIS field, it is really just getting started. In a significant evolutionary step, GIS now is adopting concepts and practices more commonly found in mainstream information technology (IT). Such adoptions are consistent with the growing importance of GIS to the organization. As the technology moves from the workgroup to the enterprise, it presents both greater benefits and higher risks. IT practices are designed to limit risk, not only from hardware and software issues, but also from data defects.

Many of the ideas presented in *Designing Geodatabases for Transportation* are derived from common IT practices. Concepts like normalization and record-level metadata may be fairly unknown in geodatabase design, but they are the bread and butter of traditional transactional database design. So, while this book may be targeted to the transportation theme, it serves as a general guide for designing geodatabases of all themes.

The GIS-T field has been held back by two common issues. First, the field has sought to develop an industry data model when such a model really has no application. This is because almost all organizations that need a transportation spatial database already have one, which is the second issue. Exacerbating these two issues is the general rule that the larger the organization, the more likely it has developed homegrown solutions to meet its information needs and these solutions tend to be poorly documented and difficult to change. As a result, it is almost impossible for an organization to abandon its long-held data structures with a wide variety of homegrown applications constructed to use them. The larger the organization, the more difficult it is to change. *Designing Geodatabases for Transportation* shows how to solve the many problems existing today as a result of haphazard database design over many years. Technical problems are not the issue. The real issue is whether the organization can survive the changes that go with implementing the solution.

This implementation issue is addressed in two ways. First, the book focuses on the data editing part of the geodatabase design problem so that any changes would affect a small part of a larger organization, limited to the people who edit data for others to use. This is done by separating the editing database—which should be normalized to preclude data redundancy and its resulting data quality effects—from the published database—which should be defined by the needs of applications that use it. This is the heart of the "measure once, cut twice" philosophy. Rather then separately and duplicatively maintaining all the parallel, application-specific databases a large organization may need, you create an editing environment that feeds information to those applications. Not only does this approach allow the applications and the supporting editing process to evolve independently, it reduces the impact of change in geodatabase design to the data-maintenance portion of the organization.

This is the first book to show the ArcGIS user how IT database design concepts can be employed by GIS data of all themes. When you follow these concepts, you are reducing risk to your organization and facilitating the migration of spatial data previously reserved for a few GIS users to the enterprise. You are also reducing risk for yourself, while increasing your value to the organization.

The second way *Designing Geodatabases for Transportation* attempts to address the difficulty of change is by presenting design ideas as a cafeteria of choices rather than as a fixed-price menu. Change can be implemented gradually. There is no need for an all-or-nothing decision. Certainly, you must set a goal, but you can reach it in small steps. Each step along the way provides additional functionality and/or solves an existing problem. As long as you know where you are going, you can get there with the solutions offered in this book.

Some of the larger graphics in the book are available for easier viewing at ESRI's transportation industries Web site:

http://www.esri.com/industries/transport/resources/data_model.html.

Preface

Whether you are interested in transportation geodatabase design for a city, county, state, province, or country, *Designing Geodatabases for Transportation* provides insight into the problems you face and the various ways they can be solved. For those agencies that share data with others or receive their data from an external source, this book illustrates how to integrate data using various structures. As the road and bridge building boom of last century has faded, transportation agencies have shifted their focus to maintenance and operation. Such a change in mission underscores the need for multimodal geodatabases in transportation agencies. *Designing Geodatabases for Transportation* shows how to meet the needs of each mode within the context of a multimodal editing environment.

J. Allison Butler
Orlando, Florida

Acknowledgments

No book of this magnitude makes it to publication without a host of people providing critical support along the way. Principal among these is Dr. Kenneth J. Dueker, professor emeritus of Portland State University's Center for Urban Studies. Although he is now officially retired, Ken continues to be active in his newly adopted community of Kirkland, Washington, where he and his wife, Donna, care for their grandchildren. Ken and I first started talking back in 1994, when we were both reviewers of the work by the National Cooperative Highway Research Program (NCHRP) 20-27(2), which developed the first industry data model for transportation geographic information systems. Within a year, Ken and I were co-authoring papers on the subject. Ken has been a mentor, collaborator, and good friend ever since. He was the first person to read the entire final draft. I am grateful that he has provided the foreword to top off his many contributions to this book. *Designing Geodatabases for Transportation* would not exist without Ken's guidance over the past 13 years.

In many ways, this book has been written over a 22-year period with a wide number of "co-authors." It began in 1986 with Jim Smith and me developing some prototype relational databases for work program administration at the Florida DOT. This effort then grew through Jesse Day, Gordon Morgan, and others to become the Traffic and Roadway Characteristics Reporting system, which resulted in the first statewide GIS in Florida. Switching to local government in 1997, Mayor Claude Ramsey and Assessor of Property Bill Bennett in Hamilton County Tennessee provided the support necessary to expand the state DOT work to meet the needs of local and regional users. Alan Voss at the Tennessee Valley Authority provided considerable professional guidance for the work at Hamilton County. That effort then led to my projects with Mike Wierzbinski and Bill Campbell at Farragut Systems, such as the data model development program at the Colorado DOT, where Tammy Lang, Marvin Koleis, Lou Henefeld, Paul Tessar, and many others both refined the ideas and provided a real-world test of most of the major ideas presented in this book. John Sutton provided the opportunity to work on a very different data structure with Nancy Armentrout at the Maine DOT, which tested the flexibility of the design.

Along the way, there were others who have contributed ideas and support. Alan P. Vonderohe, Teresa M. Adams, and Nicholas Koncz have to be at the top of that list with their work on the NCHRP 20-27 project. Al Vonderohe also worked with Todd Hepworth to develop the concepts of a linear datum. Bruce Spear and Mark Bradford at the U.S. DOT sponsored a number of efforts that generated ideas on feature identification. Harvey J. Miller and Shih-Lung Shaw, who wrote the very first scholarly book on GIS-T, set the stage for my subsequent work. Even the people who disagreed with some of my ideas, like David Fletcher, Paul Scarponcini, Val Noronha, and Zhong-Ren Peng, helped improve the content of this book through their challenging critiques of earlier work.

Equally responsible for this book's publication is the project's sponsor, Terry Bills, the transportation industry manager at ESRI. Shortly after Terry took the job, he came to me and said, "We need to do something." Do something, we have. Terry put his money where his mouth is, resulting in this book. His guiding hand was instrumental in setting the direction for the book and making sure that it both met the reader's needs and was technically correct.

Aiding Terry and me in that last task were Adrien Litton, Heather McCracken, Eric Floss, and Laurie Cooper of ESRI; and Ron Cihon of the Washington State Department of Transportation. They each spent days reading the draft chapters and offering constructive criticisms. All of these people had to make time in their schedules to get this work done, but no one else could have provided the technical insight and user perspective.

The work of all of these people would never have reached your hands without the hard work of people at ESRI Press. My editor was Mike Kataoka, who worked with designers Donna Celso and Jennifer Hasselbeck, graphics editor Jay Loteria, and copy editor Julia Nelson to produce the book under the guidance of Judy Hawkins, former ESRI Press manager, and Peter Adams, current manager. Stacks of written pages and scores of figures have been transformed into this attractive book through their hands.

I also thank David Arctur and Michael Zeiler, authors of the ESRI Press publication, *Designing Geodatabases*, for setting the standard for geodatabase design books, and Steve Grisé, chief data modeler for ESRI, for providing much of the data model development guidance contained in the book.

Of course, all the folks at ESRI and ESRI Press work for Jack and Laura Dangermond, whose financial support of this book was critical. But their support goes further than the ESRI checkbook. ESRI, under Jack's leadership, has been pursuing ways to open up the transportation world to GIS for 20 years with product functions like dynamic segmentation and support of industry data-modeling efforts like the Unified Network for Transportation (UNETRANS) effort at the University of California at Santa Barbara.

The people who were especially patient with me while I wrote the book are those who allowed me to make time for this extra task. Most notable in this list of friends and family is my wife, Robin, who endured many lonely nights while I sat in the home office typing away.

David A. Wheeler, city engineer for the city of Ocoee, Florida, also provided patient support by allowing me as much as two months at a time away from contract work in his office to write the book. They and numerous others had to experience my frequent response to a request to do something for them: "I can't right now, I have to work on the book."

Well, that task is finished. Only you can decide how well it was done. If you can find something in this publication that helps you solve a problem with transportation geodatabase design, then I guess it was worth it.

About the author

J. Allison "Al" Butler has worked for state, regional, and local transportation agencies during a 30-year career, focusing on the built environment and information technology. He has been an innovator in spatial database design, traffic engineering, planning policy, public utilities, and economic development, and has authored more than 75 published works in those fields. Mr. Butler played a key role in developing the GIS professional certification program and, with Dr. Kenneth J. Dueker, co-developed the Enterprise GIS-T database design. He is a frequent conference speaker and workshop instructor on a variety of topics, including highway safety, geodatabase design, agency management, and land use planning. Al Butler is a licensed contractor and a certified planner, mapping scientist, and GIS professional.

chapter one

Introduction

- Transport databases
- Data models
- Agile methods
- Building the agile geodatabase
- Book organization

Modes of travel can be quite different but all follow a conceptual structure consisting of an origin, a destination, a path between the two, and a conveyance to move along the path. *Designing Geodatabases for Transportation* tells how to design a geospatial information system to manage data about transportation facilities and services.

An enterprise geodatabase can help solve two common transportation challenges: the many origins, destinations, paths, and conveyances that may be present; and the need to specify locations along the facility. There is also the matter of facility suppliers usually being different from facility users. The facility user focuses on origins and destinations. The facility supplier is concerned about the many paths over which a conveyance travels and generally not about specific trips. A transportation agency provides the facilities to support travel. Shippers—those who have goods to deliver—define the origin and destination for their shipment. A shipping company supplies the conveyances and selects the paths to move the shipper's goods from origin to destination. Each origin and destination must be accessible from a transport facility. A shipping company can select a path only where a corresponding facility exists. Railroads perhaps can be viewed as being both suppliers and users of transport capacity although some may operate over facilities they do not own.

Transport databases

Transportation data is often specific to the various modes of transport. *Designing Geodatabases for Transportation* addresses six modes: walking, bicycling, motor highways, public transit (buses and commuter rail), railroads, and navigable waterways.[1] All six modes involve linear facilities supporting point-to-point travel for people and material goods. The nature of the facility supporting travel and the way it is used differs with each mode. Some facilities support multiple modes of travel. Highways and roads accommodate motor vehicles, pedestrians, and bicycles. Railroads support commuter trains, long-distance passenger travel, and freight movement. Ships travel navigable waterways that flow under highway and railroad bridges.

Points of modal connection are commonly known as terminals, depots, stations, stops, and crossings. The name "intersection" is usually applied to highways, but conceptually includes other points where facilities cross and interact, such as railroad switches, crossovers, and diamonds; rail-highway grade crossings; and limited-access highway interchanges. Transport systems also include places where facilities cross but do not intersect, such as at bridges, viaducts, and tunnels.

Geographic information systems (GIS) for transportation—GIS-T in industry shorthand—routinely deal with mode- and function-specific applications, each with its own geodatabase design. What is rare is a GIS-T geodatabase that goes beyond serving the needs of a single application. Such a geodatabase must accommodate the many segmentation schemes employed and the various linear and coordinate referencing systems available to show where the elements, conveyances, and characteristics of transportation systems are located. *Designing Geodatabases for Transportation* shows you how to construct such an enterprise multimodal geodatabase, although the ideas presented in this book can be implemented for a single mode.

A transportation geodatabase addresses concerns beyond facilities and the services that use them. For example, facility elements and characteristics are affected by projects and activities that construct, maintain, and remove elements of the transportation system. There are also traffic crashes, bus routes, train schedules, and shipping manifests to consider.

Transportation applications are much too diverse for this book to present you with a complete transportation database design. That task is up to you. ESRI has successfully worked with user groups to develop a number of industry-specific data models. That approach will not work with transportation, which lacks a single, all-encompassing view of the industry due to its diversity. Not only are there modal differences to consider, but there are also differences in detail and abstraction. A trucking company, a city public-works division, and a state department of transportation (DOT) all need data about highways, but for their own purposes that require them to adopt very different data models. Even within a single transport agency there may be several different application-specific data models.

What this book offers is a collection of ideas and geodatabase design components to help you construct a model that serves your agency and its unique set of applications. It shows you a variety of ways to handle a specific data need, describing the pros and cons of each choice. In this way, *Designing Geodatabases for Transportation* provides a cafeteria of design options rather than a fixed menu to solve the broad range of transportation spatial data requirements.

Data models

Geodatabase design is normally expressed through a data model, which is a graphical way of describing a database. A data model is essentially a set of construction plans for a database. Some data models are very conceptual, others extremely detailed. Fortunately, all data models use a few very simple symbols. You need no prior experience with data models or geodatabase design to understand and apply the suggestions in this book.

All geodatabases form a set of abstract representations of things in the real world. The process of abstraction is called modeling. You may geometrically represent linear transport facilities in your abstract geodatabase world as lines. In the real world, of course, transport facilities are areas, which may be less abstractly represented by polygons. Unfortunately, many of the analytical techniques you will want to employ, such as pathfinding, do not exist in the polygon world. This is not really a problem, though, because the central aspect of a linear transport facility is its linearity, so points and lines form the abstract world of most transport geodatabases.

Whether you formally draft a data model or not, one exists inside each geodatabase. It may also exist externally as a set of requirements, a list of class properties, or some other description of the geodatabase's contents and structure. ESRI's ArcGIS software comes with tools to produce a data model from an existing geodatabase. As geodatabases grow from supporting isolated workgroups to major portions of a complete enterprise, the complexity of geodatabase design normally increases in proportion to the number of data uses. Explicit data models and other documentation become more important as the scope of a geodatabase expands or it begins to serve a critical function within the organization. As a result, many organizations have invested significant resources into developing a good data model before moving forward with geodatabase development and migration projects. More comprehensive and ambitious projects seek to deploy at the enterprise level.

Building a good enterprise geodatabase starts with an enterprise data model. Accordingly, this book expands and enhances the previous ESRI transportation data model, UNETRANS (Unified Network for Transportation), developed by the University of California at Santa Barbara (UCSB) with financial support from ESRI. The "new and improved" UNETRANS is designed to provide a full structure that embraces all of a large organization's data and access mechanisms. This enterprise data model also takes advantage of technological advances in the ArcGIS family since the original UNETRANS was developed several years ago.

However, developing an enterprise data model need not be the first step in creating an enterprise geodatabase. Your organization may first want to construct a workgroup prototype to gain experience with the geodatabase structure and how to use it. Effective geodatabase design and deployment is an ongoing process. The needs of the organization and the capabilities of the technology both evolve. You will need an enterprise architecture that describes the computing environment and its business rules. Everything else will be somewhat reactive because you cannot possibly anticipate all data uses. You must be agile.

Agile methods

Since modern geodatabase design should be based on data modeling, this book includes a review of that process in chapter 2. But it is important to note that data modeling fits within a larger information technology process. Years ago, data modeling and application development were separate processes. Now, it is generally recognized that the speed of database access determines, to a large degree, application performance. This change is partly the result of application intelligence being built into the database management system, and partly the product of larger datasets with more data types. As a result of the stronger role for databases in application performance, data modeling is now central to application development with both following a common design process.

The agile data modeling and application design approach has been described by many practitioners, such as Scott W. Ambler, a Canadian consultant who has written several books on the subject. It is the key to surviving the changing needs of an organization and the potential solutions offered by technology. Attempting to define all the requirements up front is much riskier than following an agile methodology that can accommodate change. Thus, *Designing Geodatabases for Transportation* presents an enterprise data model only to demonstrate how all the pieces may be put together, not because the enterprise data model is a necessary first step to implementation. Again, the solutions offered here are more like the choices in a cafeteria: you can pick the ones you want. This book also guides you on making your selections. One size does not fit all.

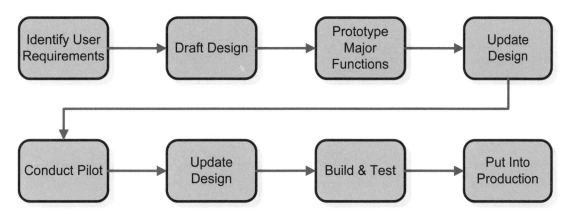

Figure 1.1 **Geodatabase design process** The agile geodatabase design process illustrated here starts with a draft design based on a set of user requirements. Design components are tested through prototypes to ensure that the overall architecture and its key parts work and then the design is revised to correct any observed problems. A pilot is tested for scalability to ensure the design's performance meets requirements under production load. (The Geodatabase Tool available as a free download from ESRI can help here.) The design is revised and tested again before the system is put into production. Although the figure stops there, the reality is that new user requirements are likely to be imposed following production deployment and the cycle begins again.

The agile method uses an iterative, team-based approach. Members of the team are not given separate assignments; all team members work together on each assignment. The ideal team member will have broad knowledge of the organization and its operational needs, the general requirements of the agile method, and expertise in one or more areas required for design development and implementation. The hallmarks of the agile method are its use of "good enough" documentation, which includes data models; frequent deliveries of solution components to get user feedback; and structured performance testing at the planned scale of deployment (number of users, transaction types and volumes, etc.). The number of and execution time for database queries and related operations determine geodatabase performance and, thus, application performance. While there may be several ways to reach the same result, you will want to use the one that provides the greatest performance. No one likes to wait. The only way to know which approach provides suitable performance is to test it under real-world conditions. Performance testing must be part of the geodatabase design process.

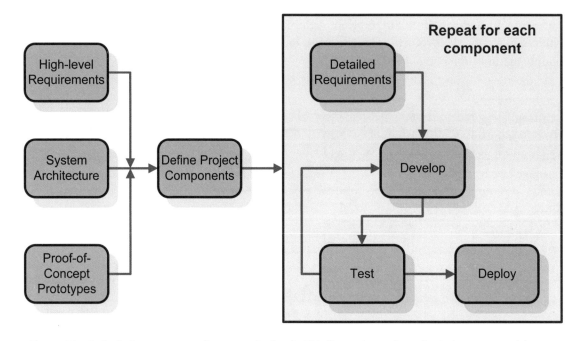

Figure 1.2 **Agile design process at the enterprise level** This figure shows the agile design process with extra structure when applied to an entire enterprise. The box on the right includes four of the eight components shown in figure 1.1. The four boxes on the left side provide the enterprise context for the agile process. High-level requirements define the overall scope. The system architecture defines the deployment environment, such as which relational database management system (RDBMS) product will be used, who will have responsibility for system maintenance, where the hardware will be located, and other constraints. Proof-of-concept prototypes may be necessary to ensure that the system architecture will perform as expected under production loads.

When you seek to apply the agile method to an entire enterprise, you need to break the project into deliverable components that fit within an overall framework. The enterprise framework allows the organization to balance the needs of various internal and external communities competing for limited development resources and to test fundamental aspects of the overall system architecture vital to major parts of the final system.

It has not been assumed that the enterprise of interest is a transport agency. This book takes into account the fact that many transportation data users do not actually care that much about transportation *per se*. These users need transportation data as a reference plane for other information, such as situs addresses tied to relative position along a transportation facility (street address). Map making is generally placed in this category. Thus, this book will show how to build transportation-oriented geodatabases that can support the editing of data needed by a variety of applications. The focus will not be solely on the more extensive data models suitable for larger organizations. The book starts with simple problems and solutions, growing in complexity in response to application requirements until it reaches the most complex problems posed by large transport agencies and transit operators. Some chapters, such as those on navigable waterways and railroads, are actually directed to users outside these industries.

Designing Geodatabases for Transportation also does not assume you are starting from scratch. The more common starting point today is the migration of a transportation dataset based on the coverage or shapefile structure to a geodatabase, often in the midst of making other changes. A common concurrent technology change is adopting a relational database management system (RDBMS) for overall data storage. Indeed, the migration to the geodatabase structure often marks the point where spatial data goes from being concentrated in workgroup datasets to becoming an enterprise resource available to the entire organization. In any event, you probably are not starting with a blank sheet of paper but rather are dealing with a number of prior decisions you need to identify in the requirements analysis and system architecture steps. You will rely on your agility to work within an existing data structure serving a number of existing applications and within a prescribed set of constraints and resources.

Building the agile geodatabase

Your task will be made much easier if you build an agile geodatabase. This starts by separating the editing geodatabase environment from the published dataset. The purpose of any GIS application is to create information. You have to put data into an application in order to get information out. No single book can show you the answer to every geodatabase design problem given the vast range in possible applications, even if the scope is restricted to a single data theme. So, instead of trying to solve all transportation application problems, *Designing Geodatabases for Transportation* seeks to solve only one: creating information out of original data sources, which is data editing. The data thus maintained is then used to populate datasets supporting other applications, each of which presents its own data structure and content requirements. The outputs of the editing process are defined by all the other applications.

Many of the applications' data requirements are likely to suggest geodatabase design characteristics that work against the data-editing process. What you really need to do is view data editing as its own application, one that creates the inputs to all the other applications, and then follows the mantra that the application determines the geodatabase design.

For example, eliminating data redundancies greatly benefits data editing by keeping each piece of information in a single location so you do not have to enter it multiple times. Thus, a geodatabase optimized for editing will eliminate data redundancies that cause extra work and increase the chance for inconsistencies.

Right about now, you may be thinking that you never want data redundancies, so why is this a big deal? The answer is that no single application may impose the need for data redundancy, but the collection of applications supported by the editing process may. For example, you could have several applications that want to know the length of a facility, some in meters and others in miles. Even if all the applications want the data in the same form, they are likely to expect the data to be stored in a data field under a specific name, such as LEN, LENGTH, or DISTANCE. Rather than store the length in all these different forms and field names within the editing database, you want to store it there once and then create the different versions needed by the various supported applications. There may also be applications that need data derived from other data. For instance, sums, averages, minimums, maximums, and counts may be employed by various applications, such as the total number of highway lane miles or the minimum length of all passing sidings located along a rail line. It is much better to have these values derived rather than enter them directly because it saves time and reduces the risk of error.

These practices mean the process of moving data from the editing to the published geodatabase will likely involve data transformations, calculation of derived fields, data replication, and other actions. But this process can be automated. In contrast, data editing is a primarily manual task. Work smarter, not harder. You will get better data with less work.

Going back to the earlier discussion of agile methods in enterprise geodatabase design, the editing environment is typically the last one to be designed and the first to be built. It is designed last because you will not know what data must be maintained—and the geodatabase design that best supports that data—until all the application inputs are defined. It is delivered first because all those using applications will not function until the inputs are provided. Assuming you cannot design and build everything at once, this chronology presents an impossible task, because the agile method assumes that an application's final requirements evolve. As a result, the geodatabase must itself be agile.

The core concept of agility is flexibility combined with robustness. Separating the editing and usage portions of the complete enterprise database allows each to evolve independently and to use a structure optimally suited to its needs. Editing involves lots of small transactions that change the geodatabase coupled with a strong need to coordinate edits made by different persons over time. In other words, maintaining database integrity. In contrast, applications involve extractions of relatively large chunks of data. Each application defines a set of data needs and imposes requirements on the geodatabase that it uses. That geodatabase should be part of the published dataset, which receives its content from the editing geodatabase. If you use the editing geodatabase directly, then your application would have to do all the heavy lifting associated with getting the data into the right form. Conversely, if you edit the application's geodatabase directly, then the editing process has to deal with the data structure the application needs. In both cases, you have editors and users churning the same data, which can often produce surprising results because of a loss of referential integrity; i.e., differences in values across the geodatabase.

This book provides detailed instructions for how to structure and process data so that it can be used to support applications without users having to separately and duplicatively maintain the data. This book is about enterprise data editing, not within a single office, but across the organization. The data it embraces is defined by other applications. Editing geodatabases evolve more frequently than do application geodatabases. The editing environment is the sum of all application data requirements. As a result, it will probably need to be modified each time any application changes or is added to the list of supported work processes.

Designing Geodatabases for Transportation describes a geodatabase design process founded on content rather than specific applications. The design of the editing application is determined by the nature of the data to be edited. Thus, the solutions presented in this book follow the general structure of, "If your user needs this kind of data, then build the editing geodatabase this way." Many of the geodatabase design principles presented here are widely applicable and need not be restricted to transportation themes. All are consistent with good data-management practices and current technology.

Book organization

This book is divided into three parts. Part 1 covers the basics of geodatabase design. Part 2 explores the various ways transportation geodatabases may be structured. Part 3 offers a variety of advanced topics on transportation geodatabase design.

As with any book intended for a wide range of readers, *Designing Geodatabases for Transportation* covers a lot of foundational concepts dealing with database design in general and geodatabases in particular. While it may tempting for a more knowledgeable reader to skip the first few chapters, even the advanced data modeler should review the content of part 1 in order to be familiar with the terms and presentation employed in this book. Similarly, you may want to explore the modal chapters in part 3 related to forms of transportation not included in your own geodatabase because there may be ideas you can use.

One of the more obvious demarcations in the book is the distinction between the segmented data structures used mainly by local governments and commercial database vendors and the route-based structures used primarily by state and provincial transport agencies. Because they are conceptually less complex, design concepts more applicable to segmented data models are generally presented in earlier chapters and those concepts with greater applicability to route-based models are covered in later chapters. Do not skip the content directed to one side of this dividing line because this distinction in application is often one of convenience. Many design techniques are applicable to both basic data structures.

Some content is targeted to a specific audience. These passages will be placed in sidebars identified by one of two icons.

The building block icon identifies basic knowledge about a fundamental aspect of the topic.

The rocket icon denotes information suitable for advanced readers that describes what is happening behind the scenes, gets into the details of a topic, or offers guidance for specific tasks.

Transportation geodatabases have been difficult to construct in the past, in large part because of a lack of basic guidance on how to address the many problems presented by this unique data and the business processes it supports. *Designing Geodatabases for Transportation* is intended to provide basic guidance on how to construct transportation geodatabases in a manner that addresses these inherent problems.

Notes

[1] Although pipelines and utility systems have a similar structure and do transport materials or energy from place to place, they are not included in this book. Other ESRI Press publications and data models address the spatial database design needs of pipelines, telecommunications, water utilities, and electric power systems.

PART I *Basic geodatabase design concepts*

chapter two

Data modeling

- Data types
- Files
- Tables
- Relationships in relational databases
- Object-relational databases
- Relationships in object-relational databases
- The data-modeling process
- Conceptual data models
- Logical data models
- Physical data models

Chapter 2: Data modeling

This chapter covers data modeling, the process of designing a dataset's structure by adopting a set of abstractions representing the real world. A dataset is a collection of facts organized around entities. An entity is a group of similar things, each of which may be referred to as an instance or a member. For example, Road could be an entity representing all roads, with State Route 50, Interstate 10, Main Street, and Simpson Highway members of that entity. You cannot store the real-world entity in the dataset, so you store a set of descriptive attributes that allow you to identify the entity and understand its characteristics. Attributes can be composed of text, numbers, geometry, images, and other forms of data. If Road is your entity, then facility ID, route number, street name, length, jurisdiction, and pavement condition could be useful attributes. When an attribute involves location, it is considered to be spatial in nature. GIS involves spatial data. Attributes, not entities, determine whether a dataset is spatial.

A database is a dataset stored in an electronic medium. A geodatabase includes spatial data. A user acts upon such a dataset through a database management system, which may also provide various security and data integrity services. A geodatabase is a collection of geographic datasets. The database management systems used for large workgroup and enterprise geodatabases are relational, which means they perform according to a number of rules, called relational algebra, that describe how to read and write information stored in the database. The language shared by relational database management system (RDBMS) products is SQL, which once stood for Structured Query Language. The RDBMS converts SQL statements entered by the user (or generated by a computer application) into relational algebra to perform operations on the data. You do not need to know about RDBMS products, relational algebra, or SQL to do data modeling. What you do need to know is included in this chapter.

Every database is a data model because a model is simply an abstract representation of the real world. A primary concern of data modeling is deciding which abstraction to use. For example, a spatial database may represent a linear transportation facility with a centerline, but the real-world facility is actually an area with one very long axis. We commonly use a centerline because it conveys the primary aspect of the facility: it has length and traverses a space. That centerline can be part of a geometric network for determining the best path between two points, or it can simply be a reference for locating other features on a map.

The information you need about the facility is determined by how the data will be used. The network pathfinding application will need information about connectivity, cost of traveling on a segment, and restrictions to travel. Any geometric representation you created for the network may be highly abstract, perhaps just a straight line between two points. In contrast, a mapping application needs just a line geometry representation, with perhaps some information for symbolizing the line. The scale of display will determine the degree of abstraction allowed for the geometry. Large-scale maps may need detailed road edgelines, while small-scale maps may need only a generalized centerline.

You can alternatively represent the linear facility as a surface, such as might be done for a digital elevation model (DEM) using a triangulated irregular network (TIN), which are both ways to represent a surface for 3D representations, or it could be a set of pixels in a raster image. You might also store the linear facility as a set of address points. You can even store the facility as a set of nonspatial attributes, employing no geometry at all. Each of these abstractions has a place within a transport agency and its variety of spatial-data applications. However, this book will concentrate on vector data forms where lines represent linear facilities, as this is the most common abstraction. Several design proposals show how to accommodate multiple geometric representations for a single entity.

Your choice of which form of abstraction to use is determined by the data's application. Since larger transportation organizations need many applications, it is likely they will need multiple abstractions. For example, a bridge might be a point feature to some, a linear feature to others, and a polygon feature to yet another group.

Data modeling is the structured process by which you examine the needs of your application and determine the most appropriate abstraction to use. It begins by understanding the application's requirements for data, which will determine the appropriate level of abstraction, the structure to use in organizing the data, the entities to be created, and the attributes assigned to each entity. In the geodatabase, an entity eventually becomes a class, which is a discrete table or feature that you will define in terms of its properties, behaviors, and attributes. A geodatabase combines data with software in an object-oriented form that takes over much of the workload needed to use and manage the data. A geodatabase is an active part of the ArcGIS platform, not a passive holder of data.

Much more about these concepts is discussed later in this chapter. What you need to know for now is that data modeling, as presented here, is founded on the capabilities and constraints of the geodatabase. However, if you are like most transportation-data users, you already have data in a variety of nongeodatabase forms, so this chapter covers other fundamental data structures along with the basic concepts of database design and data models.

Data types

When starting a data-modeling project, you first must understand the data you intend to place into your new geodatabase. In addition to the geometry you use to abstractly represent a real-world entity cartographically, you have the traditional forms of data that have always been part of transport databases.

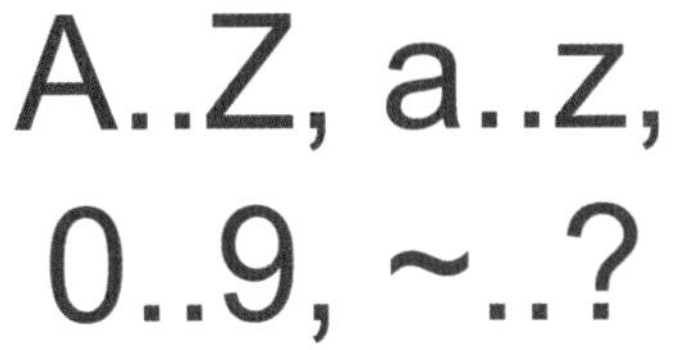

Chapter 2: Data modeling

Character strings
A character string is any value consisting of printable alphanumeric characters, such as letters, numbers, and punctuation. You specify the number of characters as part of the data type definition when you add an attribute of this type to a geodatabase class.

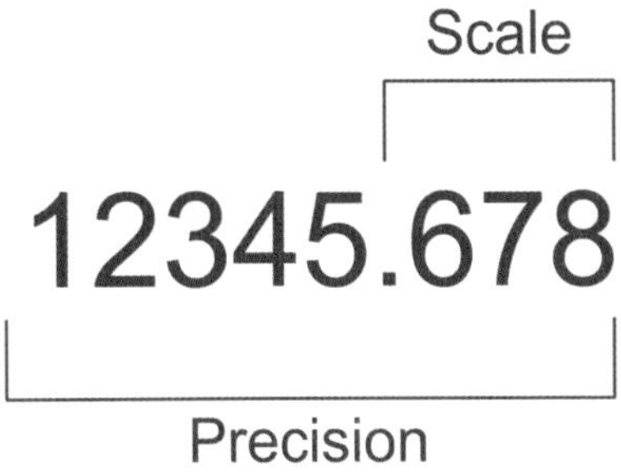

Numbers
The four numeric data types normally used in geodatabases may be defined using by specifying type, precision, and scale. Precision is the total number of digits accommodated. Scale is the maximum of digits permitted after the decimal. The actual numerical range limits for precision and scale are somewhat dependent on the RDBMS you use. Precision and scale specifications are used only when creating an ArcSDE geodatabase; only a data type choice is needed for files and personal geodatabase classes.

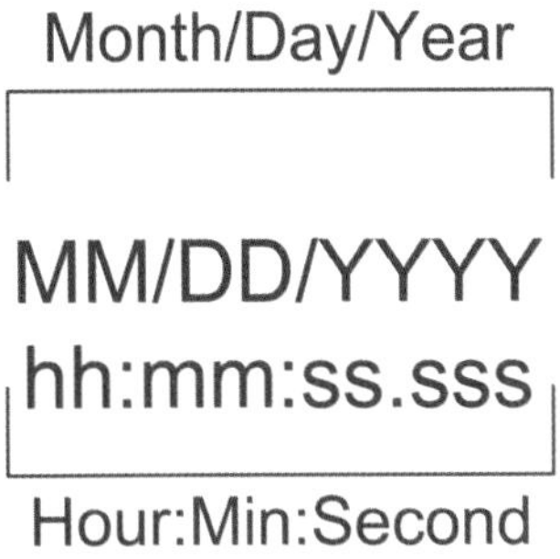

Date and time
The Date data type includes both date and time information presented as the date first, and then the time, with a resolution of thousandths of a second. Entries with only date information will have zeroes placed in the time components. Similarly, time-only entries will have zeroes in the date components.

Figure 2.1 **Data types** The ArcGIS geodatabase supports several data types for user-supplied class attributes. The primary ones you will likely use are character string, short integer, long integer, single-precision floating-point (float), double-precision floating-point (double), and date.

One of the most common kinds of data is text, which consists of a string of alphanumeric characters, like letters, numbers, and punctuation. Anything you can type on a keyboard can go into a text field. The maximum number of allowable characters defines most text fields. For example, you might see a reference like "String (30)" to define a text field with a maximum length of 30 characters.

An equally popular form of data is a number. There are many different types of number data, but to the user they all consist of a series of digits. Where they differ is how they are stored in the database. In the geodatabase, a short integer will be stored using 2 bytes of memory; a long integer requires 4 bytes. A single-precision or floating-point number is also stored using 4 bytes, while a double-precision number is stored using 8 bytes. The actual numeric range that each of these forms represents varies according to the database management system you use.

Working in concert with the type of number format you select is the way you specify it in an ArcSDE geodatabase. A number field in such a database has two characteristics that go with its type. The first is precision, which specifies the maximum number of digits that can be stored. The second characteristic is scale, which tells the database how many of those digits will fall after the decimal point.

Number type, precision, and scale interact in various ways. For example, the database will ignore scale if you specify a number type of integer, because integers consist only of whole numbers. The data type overrides the specification. Sometimes it works the other way. For instance, if you specify a floating-number (single-precision) data type but a precision of seven or more, ArcGIS will change the data type to double-precision.

Most database management systems also support date and time data types. Although stored in vendor-specific ways, ArcGIS provides a consistent representation to the user in which date and time are combined into one data type, called 'Date'. The date portion is provided as a two-digit month, a two-digit day, and a four-digit year, with the three components separated by a forward slash character. The time portion is presented as a two-digit number representing a 24-hour clock (00-23), a two-digit minute portion (00-59), and a second component with a precision of 5 and a scale of 3. The three time components are separated using colons.

Files

Relational databases were not the first kind of electronic data structure. The oldest form of database storage is the file, which consists of a block of data organized into logical groups called fields. Each position in the field is called a column. Files look like a table with their records (rows) that separate content using a special character to signify the end of a logical group of data. Everything is text. There is no inherent requirement for all the records to have the same structure. For example, the first record, often called a header, could state the number of body records or describe the fields in those records. All the intelligence needed to understand the file's content is in the application that reads and writes records.

Chapter 2: Data modeling

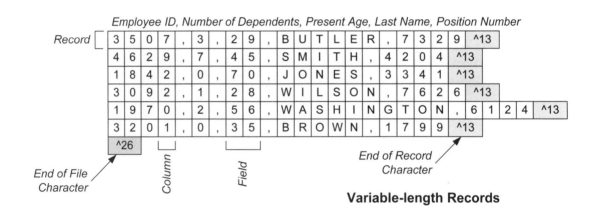

Figure 2.2 **Files** A fixed-length file uses column position to identify specific data content forming attributes. A variable-length file uses the sequential order of fields separated by a predefined special character—one that cannot appear in the data. In both cases, the application using the data must know the specific location of each piece of information.

Files come in two basic forms: fixed-length and variable-length. A fixed-length file uses the position of each character in a record to interpret its meaning. Any leftover space not needed to store the data for that record is filled with spaces, either before or after the actual data in the field. Fields in each record are identified by position. For example, a file specification may declare that record characters (columns) 1 through 47 contain an employee's name right-justified with leading spaces.

A variable-length file uses the position of a field within the record to identify its content. Variable-length records avoid space filling by using special characters to say where one field stops and another begins. You may have come across this structure when using comma- and tab-delimited text files. The commas or tab characters are the things that separate the records into fields. Usually, there is also a special end-of-file character.

The most common ArcGIS file-based data structure is the shapefile. A shapefile is a kind of spatial database structure consisting of several files. There are more than a hundred recognized shapefile component types, each with its own file extension (the three characters after the dot in a typical file name). To copy a shapefile, you must copy all the component files. The minimum components are the geometry (.shp), the nonspatial attribute data (.dbf), and the spatial index (.sbx). The structure of each component file is optimized for the information it contains. For example, the geometry file (.shp) contains a 100-byte fixed-length file header followed by variable-length records. The variable-length record is composed of an 8-byte, fixed-length record header followed by variable-length record contents. Each record defines a single geometry, with the length of the variable portion being determined by the number of vertices and whether measure (m) and elevation (z) coordinate values are included. The fixed-length record header portion provides a record number and the length of the variable portion.

Coverages, which were the original ESRI data structure, are also based on a database structure consisting of multiple files. Designed to reduce the size of a spatial database, software manipulating coverage data must manage a number of composition relationships inherent in the file structure. A special data-exchange file type was developed to be able to distribute coverages via a single file.

File data structures remain useful today and will continue to be part of GIS datasets long into the future. This book, however, will restrict itself to modeling geodatabases. What you put into and take out of a geodatabase may be a file, but the database to be modeled is a geodatabase.

Tables

The next step along the evolutionary line of database design is the table, which is a fundamental data organization unit of a relational database. A table looks very much like a file in the way it is presented to the user by the RDBMS in which it exists: a set of rows (records) and columns (fields). Rows represent members and columns represent attributes.[1] However, you cannot simply copy a table as you can a file, because each table is tightly bound to the RDBMS.

Chapter 2: Data modeling

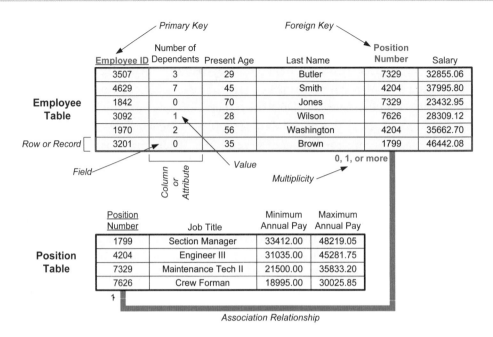

Figure 2.3 **Tables** An application seeking to use the data stored in a relational table needs to know the name of the table and the name of each attribute it seeks, but not the physical manner of data storage. That job is performed by the RDBMS. The primary key uniquely identifies each row. One or more foreign keys can be established to provide connections to other tables. In this example, the Position Number attribute serves as a foreign key to a table storing position descriptions, where Position Number is the primary key. Foreign keys express association relationships. Cardinality is the ratio of rows for two tables. The number that comprises each half of the ratio is the table's multiplicity. The cardinality of this one-to-many (1:m) association relationship says that a position number must be entered for each employee and that some position numbers may not be applicable to any employee.

Dr. Edgar Codd invented relational databases in the early 1970s at IBM, although it was several years later before a working product could be devised. Such a database management system is based on relational algebra, a kind of math that controls what can happen to data in such a storage structure. Relational algebra supports seven functions[2]:

- Retrieve (read) row
- Update (write) row
- Define virtual relations (table views)
- Create a snapshot relation
- Define and implement security rules
- Establish and meet stability requirements
- Operate under integrity rules

Relational tables are not actually stored in the row-and-column form we typically use to visualize them, but everything you need to take from this book can be accommodated with the rows-and-columns metaphor. Oracle, SQL Server, Sybase, and Informix are commonly used RDBMS platforms. Products like Microsoft Access have much of the functionality of an RDBMS but are actually database management systems that employ files.

Relationships in relational databases

The big advance offered by the relational database is its ability to represent and manage relationships between tables. Where files normally use a record's position in the file to uniquely identify each member, an RDBMS cannot impose any ordering on its member records. Thus, an RDBMS requires that at least one column be an instance identifier, called a primary key.

The relationship that relational databases are most concerned about is the association of one table to another. An association is established by placing the same column or a set of columns in both tables. This connection is called a foreign key. For example, a foreign key may link a central table storing general roadway information with other tables containing information about speed limit, traffic volume, maintenance jurisdiction, and pavement condition.

> There is considerable variety in the nature of primary keys. The duty of a primary key is to uniquely identify each row in a table, which means that there can only be one row with a given primary key value. For this reason, many database designers argue against using a primary key that is entered by the user. This guidance also means the primary key cannot have any implicit meaning other than service as the row identifier. A primary key with intrinsic meaning is called an intelligent key.
>
> Users like intelligent keys because they are easier to remember and they can serve double duty as an attribute. Database designers hate intelligent keys because they are prone to error in data entry and duplication within the database. You may want to use route number, such as SR 98, as the primary key. The problem is that you might accidentally type "RS 98," or SR 98 might be rerouted, resulting in confusion as to the version a record references. Instead, database designers populate primary keys with integer sequencers supplied by the RDBMS and large, globally unique identifiers created through various mathematical processes. These values are guaranteed to be unique within the table.
>
> All those other potential primary keys—the ones that mean something—are candidate primary keys and, thus, potential foreign keys. They could be primary keys, except for the chance that they might be duplicated within the table, which is the one thing that must never happen to a primary key. Coded values that are used as shorthand for a larger meaning, like a functional class code of 11 that means rural interstate highway, are often candidate primary keys that are chosen to serve as foreign keys. Some foreign keys may also be useful outside the database. These are called public keys, and they

> include such things as driver's license number, Social Security number, river-reach code, the three-letter airport abbreviation, the two-letter state and province abbreviation, and highway route number. All of these primary and candidate key concepts are used in this book to demonstrate specific database design solutions. Each has a number of useful applications.
>
> While on the topic of table keys, it is important to acknowledge their two varieties. A simple key consists of a single field. A complex key is composed of more than one field. For a complex key, it is the arrangement of key values that must be unique, not each individual field's value. Complex keys are useful when a combination of things is required to identify a single member. For example, instead of using a single functional class field to indicate rural/urban location and the type of roadway, you could split them into two fields, one for each aspect of highway functional class. A facility identifier in combination with a date field, such as to indicate the version of SR 98 that opened to traffic in July 2007, is another possible example you may find useful.

The two tables involved in an association relationship are called the origin and the destination. Both contain a field with the same data in the same form, although the number of instances with the same value may differ. The foreign key in the origin table is usually the primary key or a candidate primary key in the destination table. Association relationships are typically described as a ratio of the number of rows that can exist at each end of the relationship. Each number is called a multiplicity and the combination of the two multiplicities is called the relationship's cardinality.

Multiplicity can be classified as one or many. Thus, cardinality can be the various combinations of these two values: one-to-one (1:1), one-to-many (1:m), and many-to-many (m:n). When the presence of rows at one end of the relationship is optional—in other words, the association doesn't always happen—multiplicity can be zero, but that does not affect the cardinality. For example, if you designed a rail station database that contained a County table and a Station table, you must allow the number of Station table rows required for a given county to be zero, one, or more. It is, nevertheless, a one-to-many relationship because one county may have zero, one, or more rail stations. The upper bound in the multiplicity determines the cardinality.

In a one-to-one association, each row in one table may be related to one and only one row in the other table. This relationship is relatively rare because putting all the attributes in one table can often eliminate it. However, there are times when it is useful to split attributes of an entity into multiple tables. For instance, there may be a set of attributes that exist for only a small subset of entities or you may want to do different things with each subset of attributes.

The most common cardinality is one to many. In this case, one row in the origin table points to many rows in the destination table. The foreign key goes in the destination table and points to the origin table. For example, if you use coded values in your geodatabase, you will often provide a domain class that lists the range of valid values and ties each value to its

One to One (1:1)

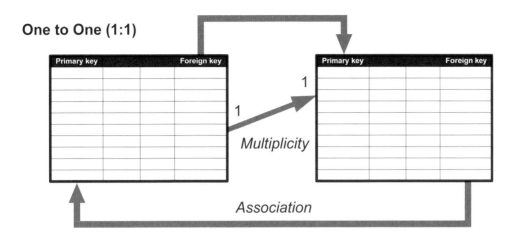

One to Many (1:m)

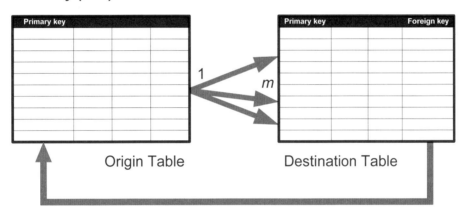

Figure 2.4 The foreign key can go in either or both tables in a one-to-one relationship. Association is a connection that shows which tables participate in the relationship. Multiplicity expresses the cardinality of the relationship; i.e., the number of rows in each table that may participate in the relationship. The foreign key goes in the "many" end of a one-to-many relationship because that is the end with a single possible value. The foreign key in the destination table stores the primary key of the origin table.

meaning. The table with value meanings is the origin and the table where those values are used is the destination. Many rows in the destination table can have the same value, all of which point to one row in the domain (origin) class.

The toughest cardinality to address in database design is many to many. This is because you cannot accommodate such an $m:n$ cardinality by simply using a foreign key. A given row in either related table may need to point to an unknown number of rows in the other table, and a column in a relational database can only have one value.

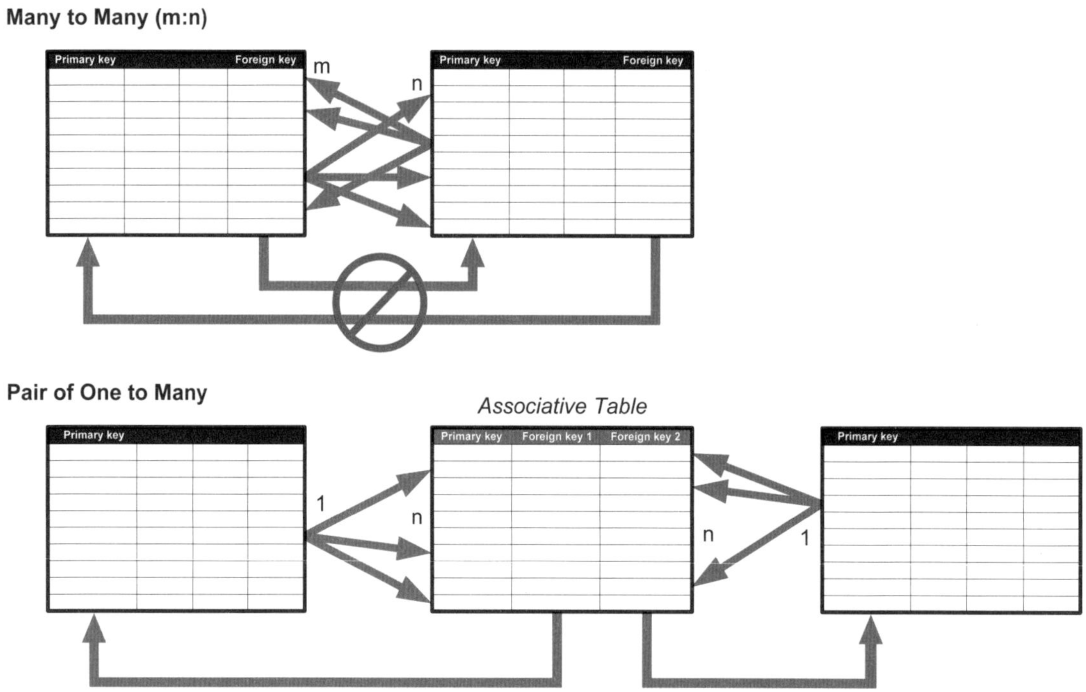

Chapter 2: Data modeling

Many to Many (m:n)

Pair of One to Many

Associative Table

Figure 2.5 Use an associative table to store the many possible relationships and give each one-to-many relationship its own foreign key. In a geodatabase, an associative table is called an attributed relationship class and can include user-defined columns.

To accommodate a many-to-many association, you have to turn the relationship into a table. Such an associative table will contain the primary keys of both related tables. The result is that each end of a many-to-many relationship can be listed in any number of rows in the associative table, thereby resolving the many-to-many relationship as a pair of one-to-many relationships. You can also store other information about the relationship.

Transportation databases are full of many-to-many relationships. For example, a work program project may affect several roads, and a given road may be affected by several projects over several years. Thus, the relationship between roads and projects is many-to-many. You will need a Road-Project table to store the relationship as a set of one-to-many relationships tying each project to one of the facilities it affected. Each row would have one road identifier and one project identifier. A given road or project identifier may occur in the associative table many times, but never more than once in the same combination.

Object-relational databases

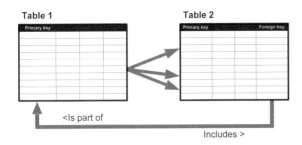

Figure 2.6   Relationships can be given names that express their role. This example shows a one-to-many relationship, where each row in table 1 relates to a collection of rows in table 2. The caret points in the applicable direction along the association. For example, objects in table 2 are part of objects in table 1.

Association relationships are usually obvious in their meaning, but naming relationships can help eliminate ambiguity. For example, you could say that an address includes a street name and that a street name is part of an address, or that an engine is part of a vehicle that may include many other components. We place relationship names next to the association connector and symbolize them with a caret that indicates the direction for which it applies.

Object-relational databases

The ArcGIS geodatabase is an object-relational design. Object-oriented software encapsulates data and the software that uses the data into an object class. The geodatabase consists of object classes that use a relational database approach to storing data. The kinds of object classes we will discuss are ArcObjects classes, which are the components of ArcGIS.

A class describes a set of data along with the functions that operate on that data. Class encapsulation is absolute. You cannot see inside the object class to view its data structure. You can only communicate with the object through the interfaces it supplies. An interface is a contract between the class and the outside world. Once an interface is declared for a class, it must always be supplied in every subsequent generation of that class and any classes that are based on that class.

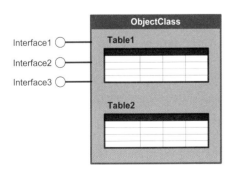

Figure 2.7   **Object relational**   The ArcGIS geodatabase is object relational, which allows ArcGIS to evolve its internal class workings without affecting how you view the data. By convention, we treat geodatabase object classes as if they were composed of a single relational table, but they are much more. **Appearance versus reality**   With a geodatabase, ArcGIS only gives you the appearance of a relational database through its class interfaces and wizards. You are not actually seeing the internal data structure. A fundamental principle of object-oriented programming, called "encapsulation," means that you can never see the internal structure.

23

Class interfaces provide a reliable way for software to be developed and used by different programmers. Each interface has a name. By convention, interfaces' names start with a capital I with the subsequent letter also capitalized. For example, ITable, IClass, and ITableCapabilities are the three interfaces added by the Table class that is part of ArcGIS. Most interfaces build on others. The parental relationship would be expressed as child : parent, as in "IObjectClass : IClass," which means that the IObjectClass interface offers additional functions for the IClass interface.

Object classes communicate with each other through messages conveyed between interfaces. You also communicate with an object class by sending, through an interface, arguments containing input data and work assignments to the class. The interface returns a result after the class's software—called variously by the names of methods, behaviors, procedures, or operations—does its thing. ArcObjects programming involves the use of class interfaces. You cannot actually change the class itself because it is not possible to see inside the class due to encapsulation. As a result, ArcObjects documentation discusses only the interfaces and how to use them.

All the software you need to manage a geodatabase is not contained in one object or feature class but in dozens of ArcObjects classes that work together to provide the performance you need in a GIS platform. Drawing-layer symbology, attribute domains, relationships, and rules are also contained in separate ArcObjects classes and are just as much a part of the geodatabase as the object classes and feature classes you create. Data models do not generally include all the things you can specify for a geodatabase in ArcGIS. They are normally restricted to tables, feature classes, relationship classes, and domain classes. Things like how a feature class is displayed as a map layer are usually omitted.

To use any of the data models shown in this book, all you need to do is add fields to a class. No programming is required. You also do not need to know about or have experience with ArcObjects. In fact, this is the only place in this book where interfaces are discussed. The intent is to make you aware that a lot of things are going on when you use a geodatabase and to help you understand what you might see in ArcGIS documentation.

Unless you plan to do some ArcObjects computer programming, all you need to know to design a geodatabase is that you will define object and feature classes using a class template supplied by ArcGIS and add attributes, define rules, establish domains, and create relationships. You need not be concerned with how ArcGIS internally handles these parts of the geodatabase. However, because of the encapsulation of software and data within the geodatabase object-relational structure, you can only exchange data by sending the entire geodatabase to another user. You cannot just select one feature class and copy it. The behavior of that class depends on the contents of several other classes. A feature or object class functions only within the context of the geodatabase.

Just as software contained in an object class may be known by many names, attributes are also called by many names: columns, properties, and fields. There are more names for each discrete "thing" contained in a class: object, member, instance, row, and record. In general, ESRI restricts the use of *object* class to mean a type of table that stores nonspatial objects. This definition does not mean that the thing the object represents is nonspatial. A dam on a navigable waterway is certainly a spatial entity. If you choose to include it in a geodatabase using an object class, you are only making the decision that this particular abstract representation does not include geometry.

In ArcGIS, an object that includes a shape attribute (geometry) is called a feature; i.e., it is a geometry object. In practical terms, a feature class adds the additional software and data structures needed to store and retrieve geometry to the software supplied by the object class. More specifically, a feature class is a table with a geometry column stored as binary long object (BLOB) pages in a relational database. A feature class contains geometric elements (simple features) or network elements (topological features) in a coverage, shapefile, or geodatabase structure. ArcGIS displays it to you as a table with a SHAPE column, but there is a lot more going on behind the scene.[3]

Relationships in object-relational databases

Our earlier discussion of relational databases introduced association relationships. Object-relational data models will explicitly display their multiplicity. Frequently encountered examples include "0..1" to mean zero or one may exist; "1" or no notation to show that one must exist (a default value); "*" to mean more than one must exist; "0..*" to represent zero, one, or more instances may exist; and "1..*" to mean at least one must exist, although more may be included. Association is represented in our models as a medium-width gray line and may carry role names.

The next most common and important relationship in an object-relational database is inheritance, which is a parent-child relationship. Inheritance is shown in a data model using a thin solid-black line with a generalization arrow pointing to the parent class. (You can think of the arrow as pointing in the direction to look for additional attributes.) The end symbol is called generalization because it shows the parent class to be a more general form of the child class, or, conversely, the child class is a specialization of the parent.

In addition to showing a parental relationship, inheritance simplifies logical data models by allowing you to omit repeating the parent class's attributes in the child class. In our data models, the child class will include only attributes that have been redefined or added to the parent's. Parent classes are often called stereotypes. Many stereotypes are abstract, which means that their purpose is solely to serve as a class template; objects of an abstract stereotype class cannot be created. An abstract class's name will be shown in italics. A class that can produce objects conforming to the class specification is said to be instantiable.

The other primary relationships you will see in data models are dependency, composition, and aggregation. A dependency relationship, also called navigability, shows which classes depend on other classes for their existence. Another way to define this relationship is to say that one class instantiates the other, meaning a function of one class creates objects of the

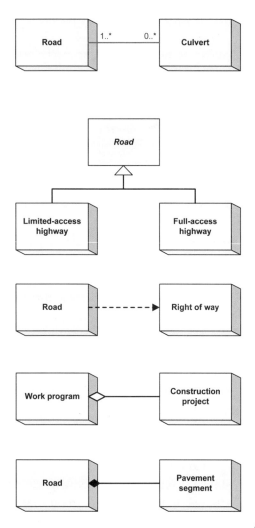

Association Association is a cardinality relationship between two classes that expresses the numeric ratio of how many of one class can exist relative to the other. Each end of the relationship includes notation for its multiplicity, except that convention omits the multiplicity of 1 as a default value that need not be written. The three basic cardinalities are one to one (1:1), one to many (1:m), and many to many (m:n). This example says that a culvert cannot exist in the absence of a related road, but that a road without culverts is possible.

Type inheritance Perhaps you need to create a class with an extra attribute or two, or different implementation rules. The stereotype (Road) serves as a model for building two subclasses (Limited-access highway and Full-access highway) that share the attributes and methods included in the stereotype. In this example, the abstract Road class, which you will never instantiate, contains the attributes and methods that will be in both subclasses. The arrow-like endpoint symbol "points" to the superclass stereotype. Type inheritance makes logical models easier to read by reducing duplication.

Instantiation Some classes can create instances of other classes through instantiation. For example, a Road object might be able to create a Right-of-way object. A data model that consists only of standard feature, table, and relationship classes will not include instantiation, as ArcGIS handles those duties.

Aggregation Aggregation is when an instance of one class (the whole) represents a collection of instances in another class (the part). A DOT work program, for example, could be viewed as a collection of construction projects. The project exists independently of the work program.

Composition In contrast to aggregation, where both classes involved in the relationship can continue to exist in the absence of the other, composition means that the "whole" class controls the existence of the "part" class. Here, if you delete a Road class instance, all the related Pavement segment instances will also be deleted.

Figure 2.8 **Relationship types** Data models at all levels of abstraction include relationships. This minitutorial explains the five primary types that you see in ArcGIS: association, type inheritance, instantiation, aggregation, and composition. All relationships are illustrated using a line and endpoint symbols.

other class. Dependency is shown using a dashed black line with an arrow pointing to the dependent class. None of the data models presented in this book will include dependency relationships, although they do frequently appear in ArcGIS documentation of the geodatabase and you may need them in your own data models.

Composition and aggregation are similar to each other with one important difference. A composition relationship is created when one class is composed of one or more instances of other classes. For example, a building may be seen as being composed of at least three walls, one floor, and one roof. Remove the building and its components cease to exist, at least as far as the database is concerned. Thus, a composition relationship tells you that when the sum of the parts is deleted from the database, you will also need to delete the objects of which it is composed. A thin black line with a solid-black diamond at the end adjacent to the composite class represents this kind of relationship.

Aggregation is not so particular. An aggregation relationship specifies that a class is a collection of other classes. For example, a baseball team may be a collection of players. If you express this relationship through aggregation, then you are saying the players will continue to exist in the database even when they do not belong to a team. If you instead use composition, then the players must be deleted from the database when the team is dissolved. Aggregation symbology is similar to that for composition, except that the diamond is outlined in black and white filled.

The data-modeling process

You need a good data model to produce a good geodatabase design. Developing a geodatabase design is a six-step process that follows the flow of the agile methods discussed in chapter 1:

Step 1—Define user requirements. First, you need to know the purpose of the data, the application requirements to be supported. Many users attempt to develop a complete set of requirements as the first step, but that cannot be done. Even for a small project, the agile method instead encourages you to create a good first effort that has the primary objective of identifying the major components. For an enterprise geodatabase that will support a wide variety of existing and yet-to-be-created applications, seeking a complete set of requirements as the first step is an impossible goal. No, your task here is to identify the general requirements.

Step 2—Develop conceptual data model. Once you have specified the general requirements for the final product, you will need to identify the basic elements of a geodatabase that meets the requirements. Such elements consist of entities and their relationships. An entity may eventually be reflected in a class, but at this point in the process, you cannot establish a one-to-one equivalency between entities and classes.

Step 3—Develop a logical data model. Once the general structure of the database—the skeleton—is established, the next step is to add some meat to the bones by specifying attributes for the geodatabase. Entities may change at this point, as attributes are assigned and new relationships discovered. The logical model is independent of the planned implementation platform.

Step 4—Develop a physical data model. Here is where entities become classes and the implementation platform makes a difference. Your RDBMS, network structure, and organizational behaviors will influence the way you translate the logical design into a physical implementation. The added benefits of the geodatabase and ArcGIS will become apparent at this stage. The geodatabase can perform many functions that would normally have to be handled by user-developed software. For many geodatabase projects, the first task will be to split entities into tables and feature classes. You will also need to decide which fields can be supported by domain classes and which relationships need to be instantiated as a relationship class, not just an implicit relationship established by foreign keys you use when you desire. You need to be alert to the difference between layers on a map and components of the geodatabase. The next task is to specify the details of each class and create the domains, rules, and other elements of a geodatabase. The physical data model specifies the data type, default value, domain, and other characteristics of each attribute. The logical data model tells you about the classes and their attributes, although you may not implement the whole model at one time, and tests may motivate changes to get the desired performance.

Step 5—Test the data model. Next, you can load the physical data model into ArcGIS and generate a prototype database for testing. Many central elements of a transport geodatabase can be implemented in more than one way. Testing the prototype before you put it into production is a good way to evaluate the efficiency of the implementation choices you made. Testing should include typical editing operations and involve a sample dataset equivalent in size to the one you will use. If the design does not pass this test, it may be necessary to go back to step 3 or 4 to make other choices, but it is much better to find out now than after you put it into production use.

Step 6—Production implementation. Now you can reap the rewards of your work. Load the geodatabase and create the default version. It is time to put everyone else to work.

These steps are generally sequential but you may move backward whenever necessary to redesign a portion of the geodatabase. You may also choose to prototype parts of the design at points well before step 5 so as to test key components. What works great for one agency may be a bad idea at another because of an organizational difference or the combination of applications to be supported. It is much cheaper to debug a paper design than an implemented geodatabase. Modeling will not eliminate all chance of error, but it certainly improves the odds of success. The balance of this chapter will explore the differences between conceptual, logical, and physical data models at one time, and tests may motivate changes to get the desired performance.

> The help section of ArcGIS online presents an 11-step geodatabase design process rather than the six shown here. The difference is due to how transportation datasets are structured. The 11-step process is oriented toward feature classes and map display. It starts with a discussion of the key thematic layers and selection of geometric abstractions. In contrast, transportation datasets are typically oriented toward object classes (tables), with geometry being a secondary consideration. While map outputs may be useful, most people editing and using a transportation dataset do so outside the map interface typically associated with GIS.
>
> The six steps shown here are in the 11 steps of the traditional geodatabase design method, which also includes such tasks as specifying the scale range and spatial representation of each data theme at each scale; designing edit workflows and selecting map display properties; and documenting your geodatabase design. You may, indeed, want to use some of these additional steps at points in the process, except documenting the design, which you'd better be doing continuously! You will certainly want to make sure that someone is assigned to putting every piece of data into the geodatabase and keeping it current. You can add spatial-display details when you decide how to geometrically represent some of the entities in your data model, but this not required until you get to the physical data model.

Conceptual data models

Data models typically go through three phases of increasing specificity, starting with conceptual modeling. This phase is primarily concerned with identifying the entities about which you will need to retain data and the relationships between those entities, including some that will also need descriptive data. Conceptual modeling considers the application the data is to support and defines database terms. For example, if you are developing a geodatabase for a transportation agency, you will need unambiguous definitions for terms that will appear in the model. This set of definitions is called an ontology.

Here is where you should expect your first philosophical debate. You'll find such common entities as *Road, Railroad Track, Bridge,* and *Airport* will often have very different meanings throughout the agency. What do we mean by *Route*? Is it the continuous piece of pavement that winds through many states, each one assigning its own name to it? Or does the name itself define the extent of a route? If the latter, what happens when the name is changed or the route takes a different path due to construction? How about if the road is realigned in some way so that the length changes? Does *Road* include *Right of Way* as an element, or is *Road* an element of *Right of Way*? Is *Airport* a piece of land, a terminal, a collection of runways, or an airspace? Is *Railroad Track* one set of rails, with a section of double-tracked mainline being two *Railroad Track* members, or is *Railroad Track* like *Road*, where each track is equivalent to a lane of traffic and the number of tracks is an attribute? Is a *Bridge* across an interstate highway part of the interstate or the road that crosses the Interstate? Answering such questions is intense, emotional, and necessary. You will quickly discover that the most important relationships are those in the room, not those shown in the data model.

A conceptual data model shows entities and their relationships. It does not include attributes. A conceptual data model expresses central concepts, illustrates data structures, and describes components of the ArcGIS object model. You will use conceptual data models to translate user requirements into data structures. Creating the data model usually begins the process of developing the application ontology, which includes formal definitions for all the entities, attributes, and operations that will be part of the final design.

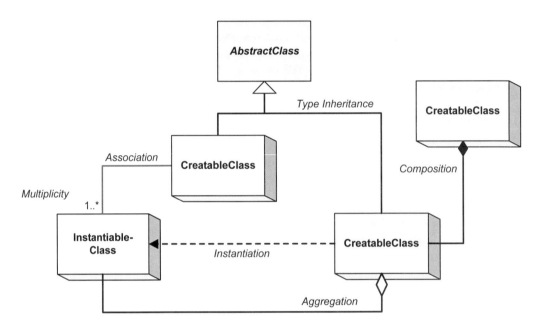

Figure 2.9 **Conceptual data models** The intent of a conceptual data model is to express the entities and relationships in a highly abstract manner. Attributes and methods are not included in a conceptual model, so the complex notation of UML is not required. Indeed, it may serve to obscure the model's meaning.

Figure 2.9 illustrates the simple 2D and 3D boxes used for conceptual data models. This is the same graphical standard used for many ArcObjects diagrams contained in ESRI documentation. For our purposes, conceptual data models consist of entities, not classes, and no one-for-one equivalency should be assumed; however, ArcObjects models presented with the same symbology do have equivalencies between entities and classes. In both cases, entities will be shown as one of three types. An abstract entity will be shown with a 2D rectangle and the name in italics. Instances of an abstract entity will not be implemented. Abstract entities form stereotypes for other entities that can be implemented.

Entities that are not abstract will be shown using a 3D cube, with a slight difference in face color between instantiable and creatable entities. This distinction really only applies to conceptual ArcObjects models representing a class structure. All non-abstract ArcObjects classes are instantiable in that members of each class (objects) can be generated. An instantiable class, as that term is used in this context, is one that is creatable only by other ArcObjects classes. Members of a creatable class can be instantiated by the user directly through ArcGIS. True conceptual data models will include only creatable entities, because users cannot generate instantiable entities directly.

The figure does not include notes and callouts, two of the most useful parts of the conceptual data model. ESRI's standard notation shows a finished product. What you are creating is a work in progress. You need to add notes that explain what the model says and callouts to describe specific entities and relationships. Business rules and definitions are not generally part of published conceptual models, but you will need them. The only important consideration is that the team members developing the model understand the model. Do not try to adhere to a particular external documentation standard for everything. This is not the time to try to teach everyone about the details of data modeling. There is no extra credit for pretty pictures. Use what works best for you. This book shows you the part that probably should be fairly uniform across teams and will be consistently used in this book.

Logical data models

Logical data models presented in this book use a simplified version of the iconic notation typical of Unified Modeling Language (UML) steady-state diagrams. These diagrams present classes and their relationships. You have already seen the UML relationship notation. You also know a class is the encapsulation of software and the data it needs. We only want to model data. ArcGIS and the geodatabase take care of the software part.

Figure 2.10 **Object classes** This is the normal form of UML steady-state data models. Visibility defines the degree to which class attributes and methods are accessible by other classes. This standard symbology is modified to create the graphical standards for logical data models. The two primary changes were omitting methods, which the ArcObjects we might include in the database design already provide, and assuming that all attributes (properties) are open for anyone to see (public). All attributes added to geodatabase tables and feature classes are also public.

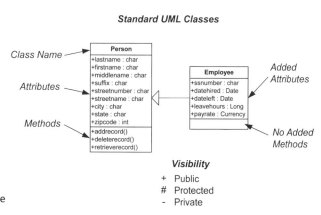

Our data model graphical standards represent a compromise between graphical consistency with UML and other ArcGIS documentation standards. UML is actually used by computer programmers to design their software. The notation is adapted here for use as a data modeling language.

Although it is common to do so, it is really a bit of a stretch and mismatch to use UML steady-state diagrams as a data model. UML is really for application design. Since the data and software that work on the data are tightly bound through encapsulation, steady-state diagrams do show the data, but with a limited view.

Normally, a class symbol is a rectangular box subdivided into three parts. The upper part of the class box holds the name of the class. The center part of the box holds the various class properties (data). The bottom part holds class methods (software). However, the bottom part is not required since ArcObjects classes already include the operations needed to implement a geodatabase. So, logical data models for geodatabase designs use a rectangular, two-part box that omits the bottom methods section. All class properties that you add will be public, so the visibility indicator is not required. Class properties can be referred to as attributes, fields, or columns, and such terms as class, table, and feature can be used to refer to the entities in logical data models.

OK, here is the truth: the other sidebar is lying. You do not actually add properties, in the UML sense of the word, to any ArcObjects class when you create a geodatabase. What you are really doing is using properties and methods that are already in the class. The ArcGIS user interface allows you to access those properties and methods by using wizards and other tools to customize classes so they serve your purposes, but you are not really changing those classes. UML steady-state diagrams are for creating software. You are not creating software; you are designing a geodatabase that is constructed of classes provided by ESRI and, perhaps, one or more of its business partners.

Think of it like using spreadsheet software. When you get started, all you see is a bunch of little boxes into which you can type numbers, text, and formulas. Anything that happens with the contents of a box is already in the spreadsheet software. You are not creating the spreadsheet software. It is the same with the geodatabase you are designing. The geodatabase is already in ArcGIS. You are just telling it which of its capabilities to use and what the inputs and outputs should be.

So, UML is really a poor way to create data models, but it is the one we have. If we were starting from scratch to create a new language, we could definitely think of something better than English, but then we would not have anyone to talk to. It's the same way with using UML to create data models. It is a language many people already understand.

In all cases, the name of a class that can be created will be shown in normal type, while the name of a class that is abstract in nature (a stereotype that cannot be created but serves as a template for defining a creatable class) will be shown in italics. Attribute names will be stated using a concatenated mnemonic name in a Roman sans serif font with an initial upper-case letter and intermediate capitals to assist in understanding, for example, FirstName.

Besides the entities, a logical data model includes relationships. Relationships are shown as a line with end symbology. For example, the UML diagram in figure 2.10 shows an inheritance relationship, which means that the Employee class is based on the Person class. By convention, UML only shows the new properties and methods added in the Employee class, with all the Person class properties and methods included by the inheritance relationship. Person is thus the parent stereotype or supertype of the Employee class.

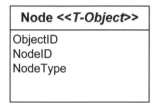

Figure 2.11 **Stereotypes** Inheritance is normally shown through an explicit relationship, but it can also be indicated by placing the name of the superclass in the class name space within double carets. This convention is normally used when the superclass and its inheritance relationship are not shown.

Sometimes the supertype class is not shown. By UML convention, the name of the superclass from which the class inherits its base attributes in the class name space can be included in the entity name space. Some classes may include subtypes listed below the normal class specification. All subtypes have the same attributes.

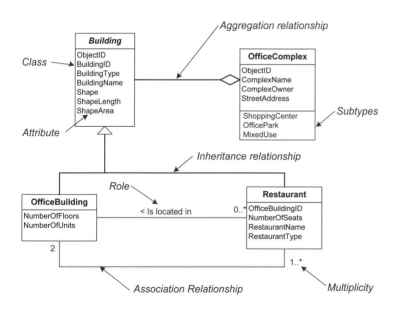

Figure 2.12 **Logical data model** A logical data model fleshes out the entities of the conceptual model by adding attributes and resolving many-to-many relationships. Since an ArcGIS geodatabase consists only of predefined classes with mandatory and user-defined attributes, methods will not be included. The result is an abbreviated form of the traditional UML notation and the usual reference to the class properties as attributes. Each ArcGIS class will determine the manner in which attributes are converted to properties.

Relationship notation continues unchanged from that of conceptual data models, but there are differences. One change you should notice is that there are no many-to-many relationships in a logical data model. They have to be resolved during the transition from conceptual to logical form. Otherwise, a logical data model will look much like a conceptual data model with attributes added. A logical data model may also include an enumeration of values that help express the domain of one or more attributes. An enumeration is an example list of values for a domain. The complete domain does not have to be specified until you create the physical data model. The enumeration may become a domain class in the physical data model.

Physical data models

The most complete version of your geodatabase design is the physical data model, which includes many of the bells and whistles a geodatabase can supply. Classes are more precisely specified, as are their attributes. As with the transition from conceptual model to logical model, changes in the design may occur as you construct or test the physical data model.

The core of any physical geodatabase model will be object and feature classes. Relationships may be implied by foreign keys or explicitly included as relationship classes. An implicit relationship is called a join relationship and represents association. Explicit relationships may include attributes or merely enforce cardinality rules.

Domain classes may be added to control data entry by limiting the available choices to a defined set. An enumeration of representative values included in a logical data model must be converted to a fully defined list of values for the physical data model if it is to be reflected in a domain class.

The geodatabase has rich capabilities that ease the transitional leap from a conceptual to a physical data model. In the past, you would have been required to break down entities in the conceptual model into component parts when you made the transition. The implementation environments for which the physical data model is designed required you to provide the behaviors and data structures necessary to express the full range of characteristics and actions embodied in each entity. For example, with a relational database implementation, you have to create lookup tables for domain control and manage association relationships with software you write. In contrast, the underlying geodatabase data model allows you to implement such behaviors by simply declaring the domain of valid values and stating the rules you want to enforce, all without writing any code at all. At the end of the day, the geodatabase classes you define in the physical data model for a geodatabase implementation look very much like the entities in the original conceptual data model.

Physical data models

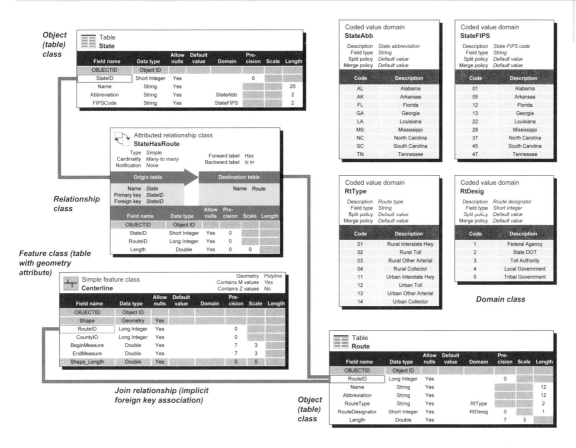

Figure 2.13 **Physical data model**   The physical data model exists to embrace the implementation environment and mold it to the form required by the logical data model. This example is for an ArcSDE geodatabase that includes road centerlines in a polyline feature class, plus states and the routes they contain in two tables. An attributed relationship class handles the many-to-many relationship between State and Route: a state can contain many routes and given route can traverse many states. Four coded-value domains have been included to manage data inputs.

The next chapter describes the geodatabase and how it works. It also presents some basic techniques of geodatabase design.

All the data models included in this book were created using Microsoft Visio, which is also supported by ArcGIS for loading database designs into ArcCatalog so as to create classes automatically. Instructions for how to do this are contained in the online ESRI Support Center.

There are many tools you can use to create data models. If you have a copy of Visio, you will notice that it contains many templates for software and database design following a wide variety of published "standards." Use the ones that work for you.

Notes

[1] A computer scientist will tell you that these structures are called relations, not tables, and they consist of tuples (rows) and vectors (columns). These terms were chosen to avoid the impression that the data is physically stored as rows and columns in a separate piece of the database. In most RDBMS implementations, all the records in all the tables are stored in one big file. From a database design perspective, it is much better to work with tables containing rows and columns than the more ephemeral concepts of tuples and vectors contained in a table space.

[2] Date, C.J. 1995. *An introduction to database systems*. Reading, MA: Addison-Wesley Publishing Co.

[3] See, e.g., Zeiler, Michael. 1999. *Modeling our world: The ESRI guide to geodatabase design,* pages 81 and 98-99. Redlands, CA: ESRI Press.

chapter three

Geodatabases

- The geodatabase
- The geodatabase framework
- The data dictionary view
- Geodatabase field types
- Attribute domains
- Valid value tables
- Subtypes
- Relationship classes
- Origin and destination tables
- Normalization
- Tracking events

This book shows how to build geodatabases that can be easily edited. In chapter 2, the geodatabase was shown to be an object-relational database that the user customized to provide the data and behavior desired. The object-oriented foundation of a geodatabase offers advantages, such as extra functionality, over a purely relational database. Many of the functions you would normally have to program, like domain checking and rule enforcement, are already part of the geodatabase.

Other geodatabase benefits that are especially useful for transportation datasets are:

- *The dataset is continuous.* In the past, traditional spatial data structures needed to subdivide their spatial data into tiles of relatively uniform density. Transportation datasets can be quite large. The geodatabase is continuous regardless of database size so a tiling scheme is unnecessary to efficiently manage data.

- *The data is uniformly managed.* Data with and without geometry can be managed in a single repository, assuring that both kinds of data are integrated and internally consistent. This book will show you how to asynchronously edit feature and object classes to preserve referential integrity.

- *Many users can edit the data simultaneously.* In larger transport agencies, several workgroups likely would be responsible for editing various portions of the enterprise dataset. The geodatabase can handle simultaneous editing cycles located at multiple sites.

- *Data entry quality assurance is built in.* One of the critical concerns for maintaining large datasets is assuring that only good data is put in. The geodatabase provides mechanisms to control what is entered into the dataset.

- *Users work with more intuitive data objects.* A properly designed geodatabase presents the user with multifaceted objects that combine spatial and nonspatial data into a single composite. Spatial features are no longer "dumb" points and lines but become roads, train stations, and bus stops. Nor is each type of object isolated from the others. By using an integrated geodatabase, you can maintain the relationships between objects. You can even use them during editing to specify what happens to other features when you move or modify a different feature.

Designing a geodatabase is a process of deciding which capabilities to use. You are not starting from a blank sheet of paper. The basic framework for an efficient enterprise design is already in place inside the geodatabase.

The geodatabase

The geodatabase is a collection of ArcObjects, but that simple definition does not really tell you what it does. ArcObjects interact with each other. If you establish a topological relationship between two lines, a change in one line can produce changes in the other line. Or you could create a connectivity rule that says a surface street may not connect directly to a limited-access highway, only to an exit ramp. You can control the direction of one-way streets or ensure that water always flows downhill. You can specify a particular map appearance for a

The geodatabase

 Tables
A geodatabase table contains objects listed as rows with a common set of columns. ArcGIS provides the real primary key, **ObjectID**, but most users will want to add their own candidate key to support row selection and table linking as part of the user-defined columns a table can contain.

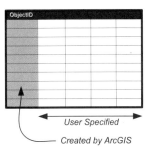

   Feature class
A feature class is a geodatabase table that contains a link to the geometric shape that represents the object. The **Shape** column, added by ArcGIS, stores the link to a geodatabase object that contains the actual geometry. Each class can contain one kind of geometry (point, multipoint, polyline, and polygon).

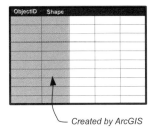

 Geometric network
A special geodatabase feature class that contains feature attributes, representative geometry, and network topology. Transportation networks consist of point features (junctions) and polyline features (edges), and are required for pathfinding applications. ArcGIS supports nonplanar networks.

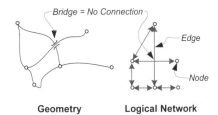

 Relationship classes
A relationship class implements association rules between features and tables. Relationship classes in the geodatabase participate in editing and versioning. There are two kinds of relationship classes. Simple relationships manage one-to-many relationships. Attributed relationships resolve many-to-many relationships. You can define attributes for both kinds. Relationships can also be established through the use of on-the-fly relates and joins, although these methods do not provide referential integrity or support for many-to-many cardinality relationships.

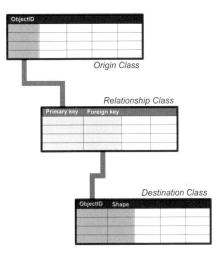

Figure 3.1 **ArcGIS geodatabase feature dataset** An ArcGIS geodatabase feature dataset can contain different class types. The more common types are discussed here.

given situation, such as label placement that does not overlap another feature. You can click on a feature and have a form appear. All of these functions are possible because the components of a geodatabase interact with each other and the user.

A geodatabase consists of feature datasets, domains, validation rules, raster datasets, TIN datasets, survey datasets, metadata documents, and locators. A feature dataset can have a spatial reference that is shared by all the feature classes in the dataset. Transportation networks, geometric networks, and planar topology rules can be in a feature dataset.

Transportation data users are most concerned with four geodatabase class types: tables, features, relationships, and transport networks. ArcGIS supplies a uniform appearance for all these class types regardless of the database management system environment you implement.

A table is a class for which you can specify attributes and behaviors to represent an entity in the real world. It looks just like a relational table. ArcGIS even creates a primary key field, called **OBJECTID**, for you. Of course, you can create another field that can also uniquely identify each row in the table in order to support association relationships.

> Although the **OBJECTID** field supplied by ArcGIS for all tables and feature classes is the primary key used by the software, it is not really a good foreign key for you to rely upon. The problem with **OBJECTID** is that it has values that can change as a result of certain database operations. For example, if you merge members of two tables into one, the **OBJECTID** values can change. Any foreign key relationships that you had constructed using **OBJECTID** no longer would work because those **OBJECTID** values are now gone or (worse) apply to other rows.
>
> You create a candidate primary key in each of the classes that you will want to link to another class through a foreign key, whether it is implicit or instantiated in a relationship class. Some ways to do that are presented later.

Most ArcGIS documentation describes a feature class as a table class with a shape column. But a feature class also has all the added software required to work with geometry.

A transportation network uses feature classes as the source for information to construct a logical topology of the system and its behaviors. For example, you can define sign features to guide users who perform pathfinding analyses, or turn tables to express the temporal restrictions of movement by particular vehicle types at junctions.

Relationship classes manage associations between classes, although you can also rely upon on-the-fly relates and joins to tie classes together using foreign keys. Relationship classes are an important part of many best practices for transportation database design.

The geodatabase framework

The balance of this chapter is one big rocket science sidebar that will take you "under the hood" to present the internal structure of the ArcGIS geodatabase. If you don't want to get your hands dirty, you can skip to the next chapter. However, if your job is to guide the data modeling process, create or implement the physical data model, or manage the geodatabase once it is created, then you are the mechanic others will rely upon and you will benefit from knowing how the geodatabase works. Of course, transportation types tend to like to take things apart to see how they work, so feel free to read this chapter for sheer pleasure, regardless of your role in a database development project.

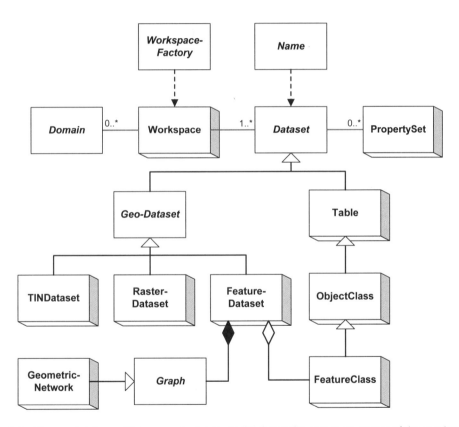

Figure 3.2 **The geodatabase** This conceptual data model shows the core components of the geodatabase support structure. The included relationships suggest the sequence of events that produce the cascading instantiation of the classes shown here. For example, a WorkspaceFactory class creates a Workspace class into which a Name class can instantiate a Dataset subclass. The figure also shows that the Table class is the mother of ObjectClass and grandmother of FeatureClass—the two primary components with which you will interact. Through this series of inheritance relationships, FeatureClass has all the interfaces of Dataset, Table, and ObjectClass.

This conceptual model (figure 3.2) shows the primary classes that comprise the geodatabase. Exploring this diagram will help you understand both how the geodatabase works and the model symbology we use.

Starting at the top, you can see that a *WorkspaceFactory* instantiates a Workspace. You invoke a workspace factory when you tell ArcGIS the directory, file name, and connection properties for the data you plan to store and use. There are many types of workspace factories, each intended to create a particular type of workspace, as determined by the database management system that applies. For example, there are workspace factories for shapefiles, personal geodatabases, enterprise geodatabases, and CAD (computer-aided design) file structures. A workspace is a geodatabase, a coverage workspace, or an operating system folder containing shapefiles.

The workspace factory creates the kind of workspace you need to manage data. There are three basic kinds of workspace: file systems (shapefiles and coverages); local databases (personal geodatabase); and remote databases (SQL Server and other RDBMS products). Each database management environment requires its own methods of communication and data operators. You use workspaces to create every other kind of class. For example, the CreateTable function of the IFeatureWorkspace interface creates a new table class when you supply the name of the table.

Domain classes are created at the workspace level and can be applied to user-supplied attribute values in any dataset that exists with the workspace. There are several domain types, each with its own creatable class. For now, we only show the abstract superclass to illustrate the central point of domains being tied to the workspace, not a particular dataset. The one-to-many cardinality for the association relationship between Workspace and *Domain* shows that no domain class is required.

A dataset is any kind of data container, and the *Dataset* class is the grand ancestor of all other data classes. *Dataset* is an abstract class that represents a named collection of data in a workspace. It provides such functions as copy, delete, and rename, and controls access to data classes based on user privileges, which is common to all data class types. When you initiate editing in a geodatabase, one thing that happens behind the scenes is that the IDatasetEdit interface of the *Dataset* class sets the IsBeingEdited property to True for each dataset that can participate in the editing session. The cardinality of the Workspace-*Dataset* relationship shows that a workspace must include at least one dataset.

PropertySet is exactly what it sounds like: a set of properties defined for each dataset. For example, it can contain the connection instructions for linking to a remote database where the dataset is stored. If you could look inside this class, you would see a list of two columns, with property names in one column and the values of those properties in the other column.

The sequence of classes on the right side of the figure—Table, ObjectClass, and FeatureClass—form the core used in geodatabase data models. The Table class provides the basic data storage mechanism of a relational table consisting of columns, referred to as fields, and an unordered collection of rows containing table members. You connect a relational table to the geodatabase through registration methods of the Table class. Since it is a child of *Dataset*, it inherits all the properties and methods of the parent class, such as knowing the name to be displayed in ArcGIS, the actual table name used in the RDBMS, and the location of the relational table in the database. The Table class uses the ITable interface to provide the methods required to read and write rows in the relational table.

ObjectClass adds the necessary properties and methods to turn a relational table into an entity in your geodatabase. For example, it knows about relationships in which the table may participate, and the domains, rules, and default values that apply to various fields. ObjectClass adds the OBJECTID primary key field to the relational table it represents. When you create a table in ArcGIS, you are customizing a copy of the ObjectClass.

You do the same thing to create a feature class, except you are working with a copy of the FeatureClass class. As you already know, a feature class is just like an object class except that its members include a geometry property and the class has methods to work with that geometry. FeatureClass contains everything that any kind of geometry will need, but ArcGIS is clever enough to show you only the parts that apply to the type of geometry you are actually using for that class. For example, the IFeatureClass interface includes properties for shape length and shape area, but ArcGIS will only show you those properties when they are appropriate for the geometry of that class. You will not, for instance, see the Shape_Length field for a class with point geometry in the data dictionary view of a physical geodatabase model.

Figure 3.2 also shows three kinds of geometry datasets in the lower-left quadrant. There are more kinds of such datasets than we include. The one we want to emphasize here is the feature dataset, which consists of a set of one or more feature classes and may include a geometric network. Notice the difference in these relationships. Deleting a feature dataset containing a geometric network also deletes the network, while the same action with a feature dataset composed of simple feature classes merely removes the grouping, not its members.

Figure 3.3 shows how the engine works. This is really hard-core stuff, so you can skip it if you are not interested. We offer this illustration to demonstrate that such simple things as a single relational table require a large number of ArcObjects classes to support the functionality offered by the geodatabase.

ArcGIS provides five classes that help the Table class keep up with the data it manages. A table consists of columns, called fields, and unordered rows of data. All of these things are not stuffed into the Table class but are spread out over supporting classes. A Fields class stores the total number of fields in the table. Each such field is described in a Field class member. The OBJECTID field is created by ArcGIS to provide a primary key. If a feature class uses the table, one of the fields will define a type of geometry. Users can specify that an index be created on other fields. An index is always created on OBJECTID. An Indexes class manages all the indexes that exist for a table.

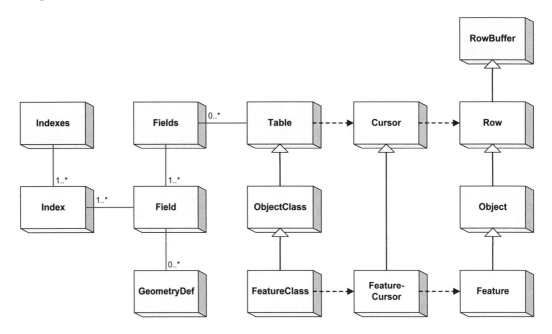

Figure 3.3 **Table classes**   In ArcGIS, a large collection of classes performs a variety of database management system functions. Classes to the left of Table store information about the format of class attributes added by the user and manages any related indexes. If a user-defined class includes geometry, the GeometryDef class stores information about the kind of geometry to expect. Classes to the right of Table allow data rows to be created and retrieved. In the geodatabase, a Table class is a way to manage a collection of Row classes. A Feature-Cursor additionally knows how to read and create rows containing a geometry attribute, but it is the Field "gene" it inherited from its Table grandmother that tells it about the geometry. We visualize and interact with object and feature classes as simple rows in a table but they are actually much more.

The Table class does not directly interact with the data container, whether it is a shapefile or an ArcSDE instance managing a relational database. The Table class works through cursors. A cursor is used to manipulate records in a file or relational table by undertaking such tasks as creating, retrieving, and deleting rows in the host database environment. The Table class instantiates the Cursor class to work with instances of the Row class. In the same way, the FeatureClass object works through a FeatureCursor class to access Feature instances. Both paths point to functions supplied by the Row class, which includes the interfaces needed to change, compare, and edit records in the data table, and to report errors. All rows in a table have the same fields. The Table class uses the Field classes to help it understand what is in each row.

The data dictionary view

ArcGIS provides the means to view a table or feature class using the physical data model perspective. This is called the data dictionary view because it displays the contents of each table or feature class. You can see the name for the class and its type in the header. If it is a feature class, you will also be able to see the type of geometry used and whether it contains measure (M) and elevation (Z) values.

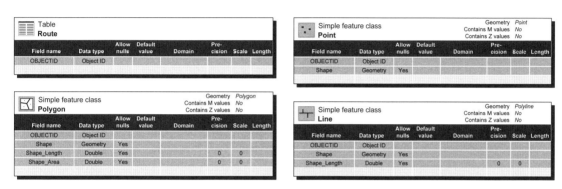

Figure 3.4 **Tables and feature classes** A geodatabase is seen as a collection of tables and feature classes. These examples show the minimal attributes such classes will contain. You may add many other attributes to any of these classes, but you cannot delete the mandatory attributes.

Below the header is a listing of each field in the class's table. The first field will always be OBJECTID. If it is a feature class, you will also see a Shape field and, depending on the type of feature, Shape_Length and Shape_Area fields. ArcGIS adds these fields and you cannot delete them or change their parameters.

When you create a field, you must specify a data type, whether null values are allowed, a default value that will be entered if the user makes no specific entry, the name of a domain to control data entries, the precision and scale of numeric fields, and the length of a string field. The user can modify field parameters that are not set to a dark gray background. At minimum, you must specify the name and data type for each field, and the length of any text field. The default value of 0 for numeric precision and scale will produce the standard form of that data type for the selected database management system. Indeed, precision and scale parameters you enter will only be used if the database management system uses ArcSDE. As we show below, your parameters for numeric fields may be only advisory, as ArcGIS can change them to optimize performance in the database management system you are using.

Chapter 3: Geodatabases

Table **Sign**							
Field name	Data type	Allow nulls	Default value	Domain	Precision	Scale	Length
OBJECTID	Object ID						
SegmentID	Long Integer	Yes			0		
SignID	Long Integer	Yes			0		
AtMeasure	Double	Yes			7	3	
MUTCDCode	String	Yes					8
Width	Short Integer	Yes			0		
Height	Short Integer	Yes			0		
Color	String	Yes		Colors			20

Simple feature class **Sign**					Geometry	Point	
					Contains M values	Yes	
					Contains Z values	No	
Field name	Data type	Allow nulls	Default value	Domain	Precision	Scale	Length
OBJECTID	Object ID						
Shape	Geometry	Yes					
SegmentID	Long Integer	Yes			0		
AtMeasure	Double	Yes			7	3	
MUTCDCode	String	Yes					8
Width	Short Integer	Yes			0		
Height	Short Integer	Yes			0		
Color	String	Yes		Colors			20

Figure 3.5 **The data dictionary view** This view reveals the descriptive information stored in the Table and Field classes. Each Field class instance is shown as a row in the data dictionary view, with the attributes of each field shown in columns. The darker gray boxes represent field attributes that cannot be changed by the user because ArcGIS controls the value that goes there or the attribute does not apply to that field. Both the table and feature class versions show the same set of field definitions, except that the feature class has an additional field called **Shape**. The feature class also has information in the upper right corner that reveals what GeometryDef knows about the class, which is also reflected in the class description and icon in the upper left corner.

Geodatabase field types

You will need to select a data type for each field you add to a table in an object or feature class. This section will discuss the most common types we will use in transportation data modeling: text (defined as a string of characters of a given maximum length), a variety of number formats, date/time, globally unique identifier (GUID), and binary large object (BLOB).

Text. A field that contains text data consists of a string of alphanumeric characters. ArcGIS uses the Unicode format to store text to facilitate the multiple languages in the geodatabase. You use text for fields that contain letters, numbers with leading zeroes, and any data where the content cannot be clearly anticipated. Once you select the String data type, you will need to specify the maximum number of characters permitted for an entry. You can control the content of a text field by using a coded-value domain when the possible values to be entered are well known and of a fairly limited number of choices. The field properties of precision and scale cannot be entered as they apply only to numbers.

Short integers. The 2 bytes needed for a field using the short-integer data type is the fewest number of bytes of any number format. The actual range of numbers available in 2 bytes depends on the database management system you are using. In most database environments, this data type can hold whole numbers in the range of -32,768 to +32,767. If you use small integer numbers to represent values in a list of choices, you can control data entry for a short integer field using a coded-value domain or a range domain. The default precision for a short integer is zero; you can supply another value as a means of ensuring that numbers entered do not exceed a general range. The valid choices depend on the database management system you are using. Oracle has a maximum precision of 4, while DB2 and Informix set a single choice for precision of 5. You must leave a digit for a negative sign but not a positive sign. You cannot enter a value for the scale parameter.

Long integers. A long-integer field uses 4 bytes to store the number. The allowable range for long integers is a whole number in the general range of -2 billion to +2 billion. A long integer has a precision of 5 to 10. As with short integers, you must include a digit to accommodate any negative sign that may be permitted. You cannot set the scale parameter.

Floating-point numbers (single-precision numbers). The Float data type is the smallest numeric format that can include fractions. Like the long-integer data type, a floating-point number field requires 4 bytes of storage. You can set both precision and scale parameters for a floating-point number field. For example, if the largest value you anticipate being entered in a floating-point field is 999.99, then you would set the precision to 5 and the scale to 2. If a user tried to enter a number with three digits after the decimal, ArcGIS would display an error message. A floating-point number can accommodate a precision of up to 6 and a scale of up to 6, with scale always being equal to or less than precision. The maximum precision for any Float number type in a file data structure or a personal geodatabase is 6. You do not need to include a digit for the decimal point when determining the precision parameter, but you do for any permitted negative sign.

Double-precision floating-point numbers. The largest numeric format in terms of storage requirements is the double-precision number, which needs 8 bytes. You can define any precision of 7 or more and any scale value. You can put some big numbers in a double-precision field. There are, however, certain limits imposed by the database management system. The practical limit for an ArcSDE database is 15 digits. You can put larger numbers into a Double field, but it will be rounded and stored in scientific notation.

As we noted earlier, ArcGIS may change the parameters of a numeric field from those you entered. For example, if you specify a data type of Double but specify a precision of 6 or less, ArcGIS will create the field as a single-precision floating-point number. Any numeric field with a precision of 10 or less and a scale of 0 should be an integer data type. Even if you expect only whole numbers to be entered into a field, any precision greater than 10 should be defined as a Double data type. Using a precision of 0 and a scale of 0 will cause ArcGIS to try to create a binary field of the type selected.

Date and time. ArcGIS uses a single data type to store date and time information, regardless of how the underlying database management system deals with dates. The format is mm/dd/yyyy hh:mm:ss.sss, with an indication as to whether the time is a.m. or p.m. You do not have to use the information stored in both parts of the Date data type, but they will be included.

Table NextID							
Field name	Data type	Allow nulls	Default value	Domain	Precision	Scale	Length
OBJECTID	Object ID						
NextFacilityID	Long Integer	Yes			0		
NextElementID	Long Integer	Yes			0		
NextCharID	Long Integer	Yes			0		
NextRelateID	Long Integer	Yes			0		
NextFeatureID	Long Integer	Yes			0		
NextEventID	Long Integer	Yes			0		
NextConveyanceID	Long Integer	Yes			0		

Next class instance identifier

Next facility object identifier
Next element object identifier
Next facility characteristic object identifier
Next relationship object identifier
Next feature identifier
Next event object identifier
Next conveyance object identifier

Supplier	ID Range
Agency A	0 – 29999
Agency B	30000 – 49999
Agency C	50000 – 63999

Agency Identifier — Sequence Number

12345-67890

Object Identifier

Figure 3.6 **Creating unique identifiers** To generate globally unique object and feature identifiers, you usually create them in a structured way. If your agency has a centralized database editing approach, then all you need is a NextID Table. Every time you create a new object or feature class instance, you read the appropriate value, add 1 to the value you read, and write the new identifier value for the next created instance to use. If you participate in a decentralized editing process, where multiple data sources may combine data, you will benefit from one of two additional actions. Should there be only a few editing agencies, you can start each NextID Table with different seed numbers so that each has a single numeric range to work with—one big enough to never extend into another agency's range. If there are several data suppliers, then you can use an identifier that consists of two parts: an agency identifier and the sequence number generated by the NextID Table approach. The latter option was proposed in the FGDC's Transportation Feature Identification Standard.

Globally unique identifier. If you want to ensure that the row identifier is always unique across the geodatabase, you can create a field of the GUID type. A GUID consists of 36 characters enclosed in curly brackets {} and is designed to be unique in the world. (When we said it was globally unique, we meant it!) ArcGIS uses GUID fields internally to manage feature replication. You can use the same technique when you expect to share data with outside users and want to ensure that your feature identifiers do not conflict with other users'. If your database management system supports the GUID data type, ArcGIS will only need 16 bytes to store the field's value. Otherwise, 38 bytes will be required.

Binary large object. A BLOB is just a big box in which to put data of any structure. It can be a scanned document image, or a video of a crash scene. It can even contain computer software. Whatever you put in a BLOB field is yours to manage. ArcGIS can put it in the field and take it out, but you must write or acquire the software needed to understand what a field with this data type contains.

Attribute domains

An attribute domain is a rule that limits users' data entries to a specific set of valid choices. A domain is a declaration of acceptable values for a field. Data quality is improved if you can keep obviously bad data out. Someone may still make a bad choice, but at least it is not something completely unexpected.

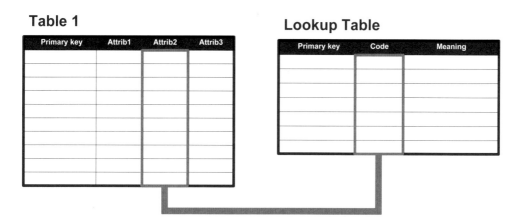

Figure 3.7 **Domain control** To ensure that only valid data gets into the database, you must control data entry. A common approach is to convert relatively limited value domains to a list of valid choices, each of which is represented by an abbreviation, called a code. The coded value can be put in a data table as a foreign key that points to a "lookup table" that stores what the value really means. (The code attribute is a candidate key, so it meets the requirements for this one-to-many relationship.) This operation requires software that can check an entry against the legitimate values stored in the look-up table. **Attrib2** contains foreign key values tied to **Lookup Table : Code**. Every person who writes a computer program to utilize the same lookup table must provide the same code.

A lookup table is the traditional way to control a domain consisting of a finite list of choices. You create a column in the entity attribute table that includes a foreign key to the lookup table that contains the meaning of various coded values. Your data entry application builds a pick list from the lookup table's rows, and users pick one of the available selections. This process requires you to create a table for each domain you want to control and write the software needed to get the input screen to go to the lookup table for the values of each field it controls.

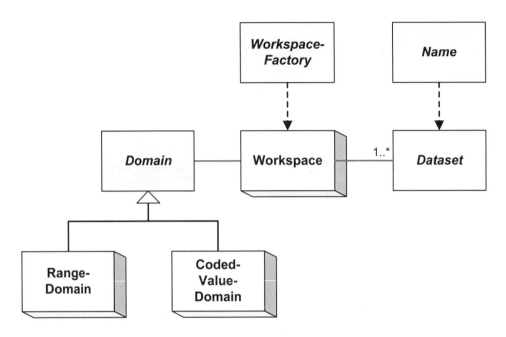

Figure 3.8 **Geodatabase domain control classes** The geodatabase supports two domain control methods: numeric range and coded values. Domains are named and apply to an entire workspace, even across datasets within that workspace.

The geodatabase greatly reduces your workload when using attribute domains. You need only tell ArcGIS what you want to do and it takes care of the work. The earlier geodatabase conceptual class model showed the abstract *Domain* class. Figures 3.8 and 3.9 show the details. There are two specializations of the *Domain* class, one for numeric range domains and one for coded-value domains. Using the domain control function in ArcGIS is fairly simple.

First, specify a domain name. You also can enter a small description for what the domain contains. You will enter the domain name in the data dictionary row describing a field each time you use this domain.

Next, specify the data type, which must be the same as any field to which you apply the domain. You can use text, short integer, long integer, float, double, and date data types in a domain class.

Then tell ArcGIS which kind of domain you want to use: numeric range or coded value. For a numeric range domain, specify the minimum and maximum values. For a coded-value domain, list the code and (optionally) its description. The data type of the code value must match that of the domain class.

The last thing you tell ArcGIS is the split and merge policies to employ. A split policy will control what value is entered in the new child objects when the parent is split. Specify one

Attribute domains

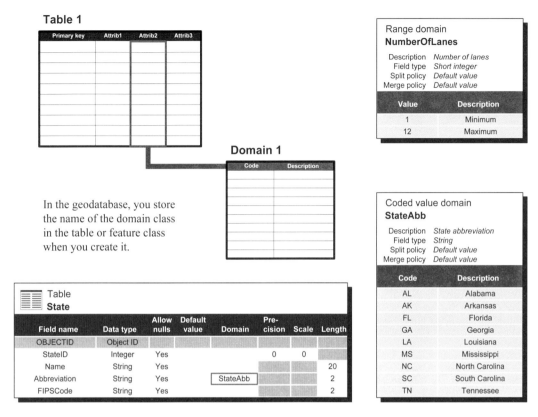

Figure 3.9 **Domain control in the geodatabase** ArcGIS saves you all the programming required for domain control using lookup tables by supplying the required software in the geodatabase itself. Coded value and numeric range domains can be defined for a geodatabase workspace. Since these domain controls are established at such a high level in the geodatabase, they will be applied globally across the dataset of tables and feature classes by reference to the domain's name.

of three choices: default value, duplicate, or geometry ratio. The default value policy means that the default value for the field specified in the data dictionary will be placed in both child objects. The general expectation of this policy is that the user will need to enter an appropriate value. The duplicate policy will place the parent's value in both child objects. The geometry ratio policy will proportionately divide the parent value among the child objects. This option only applies to feature classes. It is useful for such attributes as area, where each child feature will get the quantity determined by its size as a percentage of the parent.

The merge policy you specify will control what happens for the field controlled by the domain class when you put two features together. Merge policies are similar: default value, sum values, and geometry weighted. The default value policy works the same as for a split; the default value defined for the field is placed in the resulting combined feature. The sum values policy will add the value in this field for the two contributing features and place the total in the field for the resulting feature. The geometry weighted policy will take the weighted average of the values supplied by the two contributing features based on the ratio of the two features' geometry. For lines, the weight is determined by relative length. For polygons, the weight is determined by relative area.

Valid value tables

Sometimes using a domain class is not enough to reflect the rich mix of attribute controls you need to employ. In such cases, you can use a structure called the valid value table. Such a table includes rows that reflect the combinations of values that are legitimate.

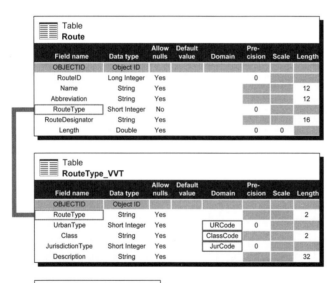
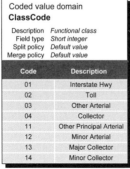
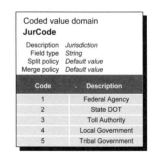

Figure 3.10 **Valid value tables**
In this illustration, we need to control **RouteType**, which is based on a combination of urban type, functional class, and jurisdiction type. Each of these three variables is separately controlled by a coded value domain, but only certain combinations are valid. For instance, "rural minor collector" is valid but "urban minor collector" is not. The RouteType_VVT table stores the legitimate combinations and can serve to control Route table entries for the **RouteType** column using custom tools.

The example in figure 3.10 is one you have may have faced in dealing with the domain of functional classes used by the Highway Performance Monitoring System (HPMS). In some states, functional classification determines jurisdiction. Higher functional classes—those that emphasize mobility over land access—are assigned to the state DOT and lower functional classes to various levels of local government. An urban type field makes the example more interesting.

Domain classes can be used to control the values placed in each of these three fields individually, but cannot control combinations of the three fields. For example, you may want to prohibit a combination consisting of interstate highway, rural land use, and local government jurisdiction. There are only certain combinations that are allowable for urban type, functional class, and jurisdiction. For example, the route type representing a combination of rural interstate highway and city jurisdiction may not be valid. The valid route types are contained in the rows of the RouteType_VVT table. By linking the RouteType field in the Route table to the same field in the RouteType_VVT table, you can check the data through an on-the-fly join to see if all the entries are a valid route type.

Subtypes

There will be instances when you need to use subtypes for a class. For example, you may want to use different line symbology for interstate highways and surface streets. But you can also use subtypes to specify a different domain or default value for a field.

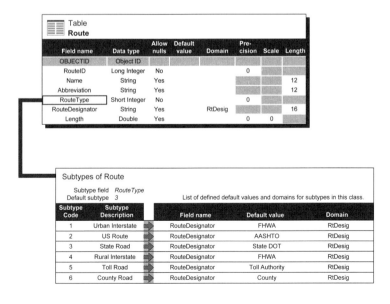

Figure 3.11 **Subtypes**
Feature class subtypes can be used to control object behavior. All subtypes share the same attributes and geometry type, but can employ different domains, default values, and rules. Subtypes can improve performance because the speed of most drawing functions is determined by the number of feature classes. Here, we use functional class subtypes to control the default value for **RouteDesignator**.

In figure 3.11, we use subtypes for a Route table based on the value of the RouteType field. Each subtype will have a different default value for the RouteDesignator field. This field stores the value representing the entity that can designate a route of that type. For example, the U.S. Route system is managed by the American Association of State Highway and Transportation Officials (AASHTO).[1] When we select a route type for a given row in the Route table, ArcGIS will insert the corresponding route designator value, which saves us a little work. Do this for a whole state, and the ArcGIS subtype function has saved us a lot of work.

Relationship classes

If the workspace and dataset ArcObjects structure is the geodatabase's engine, the relationship portion of the structure is the transmission. ArcGIS supports several kinds of association relationships with explicit geodatabase classes. The conceptual ArcObjects class model shown in figure 3.12 is somewhat misleading. By one classification method, ArcGIS supports simple and composite relationships, but there is no class called "CompositeRelationship"; only a SimpleRelationship class is included. A different classification method would split relationships into those with attributes and those without, but, again, the model only shows an AttributedRelationship class.

The AttibutedRelationship and SimpleRelationship classes are specializations of the abstract *Relationship* class. Notice that the *Relationship* class is related through association to the Object class. You know from the earlier examination of the geodatabase's "engine" that an object is like a row in a table in that it represents one member in an object class. What the association relationship says is that there is one *Relationship* class entry for each row in each table or feature class involved in a relationship. Through its composition relationship with SimpleRelationship, you can see that a RelationshipClass instance contains all the pointers that tie members of one table or feature class to members of another table or feature class. Attributed relationships work the same way, only they actually include a table to store the additional attributes you can define for such a relationship. (Most attributed relationships are used to resolve many-to-many relationships.) Simple relationships without attributes can be of the one-to-one or one-to-many type. Simple relationships of the many-to-many type require an attributed relationship.

So, where are composite relationships? A composite relationship class is one that enforces the composition relationship. A composite relationship is always of the one-to-many type. In a simple relationship, the two related classes are independent of each other. Deleting a row in the destination table has no impact on the origin table. Deleting a row in the origin table will replace the corresponding foreign key field in the affected rows of the destination table with a null value.

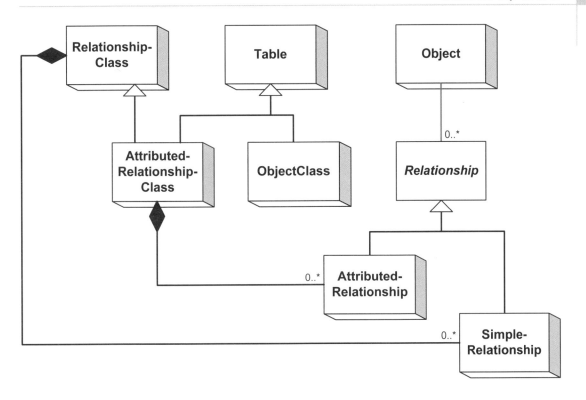

Figure 3.12 **Relationship classes** The SimpleRelationship class is related to an Object class instance. This association relationship reveals that each row in a table is tied to one or more Relationship class instances when a relationship has been formally established in the geodatabase between two tables, a table and a feature class, or two feature classes. The figure shows that each end in a cardinality relationship has its own *Relationship* class instance, because the relationship between Object and *Relationship* is one-to-many. This cardinality is required to support messaging between classes through an explicit relationship and to apply multiplicity rules, among other things. An attributed relationship includes a table to store additional attributes that go beyond merely establishing the relationship of one row to another.

If, however, you have defined a composite relationship, origin table objects are closely tied to their components in the destination table. Deleting a row in the origin table will delete the corresponding rows in the destination table. This process is called a cascade delete.

You can also go the other way in a composite relationship. The composition dependency rule can be enforced by the ArcMap Validate Features command you can run in an edit session to test referential integrity. This check will identify any instance where you create a destination table row (component) without linking it to the class of which it was a component (origin table).

Chapter 3: Geodatabases

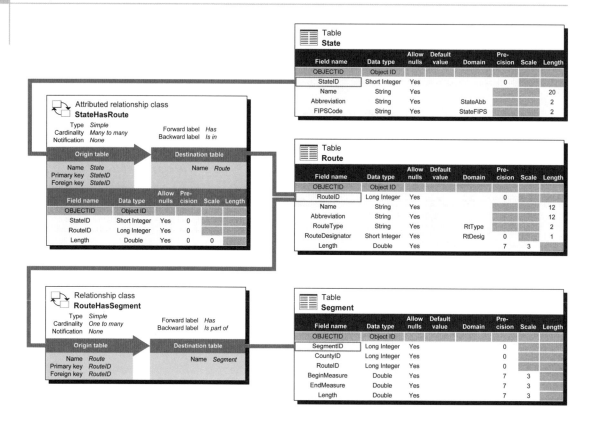

Figure 3.13 **Geodatabase relationship classes** ArcGIS offers two basic relationship classes, the difference based on whether the relationship itself has attributes. An attributed relationship is normally used to resolve a many-to-many relationship, which requires the class to store foreign keys pointing to each involved class's members. Both relationship classes support simple and composite types, and require specification of origin and destination tables and the foreign key field. A composite relationship class enforces the one-to-many composition relationship, which means that deleting an origin class row will result in a cascading deletion of all related destination class rows.

The difference between relationship and attributed relationship classes is that the data dictionary view of an attributed relationship will reveal the table it contains. The type of relationship is shown at the top in the data dictionary view. You will also notice that two field parameters are missing from an attributed relationship table: default value and domain.

Each relationship class has a name. You should use a name that tells you what the relationship represents. Below the name is a set of three relationship parameters. Type can be either simple or composite. Cardinality for a simple relationship can be one to one, one to many, or many to many. A composite relationship can only have the cardinality of one to many. The notification parameter tells ArcGIS whether it needs to send a message to a class to make an update triggered by a change in the other class. Notification sets the direction for messages between classes. A forward notification goes from the origin class to the destination class. A backward notification goes from the destination class to the origin class. You can also use "Both" and "None" as values for the notification parameter.

Although ArcGIS handles the notification, you will need to supply the software that understands what to do with the message when it is received by a class that does not participate in a composite relationship. For a composite relationship, ArcGIS provides referential integrity behaviors. Deleting an origin class will delete the destination class regardless of the notification setting. Forward notification of an action that moves or rotates a member of the origin class will move or rotate the related members of the destination class. Using "Both" as a notification setting has the same effect.

You should be aware of a few limitations on relationship classes. First, a given class cannot be the destination of more than one composition relationship. Second, if a composition relationship exists between two classes, you cannot establish any other relationship between those same classes. Third, relationship classes established between members of a geometric network can produce unexpected results during editing because both the relationship class and the requirements of a geometric network are trying to control referential integrity.

Recursive relationships where members of a class have relationships with each other are not currently supported by ArcGIS. Neither are complex primary and foreign keys. You can only use one field to define a primary or foreign key for a relationship. Custom software can be written to use complex keys. Lastly, if you put a relationship class within a feature dataset, at least one member of the relationship must be in the feature dataset.

By the way, you can have more than one simple relationship between the same two classes. This technique can be used to enforce cardinalities that differ depending on the situation. For example, the multiplicity for a node to the various links it may terminate is one to many with a minimum value of 1, but the multiplicity of a link to its terminal nodes is always 2.

Origin and destination tables

We have used the terms origin and destination to refer to the two classes participating in a relationship. It is very important that you know which is which, because it is easy to get it backward. You could lose data as a result of an identification error.

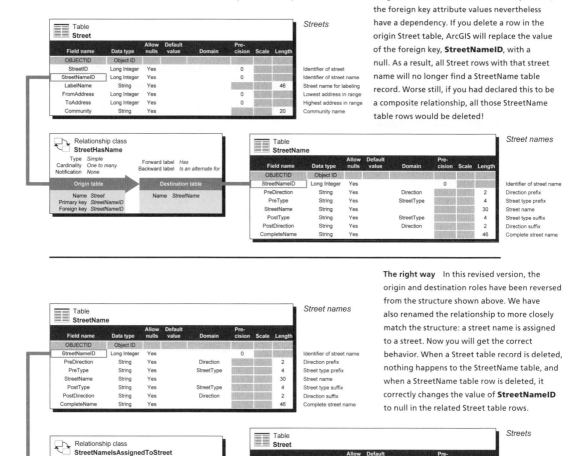

Figure 3.14 **Origin and destination tables** It may seem logical to treat the more important class as the origin, but that is usually the wrong thing to do. In this example, we seek to establish a relationship between a street class and a street name class. We have separated the names from the streets because we have a mixed-jurisdiction database that has several duplicated street names—sometimes with overlapping address ranges. The Street table is a valid combination of street name, address range, and community name. The address range and community name are stored directly in the Street table, but the street names are in a separate table. We need to create a relationship class to manage the one-to-many relationship between the Street and StreetName tables. Here are two approaches, one wrong, the other right.

In a one-to-many relationship, you will normally want the origin to be at the "one" end of the relationship. The top part of figure 3.14 shows a StreetHasName relationship between a Street table and a StreetName table. While it is certainly true that a street has a street name, this relationship is backward. The result of deleting a row in the Street (origin) table will be that the value of StreetNameID is set to null in the StreetName (destination) table. Suddenly, all Street table rows that pointed to that row no longer have a legitimate foreign key and the name disappears from all the streets that used it.

The lower part of figure 3.14 shows the right way to characterize this relationship as StreetNameIsAssignedToStreet. Now, if you delete a row in the Street able, nothing at all happens in the StreetName table. Of course, if you delete a row in the StreetName table, all the rows in the Street table that pointed to that street name no longer have a legitimate foreign key, but that is what needs to happen.

Other geodatabase classes for transportation data models

There are other geodatabase classes with specialized applications that will be discussed in later chapters. These classes include:

- Junction feature classes and edge feature classes that form a geometric network
- Turn feature classes that allow you to modify the default behavior of a transportation network

Because an understanding of these classes is necessary to use them in a data model, they are omitted from this "sidebar" chapter. Geometric networks are discussed in chapter 5. Transportation networks supported by the ArcGIS Network Analyst extension are covered in chapter 13.

Normalization

The last topic in this chapter is perhaps the toughest of all. Geodatabases that support data maintenance benefit from normalization, which is the process of removing data redundancies and dependencies. A database with data redundancies and dependencies is denormalized. You have already seen examples of normalization, such as the conversion of a many-to-many relationship to a pair of one-to-many relationships. The use of foreign keys is also part of normalization. This section will describe the breadth of normalization and why it is important.

Each table included in a geodatabase has a primary key, OBJECTID, which is used to uniquely identify each row by unambiguously identifying each member of the class. You cannot rely on record numbers as with sequential flat files or the row numbers that appear in spreadsheets because there is no ordering of rows in a relational database. Typically, each new or modified row is added to the end of the table. No two rows can have the same primary key value. Primary keys can be simple (one field) or complex (multiple fields), but the result must always be unique; no other row can have the same value. You do not want to use a manually entered attribute as a primary key. People can make mistakes and enter the same primary key value twice. You will want to use the computer to create the values in your primary key through such RDBMS functions as integer counters and date/time functions.

In addition to a primary key, a relational table can include one or more foreign keys. A foreign key in one table points to a related record in another table. It does this by having the same values in both tables. Say, for example, that you have a roadway inventory that includes a functional classification column. Rather than store the full name for each functional class, you will likely store the two-digit coded value in the inventory table, and put the coded values and their meanings in a functional class domain. Functional class is a foreign key. There will be lots of road segments with any given functional class, but each segment record must contain only one. Therefore, this is a one-to-many relationship: one functional class to many roads.

By the way, a foreign key is always a candidate primary key. This means that you do not have to pick the real primary key to enforce the relationship, just a column that could serve as the primary key. (The primary key *must* be unique. You cannot take any chances there.) Using functional class as an example, the coded value assigned to each class is unique to each record, which makes it a candidate primary key. You could store either the actual primary key or the coded value in the roadway inventory table. You will usually choose to store the coded value, as it provides information to the knowledgeable user without having to go to the functional class domain to see what it means.

Most spatial databases are normalized to the extent that they store information about discrete features, such as a parcel, a house, or a lake. Transportation data modelers are not so lucky. Linear transportation facilities are not really discrete. Agencies impose various segmentation methods to make linear facilities discrete, but in the process create denormalized data structures, such as when a street is subdivided into block-length segments. Denormalized data structures are fine, almost required, for mapping, but they are very difficult to maintain if you are trying to store a lot of data. The problem is data redundancy and the resulting need to change the same piece of data in several places. If the entire street has an asphalt surface, then you will need to repeat that information for each segment.

Data dependencies and redundancies are inefficient, but data redundancies can be downright dangerous, as they can undermine database integrity. Try to put the same information in more than one place, and you will likely find those places have different information. Put a road's speed limit in several places, and you must search for places where the speed limit changes. You also have to change them all at once. Otherwise, some users will get different answers when they read the record. Normalization is a big deal with large datasets.

Normalization eliminates redundancies and dependencies so that each piece of data is in only one place for editing. There are five cumulative forms of normalization:

- First Order (First Normal Form, or 1NF) = Attribute domains consist of only scalar values (field contains only a single value) and each row has the same columns
- Second Order (2NF) = 1NF plus every nonkey attribute is irreducibly dependent on the primary key
- Third Order (3NF) = 2NF plus every nonkey attribute is nontransitively dependent on the primary key
- Fourth Order (4NF) = 3NF plus cannot contain two or more independent multivalued facts
- Fifth Order (5NF) = 4NF plus a symmetry constraint

Do not worry about trying to figure out what all this jargon means. An example will make it simple. If you are a computer scientist, you can skip ahead to the end of the chapter.

The example involves an employee database for a national company with offices in multiple cities. The primary key is EmployeeID. The Employee table stores information about each employee, such as where they work.

It is your job to ensure a normalized database design for the Employee table. First Normal Form (1NF) is easy to do. You just have to make sure that each row includes the same columns, and that all the columns contain only scalar values. A scalar value is one with a single value that is atomic, or indivisible. It means here that you cannot just pile a bunch of different people into one row. You get one employee ID, one department, one building, and one city for each employee in the table. There are two skill fields, but they are scalar. 1NF: Mission accomplished.

employeeID	skill1	skill2	dept	building	city
101	typing		shipping	Times	New York
104	analysis	typing	personnel	Wilson	Chicago
102	computer	presentation	sales	Wilson	Chicago
106	forklift	typing	shipping	Times	New York

Table 3.1a **Employee table**

The Employee table is now in 1NF. The next step, to reach 2NF, removes nonkey dependencies, which means you must eliminate fields that depend on each other rather than the thing you are talking about. The Employee table talks about employees. Reaching 2NF means removing fields from the table that do not relate to an employee. In this case, the city in which the building is located is determined by the building, not the employee. (One city to a building.) So, you can create a lookup table for buildings that lists the building name and the city it is in.

employeeID	skill1	skill2	dept	building
101	typing		shipping	Times
104	analysis	typing	personnel	Wilson
102	computer	presentation	sales	Wilson
106	forklift	typing	shipping	Times

Table 3.1b **Employee table**

buildingID	building	city
1	Times	New York
2	Wilson	Chicago

Table 3.1c **Building table**

Achieving 3NF requires more work. The City field is an attribute that is unique to each building; i.e., building determines city. It turns out that each department is contained in one and only one building, which makes building dependent on department, reducing city to a secondary dependency. Reaching 3NF involves eliminating secondary dependencies, which is where the value for one attribute depends on an attribute that is a nonkey dependency. The Building field has a nonkey dependency in that the value of Dept, not EmployeeID, determines it. A Department table needs to be added to the design.

employeeID	skill1	skill2	dept
101	typing		shipping
104	analysis	typing	personnel
102	computer	presentation	sales
106	forklift	typing	shipping

Table 3.1d **Employee table**

deptID	dept	building
1	shipping	Times
2	personnel	Wilson
3	sales	Wilson

Table 3.1e **Department table**

buildingID	building	city
1	Times	New York
2	Wilson	Chicago

Table 3.1f **Building table**

Incidentally, you cannot simply turn the Building table into a Department table. That would cause the Building table to violate 3NF. The City field would have a secondary dependency on DeptID.

4NF removes redundant columns, like the two employee skill columns in the Employee table. The original design with two skill columns would force you to look in both columns to see if an employee had a particular skill. You need to work smarter, not harder, and move employee skills to another table where you can have multiple records for each employee, one row for each skill an employee possesses. The database design shown below is in 4NF. You now have four tables, one for assigning employees to departments, one for identifying the building in which a department is located, one for saying where a building is located, and one for describing the special skills each employee possesses. The database design is now fully normalized because there are no redundancies or secondary dependencies.

employeeID	dept
101	shipping
104	personnel
102	sales
106	shipping

Table 3.1g **Employee table**

deptID	dept	building
1	shipping	Times
2	personnel	Wilson
3	sales	Wilson

Table 3.1h **Department table**

buildingID	building	city
1	Times	New York
2	Wilson	Chicago

Table 3.1i **Building table**

empSkillID	employeeID	skill
1	102	typing
2	104	analysis
3	104	typing
4	102	computer
5	106	forklift
6	102	presentation
7	106	typing

Table 3.1j **EmployeeSkill table**

Normalization puts each piece of information in one place. If you move the Personnel Department to a new building in Chicago, updating the entire list of employee locations requires only a single change in the Department table. With the original design, you would have needed to change the record for each employee in that department.

Linear transportation facilities are segmented to provide a one-to-one relationship between attribute records and geometry. Such a segmented transportation database is not normalized in that changing a single attribute will often require that multiple records be updated. Consider the following example Street table rows, where SegmentID is the primary key.

segmentID	name	low_add	high_add	speed	lanes	pave_cond
1038	Sesame St	200	299	25	2	good
1074	Sesame St	300	399	25	2	good
1153	Sesame St	400	499	25	4	good
1076	Sesame St	500	599	30	4	fair

Table 3.2a **STREET table**

There are four rows with information about Sesame Street, which is segmented according to address block. The STREET table has redundancies. The problem this time is with rows, not columns. There are multiple rows with exactly the same values in the Name, Speed, Lanes, and Pave_Cond columns. There must be only one row for each unique piece of information for the database to be normalized. The values in Speed, Lanes, and Pave_Cond are dependent on street name and address range; i.e., they are nonkey dependencies.

Normalization

Normalizing this database will require creating separate tables for each linear attribute that may span multiple street segments.

segmentID	name	low_add	high_add
1038	Sesame St	200	299
1074	Sesame St	300	399
1153	Sesame St	400	499
1076	Sesame St	500	599

Table 3.2b **STREET table**

segmentID	name	low_add	high_add	speed
1639	Sesame St	200	499	25
1079	Sesame St	500	599	30

Table 3.2c **SPEED_LIMIT table**

segmentID	name	low_add	high_add	lanes
2174	Sesame St	200	399	2
2298	Sesame St	400	599	4

Table 3.2d **NUMBER_OF_LANES table**

segmentID	name	low_add	high_add	pave_cond
2145	Sesame St	200	499	good
1081	Sesame St	500	599	fair

Table 3.2e **PAVEMENT_CONDITION table**

There is a little more work to do. Look at the SegmentID column. Each table has its own SegmentID sequence. Imagine this database actually being composed of shapefiles or feature classes rather than relational tables. Instead of having four nonoverlapping, block-length street segments to define the four blocks of Sesame Street, there are now 10 overlapping street segments. What a maintenance nightmare!

The whole concept of linear measurement systems was created to provide some degree of normalization for transportation databases. An event table database design using linear measures would have this structure:

sectionID	...	LRMType	beginMeasure	endMeasure
1038		address	200	599

Table 3.3a **STREET table**

65

linearEventID	sectionID	linearEventType	beginMeasure	endMeasure	value
163	1038	Address Range	200	299	B
172	1038	Address Range	300	399	B
186	1038	Address Range	400	499	B
395	1038	Speed Limit	200	499	25
228	1038	Pavement Cond	200	499	good
728	1038	Address Range	500	599	B
634	1038	Pavement Cond	500	599	fair
379	1038	Speed Limit	500	599	30
631	1038	Number of Lanes	200	399	2
652	1038	Number of Lanes	400	599	4
269	1038	Name	200	599	Sesame St.

Table 3.3b **LINEAR_EVENT table**

This design is fully normalized because you only need to go to one record to change any piece of information. Of course, some changes may require you to write additional records when you subdivide an existing linear event or create a new one, but to find and change existing data is easy. Turn the SECTION table into a feature class and add dynamic segmentation and you have the traditional route-milelog method of managing data used at many state DOTs.

The design is fully normalized but the fifth form of normalization has yet to be used. 5NF cannot always be achieved. This step in normalization is actually the elimination of data altogether. You do that by using relationships to reconstruct the data you dropped.

Suppose you have assigned each customer of your business to one salesman. One customer has one salesman, and all the purchases by that customer are recorded under the name of the salesman. But you also need to know what each customer purchased so you can ship it and keep track of orders when the customer calls. This might lead you to believe that you needed to keep sales records by salesman and customer, but 5NF says that you only need to store purchases one time—the customer—you also store the name of the salesman who serves that customer. (You cannot, by the way, store the sales data only by the salesman, since each salesman has more than one customer.)

customerID	name	streetAdd	city	state	zip	salesmanID
18432	Acme Corp	386 Main St	Metro	FL	32806	32
28904	Wilson Group, Inc	4277 Elm Ave	Isnt	GA	30728	26

Table 3.4a **CUSTOMER table**

salesmanID	name	office	commRate	quota
24	Sam Jones	Tulsa	2.00%	345000
32	Steve Wilson	Atlanta	2.50%	550000
26	Jack Smith	Miami	2.25%	425000

Table 3.4b **SALESMAN table**

orderID	customerID	salesmanID	orderDate	shipDate	orderTotal
3644	18432	24	03-OCT-2005	05-OCT-2005	3472.97
4802	28904	26	12-NOV-2005	16-NOV-2005	42558.02
3290	18432	24	17-JUL-2005	20-JUL-2005	3289.45

Table 3.4c **ORDER table**

orderID	lineNumber	productID	orderQuantity	shipQuantity	backOrder	itemTotal
3644	1	0328	1	1	0	321.95
3644	2	3298	4	4	0	486.24
3644	3	1146	18	12	6	7655.48

Table 3.4d **ORDER_ITEM table**

Using this structure, you can find the items sold by a particular salesman or to a specific customer by querying the ORDER table. You can then do operations like calculate salesman commission by multiplying the total value of all sales by the commission rate. You could also track the sales of specific items by salesman for special promotional incentives. This database is in 5NF.

Tracking events

One last aspect of geodatabase design needs to be addressed before moving to the specific needs of a transportation dataset. The ArcGIS Tracking Analyst extension provides a data structure called a tracking event. This extension presents a useful way to structure data needed for some transportation applications that do not actually use the Tracking Analyst extension, and it represents an application of the normalization process.

Tracking events come in two basic types: simple and complex. A simple tracking event is dynamic in that each row in the event table stores the position of the object being tracked. One table stores information about the object and its position. A simple tracking object is one that moves.

A complex tracking event can be dynamic or stationary. A complex tracking event has one table to describe the object and another table to store its position at each moment of

Chapter 3: Geodatabases

observation. This design allows more information about the tracked object to be stored in a single row, and eliminates the redundancy of the simple tracking object table, where each row has the same descriptive data. A stationary complex tracking event is one where the object is fixed but other objects pass by it or an observed quantity changes over time. A rain gauge is an example of a sensor that could be treated as a stationary complex tracking object.

Temporal Object & Observation Table

BusID	Time	Location	Status
1	12:04:37	x1,y1	Out of service
2	12:06:29	x2,y2	At stop
3	12:06:45	x3,y3	Deadheading
2	12:07:18	x4,y4	In service
2	12:08:53	x5,y5	At stop
3	12:09:36	x6,y6	At garage

Simple tracking event This sample class is an example of a simple tracking event consisting of buses with automated vehicle location (AVL) equipment that sends a location back to the central bus terminal whenever a predefined event occurs. Everything you need to describe each event is contained in a single table.

Temporal Object Table

BusID	Route	Pattern	Status
1			Out of service
2	21	2	In service
3	48	1	In service
4	79	4	In service

Temporal Observation Table

BusID	Time	Location	Status
1	12:04:37	x1,y1	Out of service
2	12:06:29	x2,y2	At stop
3	12:06:45	x3,y3	At stop
2	12:07:18	x4,y4	Left stop
2	12:08:53	x5,y5	At stop
3	12:09:36	x6,y6	Left stop
2	12:10:05	x7,y7	Left stop
2	12:11:27	x8,y8	At time point
3	12:12:42	x9,y9	At time point
4	12:13:01	x10, y10	At stop

Complex dynamic tracking event This sample design shows a relatively static table describing the objects being tracked and a temporal observation table. The location and activity of the bus is constantly changing, but it remains on a given route and pattern combination for a comparatively long time. Other information about the bus, like bus capacity and the driver's name, could be stored in the temporal object class.

Temporal Object Table

TMSID	Route	Measure	TMSType
1	21	3.472	Directional
2	21	15.869	Lane
3	19	5.031	Lane
4	34	7.308	Bidirectional

Temporal Observation Table

TMSID	Time	Direction	Lane	Count
1	00:15:00	1	1	275
1	00:15:00	2	1	381
2	12:06:45	1	1	163
2	12:07:18	1	2	206
2	12:08:53	2	1	98
2	12:09:36	2	2	107
3	12:10:05	1	1	115
3	12:11:27	1	2	206
3	12:12:42	1	3	241
3	12:13:01	2	1	186

Complex stationary tracking event This example shows how traffic counts could be treated as temporal observations in Tracking Analyst. The location of each traffic-monitoring site is fixed, but the count of vehicles observed changes in each time period.

Figure 3.15 **Tracking events** The ArcGIS Tracking Analyst extension works with simple and complex tracking events. A simple event table contains everything needed to map the event's progress. A complex event consists of a static table that describes the tracked phenomenon and a dynamic table describing temporal observations of each tracked event. Complex events may be stationary or dynamic.

Simple and complex dynamic tracking events have obvious applications in transit system databases and fleet-management programs, where monitoring a vehicle's location is useful. For example, you could create a dynamic complex tracking event for each bus in your fleet. The temporal object table would describe buses, with one row for each bus. The temporal observation table would store the GPS-derived coordinates of each bus transmitted by an onboard automated vehicle location (AVL) unit. This function suggests applications where arrival times are forecast for metro trains at stations or online to tell users when the bus is approaching a nearby stop.

What may be less obvious is using complex stationary tracking events to store observations made over time at a single location. For example, traffic counts and crashes could be viewed as stationary complex tracking events. The temporal objects would be traffic monitoring sites and intersections. The temporal observations would be traffic volumes per unit of time and crashes that occur at random intervals.

Geodatabase performance

The number of host database queries that must be executed mainly determines the performance of a geodatabase. Performance is generally inversely proportional to the number of feature classes being used. ESRI provides a free Geodatabase Toolset that you can use when testing the geodatabase design prior to deployment. The functions available in the Toolset include:

- Editing information
- DBMS statistics
- User information
- Versioning lineage information in a state tree diagram
- Spatial index information
- Displaying versions in the geodatabase in a hierarchical tree view structure
- Relative data fetch timer
- Refresh data timer (per layer)
- Ability to start and stop DBMS tracking
- Scripting functionality

The Toolset timers will display the time required for database operations, like fetch, draw, and label, when working with feature classes. Since transportation databases involve many more tables than most geodatabase designs, you may need to use tools that came with your RDBMS to monitor geodatabase performance more fully. Of course, the ultimate measure of effectiveness is how long it takes to do useful work under production conditions. To answer this question, you will need to do performance testing with prototypes going against a populated geodatabase that is similar in size and complexity as the production environment.

Notes

[1] Please note that these subtypes and their default values are included here solely to illustrate how subtypes work, not to explain how route designations are made.

chapter four

Best practices in transportation geodatabase design

- Centerlines
- Intersections
- Realignment
- Segmentation methods
- Mapping applications
- Census files
- Accommodating multiple street names
- Emergency dispatch
- Designing a pavement management system

Chapter 4: Best practices in transportation geodatabase design

A transportation geodatabase is an abstract representation of the real world. Depending on the application, we may choose one of several possible abstractions. Whatever form is selected, the abstraction does not represent all the aspects of the real-world entities.

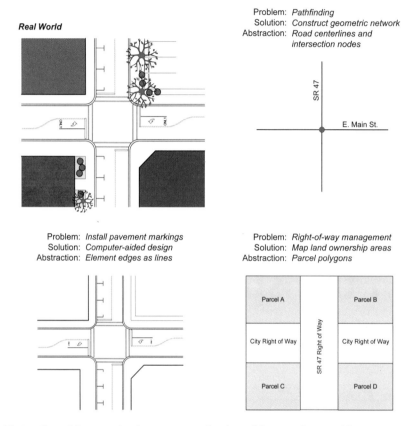

Figure 4.1 **Abstraction** Most geodatabases are a collection of feature classes with representational objects consisting of points, lines, and polygons. The abstraction process to produce a geodatabase often "loses something in the translation." While retaining the full richness present in the real world is not possible, that is not the objective of abstraction. The true purpose is to focus on real-world aspects most critical to a particular problem. Different critical phenomena, abstractions, and attributes are needed for each problem to be solved.

The key to successfully creating a usable abstraction is to represent the core aspects of the entities. Most forms of abstraction in a geodatabase revolve around the many ways you could construct geometry showing the shape and location of real-world entities. For most transportation geodatabases, the linear aspect of roads, railroads, and waterways is dominant. However, a transport agency must also consider the property—the right-of-way—on which those facilities exist. Planning and constructing changes and routine maintenance all require information about facilities. A transport agency, therefore, will likely have multiple abstractions stored in separate databases.

This chapter describes the common forms of linear facility geometry and suggests best practices to help you decide which form is best for your application. A later chapter will show several ways you can manage multiple abstractions in a single geodatabase.

Centerlines

By far, the centerline is the most common abstraction for linear facilities. There are two centerline subtypes: logical centerlines and carriageways (physical centerlines). A logical centerline represents the approximate midpoint of the facility across its width. A logical centerline does not care whether a road is divided or a rail line contains multiple tracks. The intent, after all, is to show the general location and shape of the facility at a relatively small scale. GIS is, by definition, concerned about *geographic* scales that provide relatively large spatial extents.

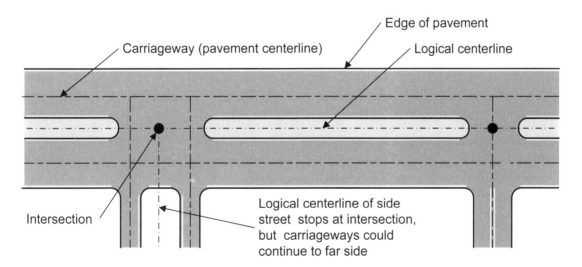

Figure 4.2 **Road geometry** A challenge with transportation-system abstraction in a geodatabase is settling on one graphical representation among so many choices. When the scale is very large, as with a CAD system for facility design, abstraction is relatively close to the real world, with lines representing the face and back of curbs, sidewalk and pavement edges, median islands, and other roadway structure details. At slightly smaller scales, many of these features disappear and carriageway centerlines represent the path followed by each linear piece of pavement. The highest level of abstraction provides a single logical centerline for each linear facility. Each of these geometric representations has a useful application, but no single choice meets all users' needs.

Sometimes, though, you need to illustrate data at larger scales, where precise position and shape are more important. Here, you can use carriageways, which are physical centerlines. A divided road or a double-track mainline will be shown using two carriageways, one for each physical path that a vehicle could travel.

At really large scales, such as for engineering design and right-of-way management, even carriageways do not provide a sufficiently "real" abstraction. At such scales we may employ facility edgelines and start to treat the linear facility as an area with both length and width.

One consistent issue is atomicity, which refers to the indivisible nature of elemental entities. Geodatabase design is usually based on a one-to-one relationship between the entity in the real world and its representative geometry. Entity attributes are used to symbolize the geometry in a way that expresses some characteristic of the entity. The geometry has to be scalar. Atomicity is always a problem for linear transport facilities. A road, rail line, or waterway is typically very long and will have many changes in attribute values along its length.

When dealing with a logical centerline, you can construct a one-to-one relationship between a facility and its representative geometry by breaking the geometry into segments. One segment has one centerline and one set of attribute values. Switch to carriageways for your geometry, and you now have one or two linear features for each facility segment, which is still manageable. Make the jump all the way to edgelines, and the number of representative edgeline features can be very large. Worse, you no longer really have a line that can be used to display common attribute values, like number of lanes and functional class. You could go with polygons, but they really look like wide centerlines and present problems at intersections.

Linear facilities are not the only things affected by the level of abstraction. Intersections and other junctions also present abstraction issues. An intersection is an atomic feature in the real world. The intersection of two divided roads is one thing; however, if you adopt a carriageway representation, then you will have four points where the physical centerlines cross. Also adopt the definition that an intersection exists wherever road lines cross, and you will find yourself with four intersections at the junction of two divided roads when you really only have one real-world intersection. We can fix this problem for pathfinding applications with internal turns at an intersection, but not for inventory and highway safety, where we need one intersection feature to illustrate entities like traffic signals and crashes. An approach that would satisfy these applications would be to treat an intersection as a multipoint feature.

To preserve atomicity, you need to organize the multiple pieces around the singular entities they represent. The easiest way is to relegate the carriageways and edgelines to geometry-only status. Use them to make maps look better at larger scales. Attach the original data to the logical centerline. Do not try to duplicate the data for each carriageway; treat them as "dumb" lines. You can still supply a number of attributes to control map symbology, as we will show later in this chapter.

If you really need to add intelligence to the carriageways for some particular application, you can use a facility identifier and a side-of-road or track identifier to move data to carriageways. But there are ways to create carriageways as offset copies of the logical centerline that will meet the needs of most larger-scale applications. (See chapter 8.)

As for the intersections, put them where the logical centerlines cross and ignore the intersections formed by dual carriageways. In addition to preserving the one-entity-to-one-geometry requirement, this practice will satisfy pathfinding applications that employ the logical centerlines. (See chapter 13 for an example.)

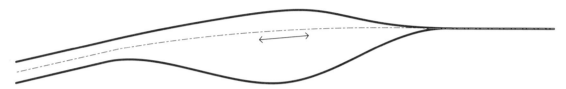

Option 1: One centerline, two carriageways

Option 2: Two centerlines where carriageway lengths differ

Figure 4.3 **Centerline choices** Although it easy to see logical centerlines on aerial photographs, they may not be so clearly discerned in the field, where data collection occurs. There is also the issue of the two carriageways having substantially different lengths, with resulting route measure issues. One solution is to store data by side of road, but this leaves the problem of unequal lengths unresolved. A more generally useful solution is to create separate directional centerlines. Adding carriageway geometry can give you the best appearance for either logical centerline choice.

There will be times when you really cannot use just one centerline, as in the example above. It is rather common to give a divided road one logical centerline and treat the fact that it is divided for all or a portion of its length as an attribute. You may even store data by the side of the road: left, right, and both. However, you will face situations where this approach does not work well, such as when each side of a road goes around an obstacle in the median. Another example is when a divided road with carriageways close together turns into a one-way pair with one or more intervening blocks full of buildings between them.

In both instances, you will want to create separate logical centerlines for each direction of travel. If necessary for the applications you intend to support, place an at-grade intersection at the junction of the single and double centerlines. Treat each centerline as its own facility. If you are using a linear referencing method (LRM) to locate positions along a centerline, reconcile the differences in length by resetting the origin of each centerline to zero. That treats each of the four centerlines as a separate facility within the database. You can readily combine all the resulting segments into a single traversal entity with one multipart centerline by selecting for route number. You will get different overall lengths depending on the direction of travel, which is appropriate for this example.

In addition to the guidance offered above, other practices can make your life easier and provide better data to the end user. You have already seen the need to make the relationship between entities and their representative geometry have a cardinality of one to one. The first task in meeting this requirement is to properly identify the entities.

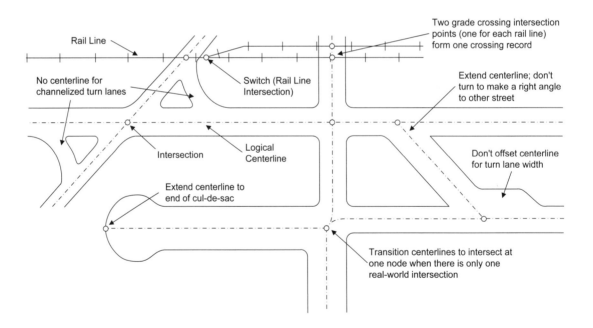

Figure 4.4 **Best practices—logical centerlines** A logical centerline represents the approximate center of a linear transportation facility. This figure illustrates the basic rules for constructing roadway centerlines. The purpose of a logical centerline is to represent a facility at a relative small scale, reflect the linear datum's rules, and support network connectivity. Appearance is a secondary consideration.

Figure 4.4 demonstrates a number of standard practices that will help you create logical centerlines for roads and railroads. Centerlines naturally go down the middle of the facility at the approximate midpoint. Do not offset the centerline when the width of the road changes asymmetrically. Extend the centerline all the way to the end of a cul-de-sac. Do not try to do anything fancy in a cul-de-sac, like shape it into a hook or circle, because that will disturb the correspondence between linear measures (addresses) and geographic location.

Put the crossing point of intersecting centerlines in the middle of the intersection, even if it means distorting the centerlines a little to accomplish that. Ignore channelized turning movements. You can give them a carriageway for large-scale mapping applications, but they are just internal details of a logical intersection. Offset intersections may be treated as one entity if that is the way traffic control works.

Use one centerline for each railroad track. You may generalize this to a single centerline for smaller scale mapping by including only the mainline track. Treat switch frogs and crossing diamonds as intersections.

Use multipoint geometry for railroad grade crossings, with one part at each intersection of a railroad track centerline and the road's logical centerline. The railroad grade crossing will receive a single federal crossing identifier regardless of the number of tracks, so you want to preserve that atomicity. Think of a railroad grade crossing as a group of road-track intersections. The situation is similar to that presented by crossing carriageways, where the resulting one to four points of line intersection are to be treated as a single entity.

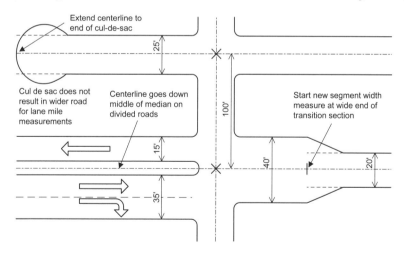

Figure 4.5 **Geometric attributes** Not all aspects of a facility's geometry are contained in its geometric representation, particularly when a line of essentially zero width is used. Many descriptive properties are included as attributes of route event tables. This figure illustrates some the common practices for determining where the facility width attribute's values may change. If you are using a predetermined segmentation design, and facility width is part of the segmentation decision, then these points of change represent segment endpoints.

Chapter 4: Best practices in transportation geodatabase design

Associated with the manner in which you construct geometric features to represent linear facilities is the manner in which you segment them. A common way to create transportation geodatabases is to create relatively short segments, such as may be defined by address blocks. In order to assure atomicity in segment attributes, you may want to consider breaking such block centerline features wherever there is an obvious change in an attribute at a midblock location. For example, if you will store segment width as an attribute for such applications as determining lane miles, then you will need to break segments wherever width changes. If there is a transition section, then make it part of the narrower segment and start the wider segment where the new full width begins. Cul-de-sac width should not affect lane mile calculations, so do not create a special cul-de-sac feature class unless you need to include cul-de-sac diameter as a network parameter to establish the cost of U-turns. (See chapter 13.)

You may also want to consider going with separate centerlines when faced with a divided road with different primary characteristics on each side of the road. An alternative would be to store attributes by side of the road.

At most locations, the logical centerline will closely approximate the location of the facility. The most common exception, save for the issue of roadway splits discussed earlier, is at limited-access highway interchanges. Most state DOTs use a form of linear location referencing, such as route-milelog. They also commonly place the point of intersection between

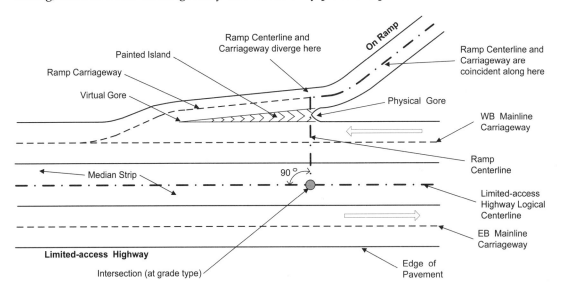

Figure 4.6 **Logical centerlines** Logical centerlines need to intersect at the location that makes sense from the perspective of connectivity and the linear datum. In this example, the agency rule says that the point of intersection, for which the route measure is defined, is the end of the physical gore; i.e., where the pavement of the ramp merges with that of the mainline. A ramp carriageway need not be constrained by that definition, since the purpose of a carriageway is to provide a good map appearance at a large scale.

a ramp and the main highway at the physical gore, as this is the location where the two pieces of pavement become one. For the ramp and mainline centerlines to intersect at this location, the ramp centerline needs to turn toward the main centerline and intersect at the appropriate milelog location at a right angle. The resulting appearance may look strange, but it ensures that the intersection will occur at the proper location from the perspective of the limited-access highway's linear measures. If you are also providing carriageways, they can continue along the path of the ramp as it merges with the highway pavement.

Intersections

An intersection is normally visualized as the location where two roads cross at grade. You may also want to look at using a number of intersection subtypes as a way of managing behaviors and symbology. For example, an interchange could be either a way to manage a collection of roads and at-grade intersections or just a big intersection with internal turns for pathfinding.

A railroad grade crossing is another intersection subtype, one between a railroad and a road. You may want to restrict the classification of this kind of intersection to those places where a track crosses a road through topology rules. Such a function can be easier to implement if you create a railroad grade crossing subtype for intersections.

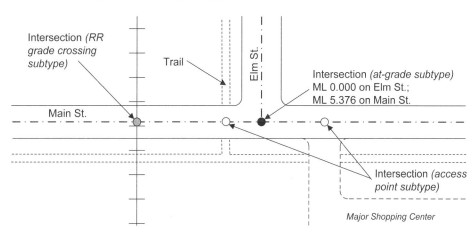

Figure 4.7 **Intersection subtypes** Three of the four basic intersection subtypes are shown here. An at-grade intersection is the typical place where two or more routes meet. A railroad grade crossing is where a route crosses a railroad. If there are multiple tracks, you may want to use a multipart point feature class. The third subtype shown here is the access point, which represents a location on a route where a nonmapped facility connects to a mapped facility. Access points include shopping center driveways, recreational trail crossings, and anywhere you have intersection-like elements with characteristics to store, such as traffic signals and crashes.

There are also places that look like intersections where only one road seems to exist. These are access point intersections, and they occur where trails cross a road or a major driveway provides access to the road system. Like at-grade intersections, you will likely want to retain information about these locations, such as traffic signals and traffic crashes.

Using intersection subtypes will provide you with the ability to check cardinality to ensure that it meets your expectations. For example, you will need at least two intersecting roads for an at-grade intersection, one road and at least one railroad track for a railroad grade crossing, and only one road for an access point. An interchange will need to include at least one road classified as a limited-access highway.

Realignment

The way you segment centerlines will affect future editing workloads. One of the more common issues you will deal with is facility realignments. A typical approach to building a geodatabase for road features is to identify the entities to be included by name and route number. You may additionally subdivide these entities at county lines or district boundaries to reduce the extent of changes or to coincide with your data maintenance organizational structure. This is all fine and good, but once the features are created, you need to break the symmetry of entities and features and treat name or route number as just another attribute. If you try to preserve the symmetry, you will have a lot of extra work when a road is realigned, particularly if you are using a linear location referencing method. The problem is that realignment causes the total length of the facility to change, resulting in a ripple effect through all the downstream locations. You either have to revise all their measure values or create some sort of reconciliation structure in the measure values to deal with the missing or duplicative values within the area of realignment. You also may lose the ability to properly map event data applicable in datasets tied to old measure values.

A better choice is to treat the new section as a new facility. Add a facility status field to the database and give it one of three values: active, retired, and replaced. (See chapter 6.) You would leave the old alignment in the dataset; just list its status as retired. If you need to display historical data on the former alignment, the feature and its linear measures are still in place to do the job. You also do not need to change any downstream data. All the data about the new section is also new, so there is no lineage issue for the facility identifier and linear measures used by it. If you want to see the present state of the system, just select all the active facilities. You can select by date and status to produce various time-stamped views of the geodatabase.

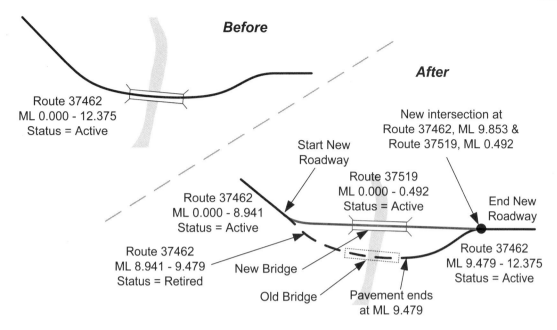

Figure 4.8 **A strategy for realignments** One of the traditionally difficult aspects of a transportation dataset based on a route-milelog linear referencing method (LRM) is facility realignments that change the length of the facility at a midpoint. Such a change, if accomplished through traditional methods, will reshape the affected geometry and recalculate the measure for downstream events. A better way that avoids modifying all the event records is to make a new feature for the realigned portion and abandon the replaced segment. A status attribute identifies active and retired route segments. This approach has the benefit of retaining the original geometry and measures for time-based analyses.

Segmentation methods

Several references have been made to the traditional transportation dataset structure of short segments with repeating attributes. This is a perfectly valid way to go if you have a manageable number of segments and relatively few changes to the system other than added facilities, which is the case for most local governments. Once built, roads are rarely removed. Subdivisions add new mileage. Annexation may change jurisdiction for a road, but it does not eliminate the road itself. Most capacity projects will widen a road, not change its location. All in all, this is a relatively stable situation where editing means adding new roads as they are opened to traffic.

The least demanding way to accommodate the needs of smaller datasets is to break roads and other linear facilities at intersections, but not to actually create intersection features. The resulting "independent" segments work fine for basic mapping needs and geocoding. The viewer will not be able to discern that a given line segment is composed of smaller pieces.

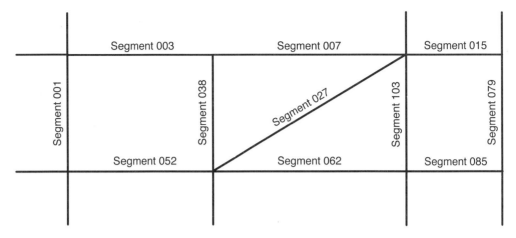

Figure 4.9 **Independent segments** The simplest approach to making a transport dataset is to create independent segments representing linear portions of the total system. A typical way to do this is to segment longer facilities at intersections. Each segment gets the same set of attributes so that values extending across segment termini are duplicated for all segments to which the value applies. You can use segment attributes to determine line symbology, or create separate map layers to represent each kind of facility.

Some database designers may argue that this simple design is inefficient. Although you will have to duplicate much data for contiguous segments of the same facility, this overhead only applies as a facility is added to the dataset. This initial workload may be offset by avoiding more complicated structures that impose their own overhead. You also do not need perfect geometry with no overlaps or gaps unless you plan to build a navigable network.

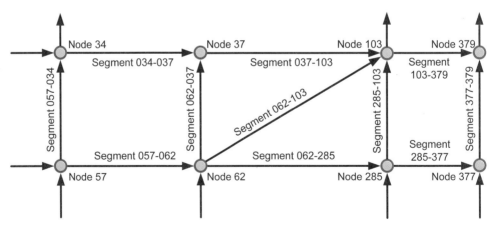

Figure 4.10 **Link-node** The link-node approach, which adds terminal nodes to the set of segment links, offers the next level of database design complexity. This design has two advantages over the independent segment approach. First, it can be used to construct a network for pathfinding. Second, it can supply features for storing intersection attributes or organizing such data as a traffic signal inventory or a crash reporting system.

If you do need a navigable network, you will likely want to add formal intersection node features to serve as junctions for segments, which become links (edges) in the network. As will be shown in chapter 13, the ArcGIS Network Analyst extension can construct junction features for you. However, there are benefits to constructing them yourself, such as providing information about the junctions to the pathfinding application. You may also want to provide different travel costs for each intersection subtype.

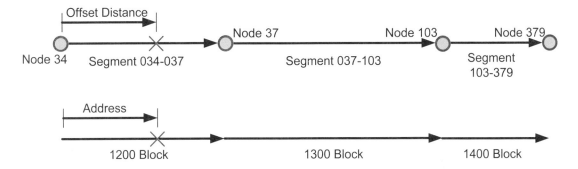

Figure 4.11 **Node offset** Sometimes you need to reference a location between nodes at a point along a link. One approach is to identify the node (origin), the link to travel down, and a distance to travel. This design is actually quite common, although the nodes are usually implied rather than expressed. We call it addressing, and it serves as the foundation for all one-dimensional (1D) location referencing systems.

One of the issues with any segmentation scheme is that you will eventually need to refer to a location within a segment. Whether you need to locate a midblock intersection, illustrate parking restrictions, or identify sign locations, it will be frustrating to find yourself limited by an indivisible segment. One approach is the node-offset linear measure. You use this approach all the time; you just call it addressing, geocoding, or project stationing. Pick a reference feature, like an intersection or a node, and a segment or link to travel down, and then specify how far to go. This is sort of like a route-milelog approach, but the routes are very short.

Mapping applications

If all you really need to do is make maps with your transportation geodatabase, then you should use a design that emphasizes the structure and attributes appropriate to such an application. Such a database would include annotation classes in addition to the feature classes of a facility inventory.

Chapter 4: Best practices in transportation geodatabase design

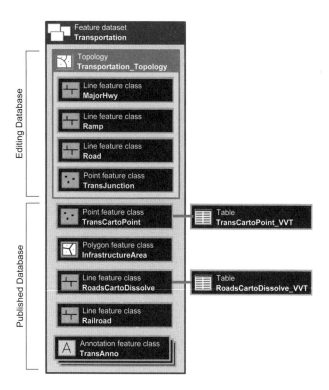

Figure 4.12 From the set of thematic layers described in chapter 8 of *Designing Geodatabases: Case Studies in GIS Data Modeling*, by David Arctur and Michael Zeiler (ESRI Press 2004).

The top part of this transportation mapping database is intended to support the editing process. It contains all the elemental pieces of the highway mode of travel like you may have in a normal segmented data structure of the type discussed above. These four feature classes exist in a transportation topology that will help ensure that clean linework is produced.

The bottom part of the database contains the classes that will be used to produce the maps. The RoadsCartoDissolve feature class contains the results of merging the various road segments in the MajorHwy, Ramp, and Road feature classes to produce the highway facility linework for our maps. This intermediary has been added to the mapping process in order to provide complex geometry. For example, you may want to outline major highways in black with a red center fill, or outline unpaved roads in black and use a white center fill. You cannot get this appearance through line styles applied to a single feature class. What you will need to do is produce a wide black line and overlay with an identical, but slightly narrower red or white line. This means printing two copies of the road features with slightly different line styles. It also means merging linear features so that intersections look outlined rather than overlapping. A valid values table (RoadsCartoDissolve_VVT) says which combinations are legitimate.

Mapping applications

Simple feature class — MajorHwy
Geometry: Polyline
Contains M values: No
Contains Z values: No

Field name	Data type	Allow nulls	Default value	Domain	Precision	Scale	Length
OBJECTID	Object ID						
Shape	Geometry	Yes					
TYPE	Long Integer	Yes			0		
TRANSID	Long Integer	Yes			0	0	
SOURCE	Long Integer	Yes			0	0	
OverpassLevel	Long Integer	Yes	0	RoadLevel	0		
Enabled	Short Integer	Yes	1	EnabledDomain	0		
Shape_Length	Double	Yes			0	0	

Limited-access highways

- Type of highway
- Permanent unique identifier
- Foreign key related to the RevisionInfo table
- Road level for drawing order
- Indicates feature's involvement in network

Simple feature class — Ramp
Geometry: Polyline
Contains M values: No
Contains Z values: No

Field name	Data type	Allow nulls	Default value	Domain	Precision	Scale	Length
OBJECTID	Object ID						
Shape	Geometry	Yes					
TYPE	Long Integer	Yes	900		0		
TRANSID	Long Integer	Yes			0	0	
SOURCE	Long Integer	Yes			0	0	
OverpassLevel	Long Integer	Yes	0	RoadLevel	0		
Enabled	Short Integer	Yes	1	EnabledDomain	0		
Shape_Length	Double	Yes			0	0	

Limited-access highway ramps

- Type of ramp
- Permanent unique identifier
- Foreign key related to the RevisionInfo table
- Road level for drawing order
- Indicates feature's involvement in network

Simple feature class — Road
Geometry: Polyline
Contains M values: No
Contains Z values: No

Field name	Data type	Allow nulls	Default value	Domain	Precision	Scale	Length
OBJECTID	Object ID						
Shape	Geometry	Yes					
TYPE	Long Integer	Yes	904		0		
CLASS	Long Integer	Yes	904		0		
TRANSID	Long Integer	Yes			0	0	
SOURCE	Long Integer	Yes			0	0	
OverpassLevel	Long Integer	Yes	0	RoadLevel	0		
Enabled	Short Integer	Yes	1	EnabledDomain	0		
Shape_Length	Double	Yes			0	0	

Roads and streets

- Type of road or street (used for symbology)
- Class of road feature (used for labeling)
- Permanent unique identifier
- Foreign key related to the RevisionInfo table
- Road level for drawing order
- Indicates feature's involvement in network

Simple feature class — TransJunction
Geometry: Point
Contains M values: No
Contains Z values: No

Field name	Data type	Allow nulls	Default value	Domain	Precision	Scale	Length
OBJECTID_1	Object ID						
Shape	Geometry	Yes					
OBJECTID	Long Integer	Yes			0		
Enabled	Short Integer	Yes			0		

Intersections
Produced by the geometric network to ensure that only ramp features connect to limited-access highways

- Indicates feature's involvement in network

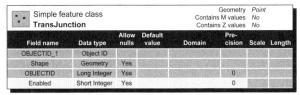

Coded value domain — EnabledDomain
Description: Boolean logic value
Field type: Short integer
Split policy: Default value
Merge policy: Default value

Code	Description
0	Disabled
1	Enabled

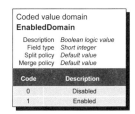

Coded value domain — RoadLevel
Description: Overpass level
Field type: Long Integer
Split policy: Default value
Merge policy: Default value

Code	Description
0	Ground level
-2	Tunnel, 2 levels down
-1	Tunnel, 1 level down
1	Overpass, 1 level up
2	Overpass, 2 levels up
3	Overpass, 3 levels up
4	Overpass, 4 levels up
5	Overpass, 5 levels up
6	Overpass, 6 levels up

Figure 4.13 **Domain control** Each polyline feature class uses two domain classes. RoadLevel specifies the drawing priority for the feature as a way of symbolizing overpasses and tunnels. EnabledDomain is a standard component of a geometric network. An enabled feature can participate in the network for pathfinding operations. A disabled feature works as a barrier to travel along a particular path.

Figure 4.13 is a close-up look at the editing part of the database. It shows the four feature classes and the two defined domains. The explanation shown next to each field in the data dictionary view offers some insight into what the attributes provide. For example, the Source field is a foreign key to feature-level metadata in a RevisioinInfo table. The domains help express the meaning of the OverpassLevel and Enabled fields. The OverpassLevel attribute controls drawing precedence, with higher values being drawn above lower values. The level information lets you properly identify which of two crossing roads can intersect so as to avoid turning a bridge into an intersection. The RoadLevel domain used to control the choices available for OverpassLevel reveals that it supports two below-ground levels and six above-ground levels.

Geometric networks use the EnabledDomain values to identify which features contained in member feature classes may participate in a network. Building a geometric network dataset also created the TransJunction class. Geometric networks and their potential use in transportation geodatabases are covered in the next chapter.

For an expanded discussion of this database design and how it can be part of a complete cartography geodatabase, please refer to *Designing Geodatabases: Case Studies in GIS Data Modeling*, by David Arctur and Michael Zeiler, available from ESRI Press.

Census files

The TIGER database that defines census blocks and tracts provides one of the more common transportation database structures used in the United States. TIGER is an acronym for Topologically Integrated Geographic Encoding and Referencing. The U.S. Census Bureau registered the acronym to ensure that users can clearly identify the source of any data they acquire. The part of interest for transportation system creation is the TIGER/Line file.

Many local governments start their mapping process by loading the census file for the jurisdiction, extracting the transportation features, and making changes to improve the cartographic appearance or to update the contents to reflect changes since the census. The census file offers many of the fundamental attributes for geocoding and other basic mapping functions, such as street name and address range, and can become the input to build a geometric network for pathfinding applications.

The file is already segmented at intersections, although the intersections used are from the coincidence of all the kinds of boundary objects used by the census, which include railroads, creeks, and political boundaries. You may want to merge segments that break at boundaries you do not include. A number of processes are in place to improve quality as part of the normal census update cycle, in part because so many people use the transportation line work for other applications.

Accommodating multiple street names

Simple feature class: **Road**								
Geometry: Polyline								
Contains M values: No								
Contains Z values: No								

Field name	Data type	Allow nulls	Default value	Domain	Precision	Scale	Length	
OBJECTID	Object ID							
Shape	Geometry	Yes						
FEDIRP	String	Yes					2	Feature direction prefix
FENAME	String	Yes					30	Feature name
FETYPE	String	Yes					4	Feature type
FEDIRS	String	Yes					2	Feature direction suffix
CFCC	String	Yes					3	Census feature class code
FRADDL	String	Yes					11	From address, left
TOADDL	String	Yes					11	To address, left
FRADDR	String	Yes					11	From address, right
TOADDR	String	Yes					11	To address, right
ZCTAL	String	Yes					5	Zip Code tabulation, left
ZCTAR	String	Yes					5	Zip Code tabulation, right
CFCC1	String	Yes					1	Census feature class code 1
CFCC2	String	Yes					2	Census feature class code 2
SOURCE	String	Yes		Sources			1	Source code
TLID	Double	Yes			7	3		TIGER line ID
Shape_Length	Double	Yes						

Coded value domain: Sources

Description: *Census source codes*
Field type: *String*
Split policy: *Default value*
Merge policy: *Default value*

Code	Description
[null]	Not documented elsewhere
A	Updated 1980 GBF/DIME file
B	USGS 1:100,000-scale DLG-3 file
C	Other USGS map
J	Pre-1990 Census updates
K	Post-1990 Census updates
L	Pre-Census 2000 local official updates
M	Pre-Census 2000 field operations
N	Pre-Census 2000 office update
O	Post-Census 2000

Figure 4.14 **TIGER files** Topologically Integrated Geographic Encoding and Referencing (TIGER) files from the U.S. Census Bureau are a common start to many transportation datasets. Out of the box, a TIGER file supplies the basic information needed for address geocoding. The first four user-defined attributes (those beginning with "FE") provide the street name. After the Census feature class code, the next six attributes supply left- and right-side address range information with ZIP Code. Roads and other linear features are used by the census to define the extent of census blocks and tracts, so you can look at the Road feature class as providing polygon boundary segments.

Accommodating multiple street names

One of the obvious design aspects of the file is that it includes left and right address ranges and ZIP Codes, but only a single street name, the latter being contained in four fields: FEDIRP, FENAME, FETYPE, and FEDIRS. This structure allows for a street name formed from a direction prefix or suffix, a base name, and a street type. Such a structure will work fine for many jurisdictions, but it will break down when there is more than one possible name for a street segment. To solve that problem, you need a more sophisticated design.

Chapter 4: Best practices in transportation geodatabase design

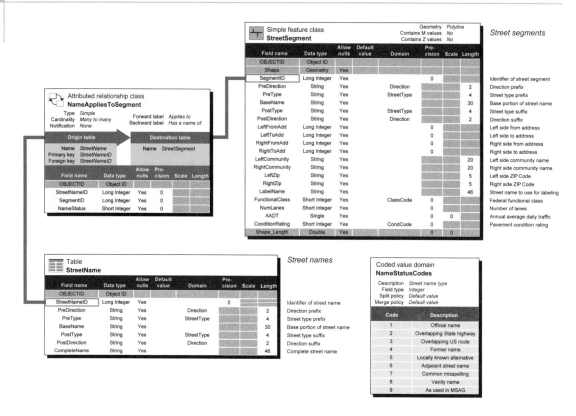

Figure 4.15 **Multiple choice** Treating the name of a street as just another attribute works fine until you run into streets that have multiple names. Complicating this situation is the normally associated corollary that a single name will apply to multiple streets. You can put the "official" name in the street segment class, but where do all those other names go? The solution suggested here is to create a separate street name table and an attributed relationship that ties each street segment to all the names by which it may be known. Some of these "unofficial" alternatives may be common misspellings.

The Census Bureau just needs to know one name to identify the boundary feature. You and your users, though, may need to know all the possible names that may be valid—or maybe even invalid. The design shown in figure 4.15 solves the problem. It also allows for more complicated street names, such as "Avenue B" or "Highway 96," by including a type prefix field. Otherwise, the StreetSegment feature class looks like the census file but with a few extra attributes thrown in. One of them, the LabelName field, stores a street name value to be used as annotation on a map.

You will put the official street name in the StreetSegment class and store the official and alternative names in the StreetName table. The StreetHasAlternateName attributed relationship class connects the two classes, which reveals this as a many-to-many relationship; i.e., a street may have many names and a name may apply to many streets. In addition to holding a foreign key from each related class, the attributed relationship also has a NameStatus field, which indicates the kind of alternate name. The NameStatusCodes domain illustrates the choices, although you cannot really use a domain to control values entered into a relationship class.

Such a geodatabase design can support applications for vehicle routing or to respond to customer calls, where someone might refer to State Road 436 rather than the official local street name of Semoran Boulevard. You may also want to accommodate various legitimate alternatives (e.g., SR 436, S.R. 436, SR436, and State Rd. 436) and common misspellings (e.g., Simoran, Semorun, and Sumoran) as a strategy to help your users be more successful in their searches.

To do so, you put all the alternative names and spellings in the StreetName table and establish the relationships between names and segments. Be sure to include the official names, too, because they may be alternatives for street segments that officially have a different name. Plus, it is a good idea to put all the names in one place. If a user queries the StreetSegment feature class looking for a street by name, and fails to find one, you could write a routine that then looks in the StreetName table and identifies any segments that use that name as an alias. Alternatively, your query tool could just go to the StreetName table and then use the relationship class to show all the possible matches. You may want to put the segments with that official name first in the result list.

Putting the segments together

One of the obvious problems of the multiname geodatabase design is the workload imposed by creating a relationship between a name and all the segments to which the name may apply, either officially or unofficially. Such associations are fuzzy, at best, so it is a case of false accuracy to establish these equivalences at the segment level. It is also a lot of work. What you really need is to correlate names to each other, but the geodatabase does not support recursive relationships. Fortunately, there is an even better idea.

This modified design inserts a new Street table between StreetName and StreetSegment. The relationship between names and the features to which they apply is now accomplished at the street level, where it can be applied to multiple segments.

A second relationship class has also been added to show that streets are a collection (aggregation) of segments. This design puts the geometry at the segment level, but you could copy it to the street level, too. Street will need to remain a table if it is used to store official street names and address ranges for facilities that do not yet exist. Such a use would allow you to manage future street names proposed for subdivisions still under development in order to avoid accidentally duplicating a name. You could link each Street row representing an existing facility to a Centerline feature.

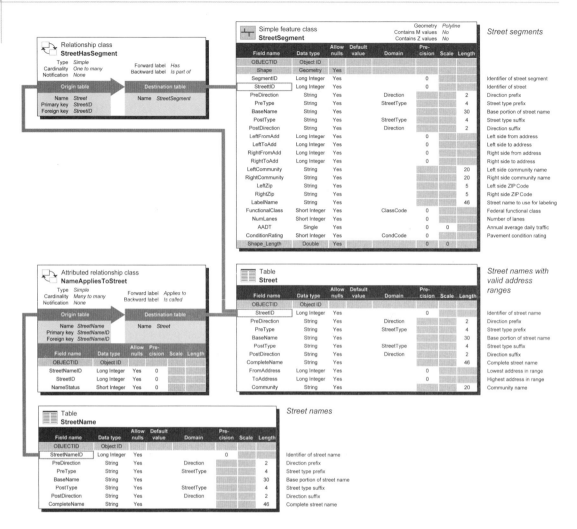

Figure 4.16 **Address range control** After dealing with multiple street names, the next step up in database design complexity is controlling the range of addresses for each similarly named street. This design aspect is often required when you want to put segments together to form logical streets based on name and continuity. You may also need to control for duplicate names and address ranges in different jurisdictions within your database. The solution shown here is a form of the valid value table design in that only certain addresses are valid for a given combination of street name and community.

It is possible to omit the five street name component fields from the StreetSegment feature class and just rely on the StreetID value to point the user to the right Street table record to find the name. However, copying the street name values to the StreetSegment class eliminates the extra step of doing a join on StreetID to see the name for a given segment. You could use the simpler design for the StreetSegment record for editing, and then load the street name data after the editing session is finished to make a denormalized version for others to use.

The Street table also provides a domain control for addresses used in the component segments in the FromAddress and ToAddress fields, which set the minimum and maximum values present on the street. To take full advantage of this design aspect, StreetID values should be assigned to only continuous street extents. You would accommodate discontinuities, such as when there are breaks in a street or when more than one street has the same name, by assigning a different street identifier to each part. Name is an attribute, not an identifier.

Emergency dispatch

A different set of changes can be made to the original design to increase its usefulness to emergency-dispatch applications. To support such applications requires us to know not only the address of an emergency call but also which responder should be contacted. This application requirement means the design must tie each street segment to an emergency service zone (ESZ). An ESZ is created for each unique combination of police agency, fire department, and EMS provider. Since a street segment may, itself, be a zone boundary, the design accommodates emergency service zones by side of street.

Figure 4.17 omits the Street table from the design because it is not central to the design objectives being discussed here, but it is compatible with the emergency-dispatch accommodation illustrated.

There are often several ways to implement a logical database design. This example is a good illustration of that principle. For example, this implementation does not simply store two ESZ identifiers for each street segment, because that may require you to change lots of segment records every time a boundary is moved. Instead, the design uses an attributed relationship between ESZ and StreetSegment.

This is really a judgment call. As mentioned earlier, streets in a local government geodatabase very likely would not really change much except to add new segments as a result of construction. The decision here is that ESZ records would change fairly often as a result of annexations. Each time a city annexes a new area, it is probable that the city will take over the provision of at least one of the emergency services that defines an ESZ.

Chapter 4: Best practices in transportation geodatabase design

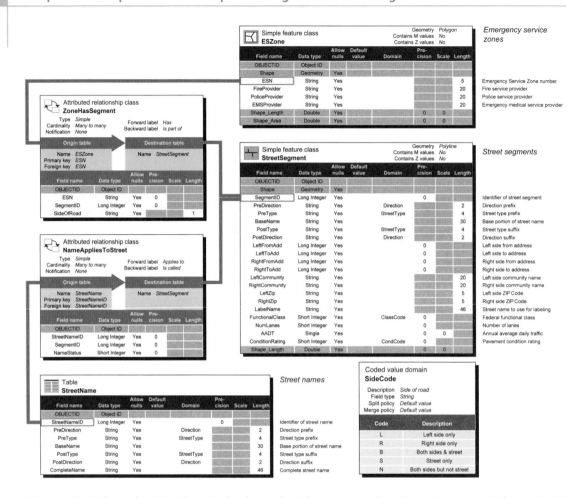

Figure 4.17 **Enhanced 9-1-1** It is quite simple to adapt the two preceding designs to support Enhanced 9-1-1 services. You create an emergency service zone (ESZ) for each combination of fire, police, and EMS provider. Street segments will have to be subdivided when they cross an ESZ boundary. A segment may itself serve as an ESZ boundary, resulting in each side of the street being in a different ESZ. There can also be linear ESZs, such as when the street right-of-way is in a different jurisdiction than the adjacent properties.

If your agency serves a jurisdiction where annexations are infrequent, you might be tempted to consider using a repeating-field design. Don't do it. Annexations can have some fairly complex interactions with addresses and ESZ boundaries. Look closely at the SideCode domain class. You will see that in addition to "Left," "Right," and "Both," there are also "Street only" and "Both sides but not street" choices. So, handling all the possible outcomes of annexation requires three fields in a denormalized StreetSegment feature class: left side, right side, and right of way. That seemed to be a lot of fields when most segments will have the same value for all three, because most streets do not form ESZ boundaries.

Designing a pavement management system

An exercise to design a local government pavement management system offers a good opportunity to use many of the design practices illustrated so far. This geodatabase will include pavement segments that are formed from one or more street segments. Each pavement segment will consist of any number of layers, arranged vertically. The geodatabase will also include public-works projects, some of which may change attributes of one or more segments.

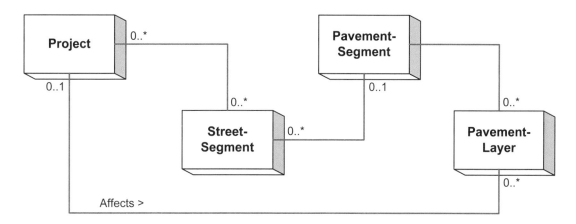

Figure 4.18 **Pavement management system**   This conceptual data model illustrates how you could put together a simple pavement management system (PMS) for a local jurisdiction that used the street-segment approach for its roadway feature dataset. Instead of adding pavement attributes to the StreetSegment feature class, this design uses a separate PavementSegment entity that describes the structure of one or more street segments, reducing the pavement data entry and maintenance workload significantly. The design also provides a Project entity to store information about public works projects that can alter the roadway structure. The pavement layer information is separated into its own object class to provide a normalized design. Pavement segments and projects can be mapped using the geometry of street segments.

The conceptual data model shows the StreetSegment entity used several times in this chapter. The model also shows three other entities: Project, PavementSegment, and PavementLayer.

The association relationships reveal a many-to-many cardinality between Project and StreetSegment and a one-to-many cardinality between Project and PavementLayer. These relationships indicate that a project may involve zero, one, or more street segments and install or modify one or more pavement layers. Because a pavement segment is an aggregation of one or more street segments and pavement layers, projects that affect either of those entities will also affect pavement segments. Pavement layers are always put in place or modified by a project; however, the design accommodates instances when you do not have information on that project.

Projects are related to street segments rather than to pavement segments because most public-works projects involve activities other than roadway pavement such as striping, signing, traffic control, or landscaping. Pavement segments may change without being part of any project through the normal deterioration caused by the environment and traffic conditions.

The potential for multiplicity to be zero shows that the relationship is aggregation, not composition. Composition would require that at least one of the components exist for the whole to exist. Using aggregation allows you to define pavement segments without having to provide all the details.

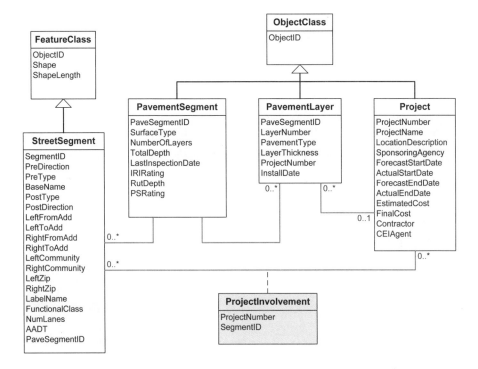

Figure 4.19 **PMS logical data model** This logical data model fleshes out the conceptual PMS data model by adding attributes and an associative relationship class to manage the many-to-many relationship between projects and the street segments they may affect. A pavement segment may be defined by a set of pavement layers, with each layer being uniform along the pavement segment. The foreign key attributes and relationship cardinality show that a pavement segment may include one or more whole street segments. The attributes of StreetSegment were modified to replace **ConditionRating** with **PaveSegmentID**. A street segment will need to be subdivided at midblock locations if pavement is not uniform along its length. A project may install one or more pavement layers on one or more street segments, and several projects may affect a given street segment.

This model is denormalized as far as pavement layers are concerned, because they are broken at each pavement segment boundary. This was done to simplify the database design and to recognize that most pavement layers never change. A typical pavement-related project will mill and replace the top layer, leaving the base layers unaffected. This practice limits the number of layer records that may change in response to a project. The alternative to denormalizing pavement layers is to manage them independently, which requires both a way to define their limits relative to some higher-order facility and a way to tie them to pavement segments through a many-to-many relationship. An example of such a design is provided in chapter 12.

The logical data model shows the geodatabase classes chosen for each entity and the attributes of each class. StreetSegment is the same feature class used earlier in this chapter, except that PaveSegmentID has replaced ConditionRating. The other classes are tables.

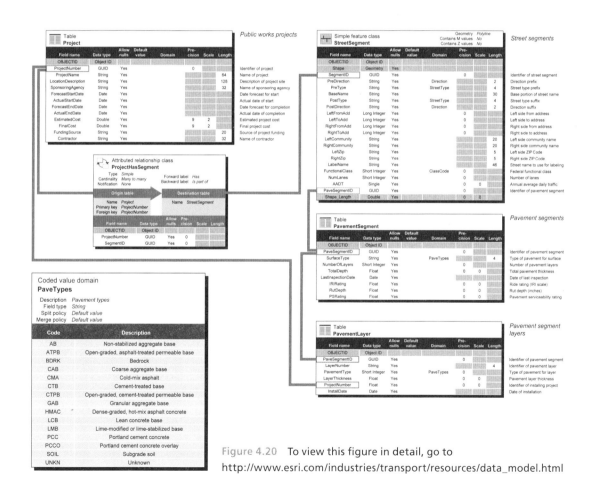

Figure 4.20   To view this figure in detail, go to http://www.esri.com/industries/transport/resources/data_model.html

The model also includes a relationship class, called ProjectInvolvement, in order to resolve the many-to-many relationship between projects and street segments. A relationship class that appears in a logical data model will eventually become an attributed relationship class in the physical data model. Other kinds of association relationships are described by the connecting lines.

The last figure reveals the physical data model, which is a data-dictionary view of the logical data model with the addition of a PaveTypes domain class.

One aspect of the data model that the figure does not illustrate well is the effect of pavement layer changes on pavement segments. For instance, if a project begins or ends in the middle of a pavement segment, that segment will need to be subdivided to reflect the differing pavement layer compositions of each resulting portion.

All of the design ideas presented in this chapter can be used together or in isolation. They involve a small number of classes constructed in the usual ways employed by local governments to manage transportation data using discrete facility segments. The concepts can be equally applied to other modes of transportation when an intersection-to-intersection segmentation schema creates the basic linear features in the geodatabase.

chapter five

Geometric networks

- Travel demand models
- Building the travel demand model
- Map geodatabase for pathfinding

Although ArcGIS Network Analyst extension provides the most versatile environment for building transportation databases, ArcGIS natively supplies geometric network capability that can support a number of transportation-related applications. This chapter will introduce geometric networks and present two common transportation applications.

A geometric network consists of features forming edges and junctions, which look like the links and nodes discussed in the previous chapter. An edge is a path between two junctions. A junction cannot exist in isolation. Edges and junctions are defined by special feature classes and are combined to form a planar graph.

Chapter 5: Geometric networks

The feature classes in a geometric network manage a set of logical network elements and the rules under which they operate. ArcGIS uses the feature geometry to infer the structure of the logical network, but the geometric features and network elements are separate entities in the geodatabase. Out of the box, ArcGIS can build and use networks consisting of simple junctions, simple edges, and complex edges. ArcGIS can also accommodate complex junctions, but their use requires construction of additional tools. For transportation networks, the most useful combination is simple junctions and complex edges.

The complete model of network features and elements is presented in figure 5.1.

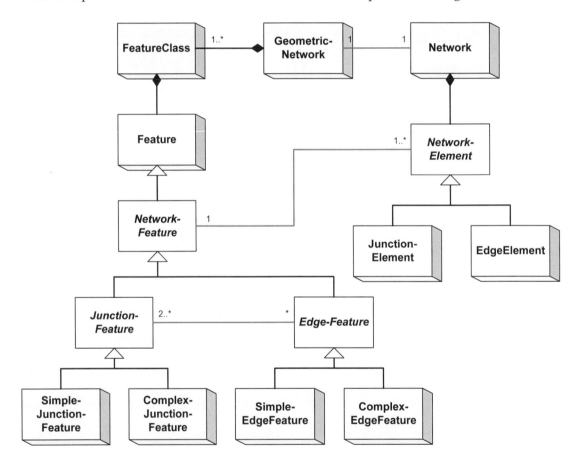

Figure 5.1 **ArcGIS geometric networks class model**

A network consists of a collection of edge and junction elements that form a schematic diagram of the system. A network element has no position except as defined by the relative location through connectivity of that element to other elements. A network element has no geometry and no dimensions. Network features are used to create and manage network elements. The geometry of network features is used to discern the network topology, which is reflected in the relationships between network elements. The relationship of network features to the elements they manage can be one to one for simple features or one to many for complex features.

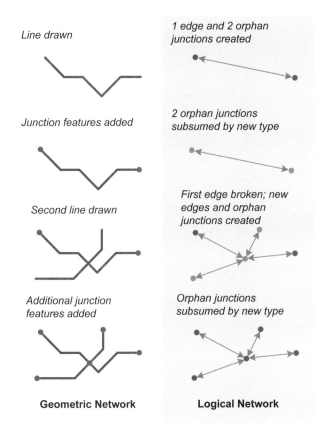

Figure 5.2 **Simple edges and junctions** Simple edges and junctions have a one-to-one correspondence with single-part geometric features. Actions in the geometric network produce changes in the logical network. When a polyline is drawn in an edge feature class, ArcGIS places an orphan junction at each end. These default junctions can be replaced with specific junction types by adding the desired junction feature at the endpoints. If a second line is constructed that crosses the first line, then both lines are broken at the point of intersection and new orphan junctions are added to the network. At this point, there are four edge features and four simple edge elements. There are also five junction features and five corresponding junction elements. The newly created orphan features may be subsumed by explicitly placed simple junction features.

A simple junction is always the end of an edge. A simple edge consists of a single line feature. Any line that crosses that feature is considered to have intersected it, resulting in the feature being subdivided into multiple edge elements. The coincident point where the two lines cross will form a new junction element. You can place explicit junction features at line termini and intersections, or simply accept the default junction ArcGIS places at these locations.

A complex junction can be a collection of internal edges and junctions. You have to create the structure and management software for complex junctions, as they are not supplied by ArcGIS. None of the network examples presented in this chapter will require complex junctions.

99

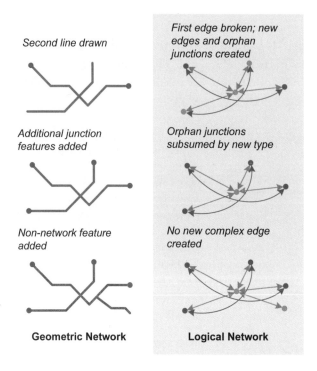

Figure 5.3 **Complex edges** Complex edge features have a one-to-many correspondence with network elements. The logical network remains a planar graph, with each edge feature intersection being represented by a junction element. Unless you explicitly place a junction feature at the intersection, however, ArcGIS network operations, like path tracing, will essentially ignore the intersection, thereby allowing the network to perform as if it were a nonplanar graph. Complex edges can include linear features that do not participate in the network, such as driveways, recreational trails, proposed facilities, and alleys. Most networks will benefit from using complex edges for a more realistic map appearance.

A complex edge consists of one or more simple edges treated as a single element of the network. A complex edge can have one or more crossing edges that will not form explicit junction features, although default junctions will be created internally by ArcGIS. Complex edges allow you to more fully model transport facilities in feature classes with members that do not participate in the network, such as driveways and railroad stub sidings. Complex edges also offer a way to deal with bridges, tunnels, and other network structures that do not represent actual intersections in the real world, although they may appear to in a planar graph.

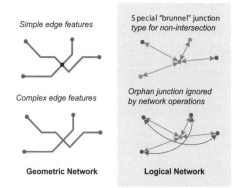

Figure 5.4 **Bridges in the network** Geometric networks in ArcGIS offer several choices for dealing with bridges, tunnels, and other parts of the transportation network where two lines cross but do not represent an actual intersection. One option is to create a special "nonintersection" simple junction feature type, which allows for simple edge features. Another is to employ complex edge features and to place no explicit junction feature at the point of overlap. You can also define connectivity rules.

Travel demand models

By using complex edge features for roadways, you can control whether a planar intersection is a point where travel can occur between facilities. If it can, put an intersection junction there, otherwise do not.

Travel demand models

A common application for geometric networks in a transportation geodatabase is to support travel demand modeling. A travel demand model examines the impact of changes in land use, the transport system, and system-user behaviors under various scenarios. Originally, a travel demand model was composed of network elements without any geometry. In its most basic form, a travel demand model is a set of tables that define the network, trip generation rates, and the probability for a given trip to exist between origins and destinations. However, if you do not use geometry, then it is difficult to see the network.

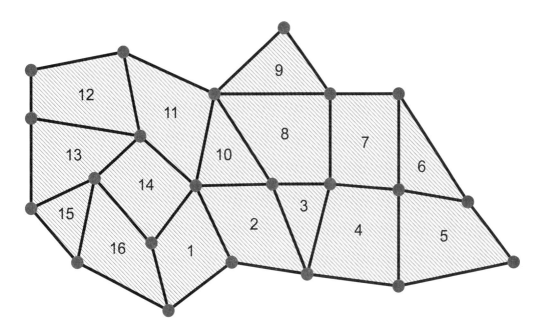

Figure 5.5 **TAZ network** A travel demand model consists of traffic analysis zones (TAZs), where trips are generated or attracted, bound by a travel network. The network consists of links and nodes. In a model with modal choice, there may be multiple networks, one for each mode of travel included. The TAZs would be common to all modes. A travel demand model is purely conceptual; link length and path do not represent anything used by the model. However, reasonably good geographic positioning of nodes will help the user visualize the network in terms of the real-world facilities it represents.

The model consists of traffic analysis zones (TAZs) bounded by transport facilities. The world outside the modeled area is itself treated as one or more external TAZs. Trips are produced by and attracted to TAZs. A trip may originate and end within a single TAZ in a process called internal capture. Trips generally originate in one TAZ and end in another. Of course, some trips will represent vehicles and people moving from the modeled system to an outside destination (internal-external trips), from outside the modeled area to a destination within the area (external-internal trips), or from one outside origin to a different outside destination (external-external or pass-through trips). Each edge bounding a TAZ has a maximum capacity and a traversal cost.

 Trips generally are classified by type, such as home to work and home to nonwork. The number of each type of trip produced by or attracted to a TAZ is determined by its land use. For example, a TAZ consisting of commercial land uses will attract home-to-work trips in the morning peak and be the source of return trips in the afternoon peak.

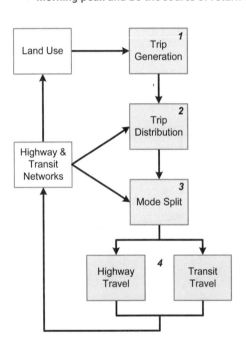

Figure 5.6 **Travel demand modeling process**
In this traditional four-step process, the first step generates person-trip travel demand based on land-use activities for each defined trip purpose within each TAZ. Access to transportation can influence land-use types and intensities. The second step estimates the probability that a given trip will end in a particular TAZ, including the one where it started or one outside the modeled area. Accessibility and availability of the means to travel influences trip distribution results. The third step involves making a choice as to how the trip will occur, usually either through private vehicle travel or a transit-supported trip. The modal choice probabilities are then assigned to the trips forecast to occur between each TAZ to produce network link-load estimates.

Using a travel demand model is traditionally a four-step process that begins by generating trips based on the land-use information supplied for each TAZ. The output of this process is the total number of trips that must be accommodated by the transport network during the modeled time period, which will typically be a peak travel time. There is some influence on land-use generation rates based on the degree of access to transportation and the size of individual units. For example, a TAZ with numerous residential subdivisions and easy access to the transport system will generate more trips than a sparsely settled area with few roads, even though both may have a residential land use.

Travel demand models

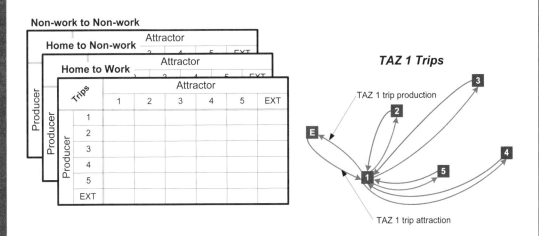

Figure 5.7 **Trip generation** It is common to use three kinds of trips: home to work, home to nonwork (school, shopping, recreation, medical, etc.), and nonwork to nonwork (usually the result of trip chaining within a home-to-work or home-to-nonwork trip). Each of these trips has different rates of production and attraction determined by TAZ land-use data. Trips are also generated by external users, some of whom may pass entirely through the modeled area on an external-to-external trip. The model must also estimate the desirability for the trip to end in each TAZ or go to an external destination. The relative attractiveness of a given TAZ as the destination of a trip is determined by the amount of appropriate land uses and relative difficulty of reaching those destinations. The result is a big matrix of trip production and attraction probabilities for each trip purpose produced over a number of balancing iterations.

The second step is to distribute trips. This involves trying to predict the TAZ destination for trips of each type originating in each TAZ. A set of probabilities based on distance and land use is used for this step, which is also influenced by the number of choices. If it is generally difficult to get to a particular destination TAZ, or it does not have much to offer, then the TAZ will not attract many trips.

The third step is to split the trips between the available modes of travel, which are typically highway travel in private cars or public travel in transit vehicles. Smaller urban areas without transit service may omit this step. The output of this step is a set of tables that represent trip volumes between origin and destination TAZs by mode of travel. These volumes are then loaded onto the travel network for the appropriate mode and routed to their destinations. Most models ignore truck travel to varying degrees.

In low-volume conditions, trips will follow the lowest-cost path. Trips must go a different path once a member edge has reached its capacity. A model is validated and calibrated by modeling existing conditions and comparing the model's traffic volumes on each edge to the volume actually being observed in the real world. Such a process is not absolute, because major roads form most edges and some edges may be a composite of multiple facilities. Travel on minor roads may be invisible to the model.

Based on what is observed in the model's results, transportation-system managers can identify those facilities where capacity improvements will provide the greatest benefits. The output can also indicate optimal locations for increases in the intensity of particular land uses and the impact of proposed changes. For example, a proposed commercial development may be expected to reroute existing trips now going to similar destinations in other TAZs, and to induce new trips from nearby locations. A travel demand model that includes the new development can be used to show its impact on the transport system and allocate capacity costs.

Building the travel demand model

Out of the box, ArcGIS can be used to build TAZs, develop networks, provide data loads into external modeling software, and display model results. Linking ArcGIS to the modeling software's database will facilitate these functions. It is also possible to extend ArcGIS to include the modeling process, although such an effort is beyond the scope of this book While conceptually a travel demand model looks like a set of polygons and their edges, the reality is a bit more complicated than that. First of all, there will likely be two networks, one for road travel and one for transit travel. These systems serve the same TAZs but follow different paths on the ground.

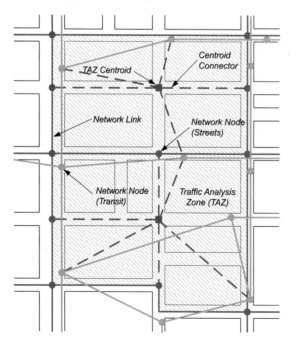

Figure 5.8 **Traffic analysis zones** TAZ trips appear to enter or leave the network at nodes where centroid connectors terminate. As this example shows, the network is generally a simplified abstraction of reality. Most models support one-way links. Trips are allocated on the network based on the difficulty of travel, which may be determined by speed, number of trips already assigned, and other considerations. This diagram shows two networks, one for highway travel and one for bus transit. The travel demand model will assign trips to both networks based on modal split probabilities.

Building the travel demand model

The problem of building a TAZ is not as bad as you might imagine. A TAZ may look like a polygon, but it really is a centroid, where all the trips are generated and attracted. The trips so generated and attracted appear on the network at junctions that end on edges leading to the centroid. A model can have different edges connecting the centroid to the various modal networks it contains. So, while the TAZ is the same for all modes of travel, each mode may be loaded at different nodes and can be described by different networks, each composed of a subset of modeled edges and junctions. The distribution of TAZ trips to the various connecting edges is usually proportional to the number of such edges. With three centroid connections, each edge would typically get a third of the total number of trips.

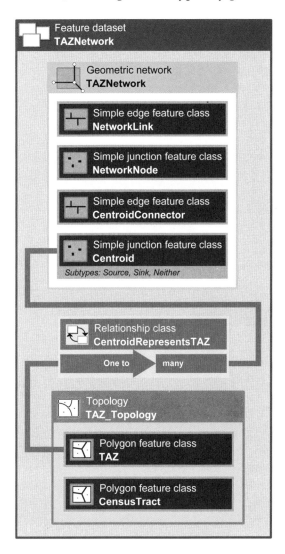

Figure 5.9 **Travel demand modeling** This example network dataset supports traffic analyses involving the production and attraction of trips between TAZs. Travel occurs along a highly abstract network. A TAZ represents an area of trip production and attraction, the travel demand data for which are often derived from census tracts and similar sources of TAZ land use. Simple edge features are adequate for this application because all network intersections represent a point of interconnection within the network. A relationship class ties each TAZ to the Centroid simple junction feature that serves as a source or a sink in the network through an option called the ancillary role. This part of the model is not necessary if you seek only to display the inputs or results of a traffic demand model. A TAZ is composed of one or more census tracts, so there should be topology rules that say: (1) census tracts must not overlap; (2) census tract must be covered by TAZ; and (3) census tracts and TAZs must cover each other.

Most users build TAZ polygons by combining census tracts. The data model reflects this approach. The classes are included in a topology group to make sure that each TAZ is composed of a whole number of census tracts and that TAZ polygons do not overlap.

The geometric network does not contain the TAZ polygons, only their centroids, which are stored in a simple junction feature class. The other class of this type stores network nodes. Two simple edge feature classes store network links and centroid connectors. Each TAZ centroid is connected to its TAZ polygon through a simple relationship class with a cardinality of one to one.

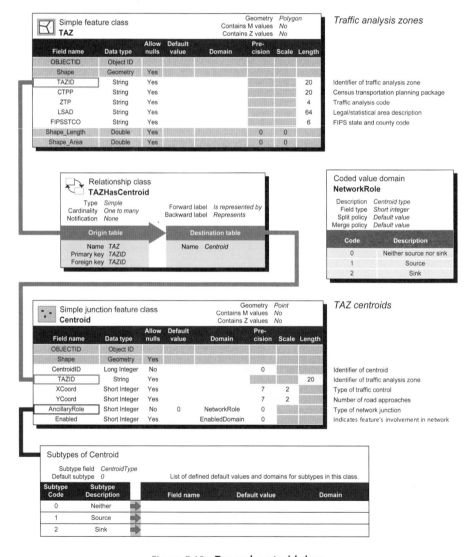

Figure 5.10 **Taz and centroid class**

Simple junction feature class **NetworkNode**				Geometry *Point* Contains M values *No* Contains Z values *No*				*Network junctions in travel demand model*
Field name	Data type	Allow nulls	Default value	Domain	Precision	Scale	Length	
OBJECTID	Object ID							
Shape	Geometry	Yes						
Node	Long Integer	Yes			0			Identifier of junction
XCoord	Double	Yes			7	2		X coordinate of node
YCoord	Double	Yes			7	2		Y coordinate of node
Enabled	Short Integer	Yes		EnabledDomain	0			Indicates feature's involvement in network

Figure 5.11 **Network node class**

This part of the physical data model shows the TAZ polygon feature class and the two simple junction feature classes. In the vernacular of ArcGIS, a junction can be a source or a sink of flows on the network, a property that is expressed in an AncillaryRole field that ArcGIS will add to a junction class when you indicate that it can serve one of these roles. The NetworkNode class does not include this field because all its members are neither sources nor sinks.

Most geometric networks manage flows that move away from sources and toward sinks. In a traditional network, such as may be constructed for an electric power grid or water system, sources and sinks are distinct. In a travel demand model, however, each TAZ centroid is both a source and a sink. ArcGIS requires us to choose which role it may serve, because a given junction cannot be both a source and a sink. As a result, we will need to define two centroid features for each TAZ and indicate through the value of the AncillaryRole field which centroid serves which role. Note that if you will only use the network to create the travel demand model's network and/or display the results of a travel demand model, then a centroid will not need to include the AncillaryRole field, nor will you need two class subtypes. You will not even need centroids. The Centroid class needs an attribute to store a foreign key pointing to the TAZ polygon it represents in all cases.

A network node has coordinate attributes to plot a general view of the network. Traditional practice says that NetworkLink geometries should be straight lines between these nodes. It would be a fairly simple matter to use the two node identifiers stored in the NetworkLink class to build the required link geometry. You do not need to manually create NetworkLink features. Instead, create a NetworkLink table with just terminal node fields, and then use it to generate NetworkLink features.

Chapter 5: Geometric networks

Simple edge feature class
NetworkLink

Geometry *Polyline*
Contains M values *No*
Contains Z values *No*

Network links in travel demand model

Field name	Data type	Allow nulls	Default value	Domain	Precision	Scale	Length	
OBJECTID	Object ID							
Shape	Geometry	Yes						
Focus	String	Yes	FocusCode			1	One-way indicator; a.k.a., focus code
ANode	Long Integer	Yes		0			Identifier of A node
BNode	Long Integer	Yes		0			Identifier of B node
Distance	Short Integer	Yes		0			Link distance
TollClass	Short Integer	Yes		0			Toll road class
FType	Short Integer	Yes	FCode	0			One-digit facility type code
AType	Short Integer	Yes	ACode	0			One-digit area type code
Const	Short Integer	Yes		0			Construction code
LUse	Short Integer	Yes		0			Land use code
Zone	Short Integer	Yes		0			Identifier of traffic analysis zone
Time	String	Yes	TimeCode			1	Indicates type of travel time/speed
Locat	Long Integer	Yes		0			Geographic location
Use	Long Integer	Yes		0			Use code
FreeSpeed	Short Integer	Yes		0			Estimated free-flow speed or time
ObsSpeed	Short Integer	Yes		0			Observed travel speed or time
Cap	Single	Yes		0	0		Hourly lane capacity
Count	Single	Yes		0	0		Hourly vehicle count
NoLn	Short Integer	Yes		0			Number of lanes
LGrp	Short Integer	Yes		0			Link group (screenline)
FType2	Short Integer	Yes	FCode	0			Two-digit facility type
AType2	Short Integer	Yes	ACode	0			Two-digit area type
Enabled	Short Integer	Yes	EnabledDomain	0			Indicates feature's involvement in network
Shape_Length	Double	Yes		0	0		

Figure 5.12 **Network Edges** This sample feature class is derived from the Florida Standard Urban Traffic Modeling System (FSUTMS). The link identifier is constructed as the combined A and B node identifiers. The Focus attribute indicates whether the link supports one-way or two-way travel (1 = one-way). If it is a one-way link, the flow is from the A node to the B node. The Time attribute indicates whether travel speed in **FreeSpeed** and **ObsSpeed** is stated as time or speed; a uniform time option is also provided. Facility type includes such classes as "31—undivided arterial unsignalized with turn bays," while area types include designations like "41—High-density outlying business district" and "43—Beach residential."

The attributes shown in this representative network link class were derived from those of the Florida Standard Urban Traffic Modeling System, but classes derived from other traditional models contain similar fields. Travel costs stored in the FreeSpeed and/or ObsSpeed fields can be expressed as speed or time. These fields can be used to calculate network element weights, which are the values used to indicate the relative ease or difficulty of moving along a particular edge. You can define multiple weights for a given edge class. Using the attributes provided, you could calculate a travel time weight as the product of Distance and FreeSpeed. This value could be modified by a volume-to-capacity ratio calculated using the Cap and Count fields or the presence of a toll imposed on the link, as indicated by the value of the TollClass field.

You can also explore various network configurations without having to produce multiple geodatabases. For instance, you can include all existing and proposed edges and defining junctions in this model and use the Enabled field to control which ones are part of a particular scenario in order to evaluate various alternatives. This works equally well for overlapping links with different capacities (widen a road or not) and for introducing entirely new links to the network (build a road or not). You can similarly include or remove centroids with various land uses and resulting trip-generation rates (build a shopping center or not).

Because of the one-to-one relationship between network feature classes and the network elements they represent, changing attribute values in the features will produce a corresponding change in the network. You do not have to rebuild the network and start over; the effect is dynamic. Adding and deleting network features or modifying the geometry of existing network features will similarly result in immediate changes to the network; however, the process used by ArcGIS to discover all the new connectivity can produce a decline in editing performance.

The impact of this processing overhead is proportional to the number of network feature classes used, as ArcGIS must make one database query for each involved network feature class. Using the map cache can allow more of this work to occur in memory and reduce the overhead workload in ArcGIS. Consideration of this performance impact during editing is one of the reasons the design uses a single Centroid class rather than two. Subtypes do not impact performance.

Map geodatabase for pathfinding

The second application we want to review as a use for geometric networks can create highly stylized maps and support pathfinding. This geodatabase is similar in design to the editing portion of the cartography geodatabase presented in chapter 4, which was itself taken from a geometric network.

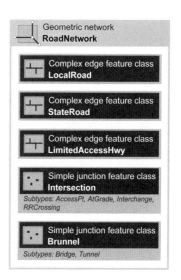

Figure 5.13 **Mapping and pathfinding** This illustrative network dataset provides three road complex edge feature classes and two simple junction feature classes. Using complex edges for road features allows them to ignore planar intersections where connections do not actually exist. Subtypes for the two simple junction feature classes are provided to manage differences in travel difficulty.

Chapter 5: Geometric networks

Complex edge feature class: LocalRoad
Geometry: Polyline
Contains M values: No
Contains Z values: No

Local road edges in network

Field name	Data type	Allow nulls	Default value	Domain	Precision	Scale	Length	Description
OBJECTID	Object ID							
Shape	Geometry	Yes						
RoadID	Long Integer	Yes			0			Identifier of road edge
PreDirection	String	Yes		Direction			2	Direction prefix
PreType	String	Yes		StreetType			4	Street type prefix
BaseName	String	Yes					30	Main portion of street name
PostType	String	Yes		StreetType			4	Street type suffix
PostDirection	String	Yes		Direction			2	Direction suffix
LeftFromAdd	Long Integer	Yes			0			Left side from address
LeftToAdd	Long Integer	Yes			0			Left side to address
RightFromAdd	Long Integer	Yes			0			Right side from address
RightToAdd	Long Integer	Yes			0			Right side to address
LeftCommunity	String	Yes					20	Left side community name
RightCommunity	String	Yes					20	Right side community name
LeftZip	String	Yes					5	Left side ZIP Code
RightZip	String	Yes					5	Right side ZIP Code
PeakSpeed	Short Integer	Yes			0			Estimated peak-hour free-flow speed
OffPeakSpeed	Short Integer	Yes			0			Estimated off-peak free-flow speed
Capacity	Single	Yes			0	0		Estimated peak hour capacity
AADT	Single	Yes			0	0		Annual average daily traffic
NumLanes	Short Integer	Yes			0			Number of lanes
FromNode	Long Integer	Yes			0	0		Identifier of from node (as addressed)
ToNode	Long Integer	Yes			0	0		Identifier of to node (as addressed)
Direction	Short Integer	Yes		TravelCode	0			Allowed directions of travel
Enabled	Short Integer	Yes		EnabledDomain	0			Indicates feature's involvement in network
Shape_Length	Double	Yes			0	0		

Complex edge feature class: StateRoad
Geometry: Polyline
Contains M values: No
Contains Z values: No

State road edges in network

Field name	Data type	Allow nulls	Default value	Domain	Precision	Scale	Length	Description
OBJECTID	Object ID							
Shape	Geometry	Yes						Identifier of road edge
RoadID	Long Integer	Yes			0			Direction prefix
PreDirection	String	Yes		Direction			2	Street type prefix
PreType	String	Yes		StreetType			4	Main portion of street name
BaseName	String	Yes					30	Street type suffix
PostType	String	Yes		StreetType			4	Direction suffix
PostDirection	String	Yes		Direction			2	Left side from address
LeftFromAdd	Long Integer	Yes			0			Left side to address
LeftToAdd	Long Integer	Yes			0			Right side from address
RightFromAdd	Long Integer	Yes			0			Right side to address
RightToAdd	Long Integer	Yes			0			Left side community name
LeftCommunity	String	Yes					20	Right side community name
RightCommunity	String	Yes					20	Left side ZIP Code
LeftZip	String	Yes					5	Right side ZIP Code
RightZip	String	Yes					5	Use code
RouteID	Long Integer	Yes			0			Identifier of route of which link is a part
BeginMeasure	Single	Yes			6	3		Beginning LRM measure for link
EndMeasure	Single	Yes			6	3		Ending LRM measure for link
PeakSpeed	Short Integer	Yes			0			Estimated peak hour free-flow speed
OffPeakSpeed	Short Integer	Yes			0			Estimated off-peak free-flow speed
Capacity	Single	Yes			0	0		Peak hour capacity
AADT	Single	Yes			0	0		Annual average daily traffic
NumLanes	Short Integer	Yes			0			Number of lanes
FromNode	Long Integer	Yes			0			Identifier of from node (as addressed)
ToNode	Long Integer	Yes			0			Identifier of to node (as addressed)
Direction	Short Integer	Yes		TravelCode	0			Allowed directions of travel
Enabled	Short Integer	Yes		EnabledDomain	0			Indicates feature's involvement in network
Shape_Length	Double	Yes			0	0		

Figure 5.14 **Network edge features** (continued on next page)

Map geodatabase for pathfinding

Complex edge feature class
LimtedAccessHwy

Geometry: *Polyline*
Contains M values: *No*
Contains Z values: *No*

Limited-access highway edges in network

Field name	Data type	Allow nulls	Default value	Domain	Precision	Scale	Length	
OBJECTID	Object ID							
Shape	Geometry	Yes						
RoadID	Long Integer	Yes			0			Identifier of road edge
RouteID	Long Integer	Yes			0			Identifier of route of which edge is a part
BeginMeasure	Single	Yes			5	2		Beginning LRM measure for edge
EndMeasure	Single	Yes			5	2		Ending LRM measure for edge
PeakSpeed	Short Integer	Yes			0			Estimated peak hour free-flow speed
OffPeakSpeed	Short Integer	Yes			0			Estimated off-peak free-flow speed
Capacity	Single	Yes			0	0		Peak hour capacity
AADT	Single	Yes			0	0		Annual average daily traffic
NumLanes	Short Integer	Yes			0			Number of lanes
FromNode	Long Integer	Yes			0			Identifier of from node (LRM direction)
ToNode	Long Integer	Yes			0			Identifier of to node (LRM direction)
Direction	Short Integer	Yes		TravelCode	0			Allowed directions of travel
TollFlag	Short Integer	Yes		TollCode	0			Indicator as to whether facility is a toll road
Enabled	Short Integer	Yes		EnabledDomain	0			Indicates feature's involvement in network
Shape_Length	Double	Yes			0	0		

Coded value domain
Direction

Description: *Street direction modifier*
Field type: *String*
Split policy: *Default value*
Merge policy: *Default value*

Code	Description
N	North
S	South
E	East
W	West
NE	Northeast
NW	Northwest
SE	Southeast
SW	Southwest
	None

Coded value domain
StreetType

Description: *Street type*
Field type: *String*
Split policy: *Default value*
Merge policy: *Default value*

Code	Description
Ave	Avenue
Blvd	Boulevard
Cir	Circle
Ct	Court
Dr	Drive
Hwy	Highway
Ln	Lane
Loop	Loop
Pkwy	Parkway
Rd	Road
St	Street
Terr	Terrace
Trl	Trail
Way	Way

Coded value domain
TravelCode

Description: *Direction of travel*
Field type: *Short integer*
Split policy: *Default value*
Merge policy: *Default value*

Code	Description
1	1-way, From > To
2	2-way
3	1-way, To > From

Coded value domain
TollCode

Description: *Toll indicator*
Field type: *Short integer*
Split policy: *Default value*
Merge policy: *Default value*

Code	Description
0	No tolls
1	Tolls

Street Types.
The domain for StreetType is abbreviated from the complete list that you may want to include. For a more detailed set of examples commonly found in U.S. addressing, refer to Appendix C of Publication 28 from the U.S. Postal Service.

Figure 5.14 **(continued) Network edge features** This sample database design features three kinds of network edges: LocalRoad, StateRoad, and LimitedAccessHwy. An edge classified as a state road or limited-access highway can be tied to a larger route that consists of several edges; local road edges are joined by a common street name. Some attributes are common to all three edge feature classes as an aid to calculating edge weights. Terminal junctions are ordered by address or, for limited-access roads, increasing LRM measures. A direction attribute expresses whether one-way or two-way operation is permitted.

The sample geodatabase includes three complex edge feature classes for different kinds of roadways and two simple junction classes. There are four Intersection subtypes and two Brunnel (a composite representing bridges and tunnels) subtypes. These subtypes support the various cardinalities that exist for these features and the edges they involve without imposing a performance penalty. The subtypes allow you to apply different weights for traveling through the various edge and junction types. For example, you might assign a greater weight for travel through an at-grade intersection than through an access point.

The figure 5.14 physical data model shows the three different sets of attributes for each kind of roadway. State roads and limited-access highways have route-milelog measure fields, which will be discussed in chapter 7, because these are common in state DOT databases. State and local roads have address fields, because they may be used for geocoding. Address fields are omitted for limited-access highways because they generally do not have addresses. You cannot get to the adjacent land from these facilities. Route-milelog fields were omitted for local road features on the expectation that they are not included in the state system.

The geometric network uses some of the included fields to manage edge elements. For instance, the Enabled field tells the network whether a particular edge should be included in a pathfinding exercise. The Direction field is used to tell the geometric network whether travel is restricted by direction of flow. The related TravelCode domain class indicates the available direction choices.

Fields were included to store from- and to-junction identifiers, which are found in the simple junction feature classes. (They are called nodes in the mnemonic field names to be consistent with traditional transportation terminology.) The geometric network does not need you to supply this information. It determines terminal junctions by looking at the features and where they touch. However, it will help you keep everything straight when you use explicit junction features by reminding you what the network connectivity may be for each feature. (You cannot view network element classes.) There is no explicit junction type for the end of a road, so not every complex edge feature class will have two terminal node entries even if you enter all the explicit Intersection and Brunnel junction class identifiers. If you wanted to include such a feature, you could create a fifth intersection subtype for the end of a road.

There are two simple junction classes in order to store different attributes for each basic type. For intersections, you need to know things like the kind of traffic control used. That information may be helpful to calculate a travel cost used to determine an element weight. The different set of attributes provided for the Brunnel class lets you store information about how large a vehicle may go through the junction. The design includes the brunnel subtypes of Bridge and Tunnel in the JunctionCode domain class. You could use separate domain classes.

Chapter 13 shows how to use the ArcGIS Network Analyst extension to create other kinds of network geodatabases that provide a richer set of functions for transportation applications.

Map geodatabase for pathfinding

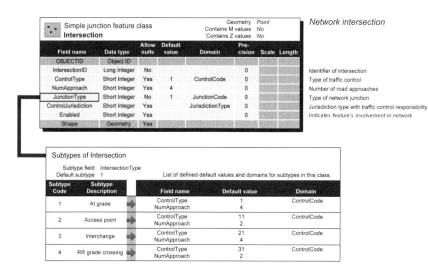

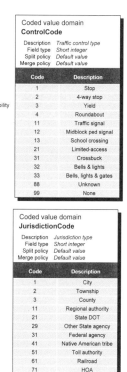

Intersection subtypes If you do not care about the different attributes that can be defined for the four intersection subclasses, then you can take advantage of the performance enhancement offered by subtypes. Although all the attributes have to be the same, you can define different topology and network rules for each subtype.

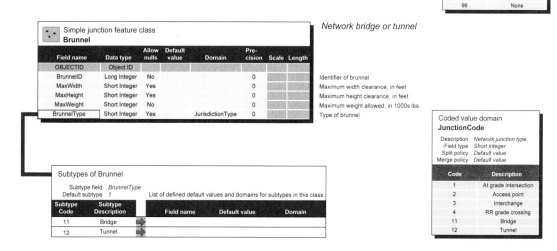

Figure 5.15 **Brunnel subtypes** This network model includes two simple junction feature classes. The "normal" version includes intersections of various types. This second, somewhat contrived Brunnel class reflects the nonintersection restrictions that may be imposed by bridges and tunnels. Overpasses may impose two Brunnel nodes, one for the road passing over the bridge, which has a maximum weight restriction, and one for the road passing under the bridge, which may be subject to both height and weight restrictions.

113

PART II *Understanding transportation geodatabase design issues*

chapter six

Data editing

- Traditional editing process
- Multiple editors
- Continuous versioning
- Preserving data dictionary history
- Separating the editing and publishing geodatabases
- Service-oriented architecture

This book is about designing geodatabases for editing transportation data of all types. This chapter will present some of the design approaches that can improve the functionality of a database for editing.

Why the focus on editing? Most transportation geodatabases are composed of data subsets supplied by various workgroups within the organization. Often, a metropolitan planning organization (MPO), local government agency, federal data resource, or other external group supplies the data. This means that transportation geodatabases are inherently multiparticipant in nature, which requires the various data suppliers and users to work somewhat independently.

In a small agency, census data from the federal government is a common starting point for street centerlines, with land data coming from a county property appraiser. MPOs receive data from their members and need to aggregate it into a regional dataset. In a larger organization, multiple workgroups supply pieces of the total dataset, like traffic data from planning, crash data from highway safety, pavement condition data from engineering, and highway inventory information from district offices.

The transportation-data user community is similarly structured with local governments and state and provincial agencies sharing data. Those agencies must supply data in a number of federally mandated inventory and condition reporting systems. Transportation data is both ubiquitous and varied in its structure.

This independent confederation of actors imposes a wide variety of data structures on the geodatabase-design process, from the fixed segmentation schemas common to local governments to the route-milelog structures normally used at state and provincial transport agencies. The editing environment that seeks to consolidate these sources is the focal point for the difficult geodatabase-design problems this book addresses. Whether one person or a several workgroups take part in the editing process, the design problems must be solved.

Traditional editing process

The editing process begins when data is initially placed in the geodatabase. The traditional approach is to compile features from aerial photography, add attributes, or load a dataset supplied by someone else. Either way, you likely will start building the initial geodatabase with a collection of lines representing linear facilities. It is very likely that these lines are not subdivided at the locations you might choose. If you work for a local government following the usual method of segmenting transport facilities, you will subdivide streets at intersections to produce block-length features. You will then assign the official street name and address range to each block. If you work for a state DOT, you might subdivide the lines at county or district boundaries, and perhaps at intersections of state roads, and add route identifiers and milelog references. Either way, the output is a collection of lines cut to the right length with identifiers and location references.

The next step is to add other attributes. Here, the process becomes more differentiated depending on whether you are using blocks and addresses or routes and milelog measures. The street block features will simply get more attributes defined and the appropriate values will be entered. You may have to subdivide a block if one of the attributes changes value along its length. Conversely, if two or more adjacent blocks have the same value for a particular attribute, then you will enter it more than once. This is a denormalized geodatabase design with the resulting data duplication. Fortunately, you only have to load the database

Traditional editing process

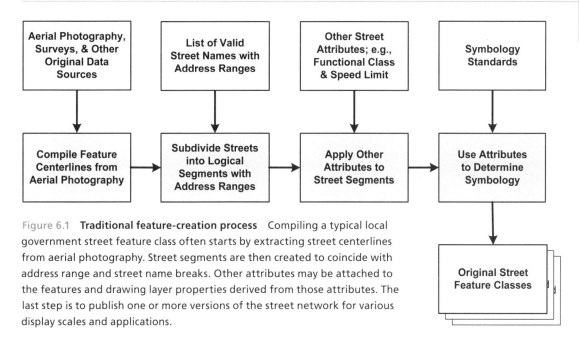

Figure 6.1 **Traditional feature-creation process** Compiling a typical local government street feature class often starts by extracting street centerlines from aerial photography. Street segments are then created to coincide with address range and street name breaks. Other attributes may be attached to the features and drawing layer properties derived from those attributes. The last step is to publish one or more versions of the street network for various display scales and applications.

once and it likely will not change much except to add new segments as growth occurs. Plus, the end result can be used directly by others to support their applications. If there is an interest in only a few of the attributes, then the user can merge adjacent segments to reduce the number of features to just those that contain unique values. One or more attributes are used to control line symbology.

A state DOT using a route-milelog reference system will create event tables to store the additional attributes; feature editing is complete. Users will select the event types of interest and use dynamic segmentation to create the features they need with unique values in each segment. Local governments combine segments to get a final database while state DOTs cut segments into the pieces they need.

After the initial dataset is created, the periodic editing process similarly follows a rather common sequence of events. Information on changes, additions, and deletions is compiled during the intervening period and applied to the dataset. Because you are working with a geodatabase, the first thing to do is edit the geometry. You can just simply delete the features that no longer exist, but that rarely happens with transport facilities. It is more likely that you will need to modify some attribute or add new features and load all the attribute data. Once the edits are made and checked, you publish a new geodatabase version for everyone to use.

The traditional dataset creation and editing process assumes that there are very few editors. It arose at a time when few people had access to GIS software and the powerful

Chapter 6: Data editing

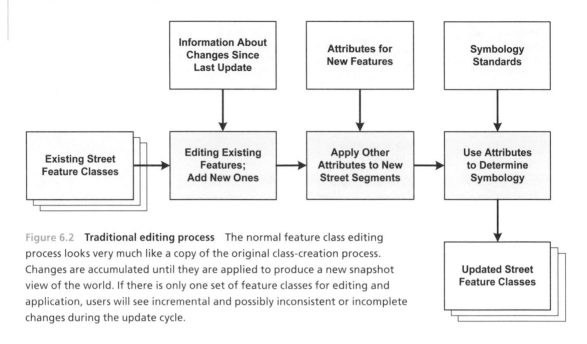

Figure 6.2 **Traditional editing process** The normal feature class editing process looks very much like a copy of the original class-creation process. Changes are accumulated until they are applied to produce a new snapshot view of the world. If there is only one set of feature classes for editing and application, users will see incremental and possibly inconsistent or incomplete changes during the update cycle.

(and expensive) computers it ran on. These conditions no longer exist. Even in a single-person editing shop, it is likely that other people use the data you produce. This chapter will focus on the needs of multiple editors and users who can benefit from ArcGIS functions and geodatabase design principles.

Multiple editors

A state DOT or a large local government will typically have internal workgroups responsible for various parts of the transport system. Each of these workgroups edits a portion of the overall geodatabase, and may make changes at different times.

Some decentralized organizations can be quite large and present challenges to

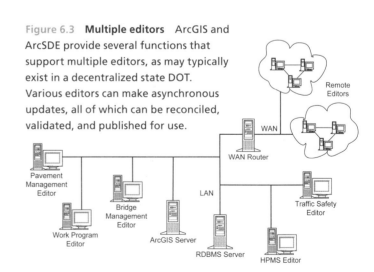

Figure 6.3 **Multiple editors** ArcGIS and ArcSDE provide several functions that support multiple editors, as may typically exist in a decentralized state DOT. Various editors can make asynchronous updates, all of which can be reconciled, validated, and published for use.

the editing process in terms of coordinating and validating changes to the geodatabase. Consider the situation of a state DOT, where various statewide offices in the headquarters city are responsible for particular pieces of the geodatabase. They are editing mainly event tables. The remote district offices also may be editing some event tables, but they alone are responsible for editing the geometry. There may also be a Web site where the public and field-office workers not on the wide-area network can access a version of the geodatabase.

One of the first things you need to consider is versioning to help manage this diverse group of editors and users. ArcSDE supports versioning by managing each class as a collection of three subsets. The base table is the initial version, what is referred to as the default version. Changes are stored in the two other tables. An Adds table has all the new rows and a Deletes table has all the removed rows. An edit action that modifies an existing row will produce one Deletes table row (the original row) and at least one Adds table row (the new row). To show a user the current version, ArcGIS will go to the Base table, remove the Deletes table rows and add the Adds table rows, and then show the end result to the user. As you might imagine, the more edits you make, the longer this process takes. You can compress the geodatabase and apply all the additions and deletions to the default version to create a new default version, which will clear the Adds and Deletes tables.

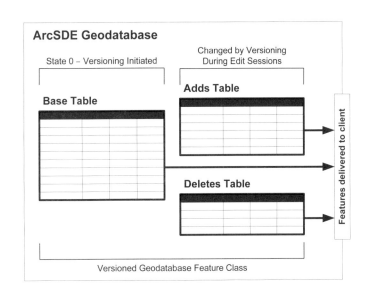

Figure 6.4   **ArcSDE versioning**

The reason this helps manage a multieditor environment is that each editor is writing new rows in the Adds and Deletes tables, not in the "real" geodatabase. This means you have a big Undo button at your command. An end-of-editing reconciliation process can be used to make sure that all the edits are compatible.

Chapter 6: Data editing

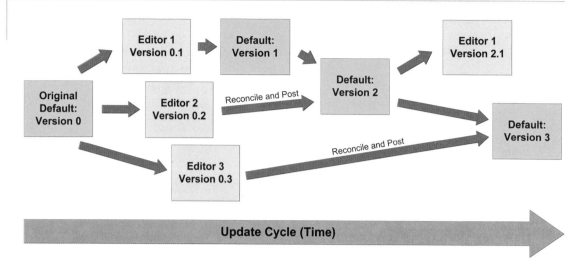

Figure 6.5 **Direct editing of default geodatabase version** The simplest way to use versioning in ArcGIS is to directly edit the default version and reconcile changes with every save. One of the primary issues with this approach is the constantly changing nature of the geodatabase. Transportation agencies may also be concerned about the lack of support for long transactions, as may be required for the evaluation of alternative designs. (Derived from **Versioning Workflows**, p. 2, an ESRI technical reference document.)

There are many ways versioning helps manage edits to a geodatabase in ArcSDE. The simplest way is to directly edit the default version. Whenever you initiate an editing session, ArcGIS will create a copy of the default version for you to work with. As each editor seeks to commit changes to the geodatabase, ArcGIS will go through a reconcile-and-post process to create a new default version. This process compares the current version of the geodatabase with the edits to make sure that there are no conflicts, like modifying a feature that was previously deleted by another editor. Each editor who subsequently commits edits to the geodatabase goes through a similar reconcile-and-post process. As long as no conflicts are detected, the process goes fairly quickly. If, however, a conflict is found, the editors are notified and the two versions must be modified before the second set of edits can be committed to the geodatabase.

As you might imagine, one of the potential problems with this style of versioning is that the geodatabase is constantly changing. The first editor to commit changes gets everything asked for. Everyone else is now working with an old copy of the geodatabase, as it existed in a previous state. Users may get different results from an analysis operation. This style of versioning also does not support long transactions, because every commit goes to the default version.

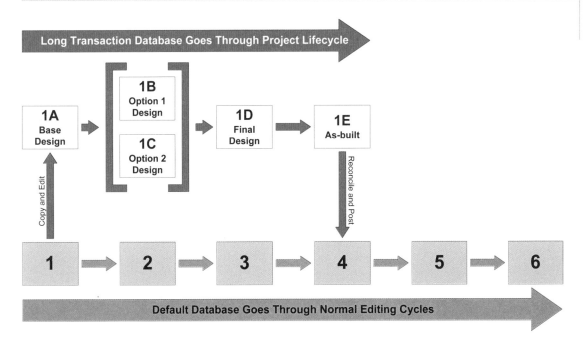

Figure 6.6 **Explicit state versioning** In this example of a multitier version tree, a project that will change the database goes through its independent lifecycle. Once the project is completed, the as-built version is reconciled to the default geodatabase—which has been going its own merry way—and a new updated version is created with all compatible edits posted. ArcSDE supports a large number of such independent editing processes in a single instance using this explicit state model. (Derived from Fig. 6-8 in **System Design Strategies**, an ESRI technical reference document.)

Long transactions depend on multiple default versions in an ArcSDE structure called a multitier version tree following an explicit-state data model. Each version can move through its life and be reconciled at a later time. Editors commit changes to each separate lineage, just as if the others did not exist.

A typical application of long transactions in transport agencies is to evaluate alternative designs. One or more editors, in isolation from the normal editing process, modify a copy of the default version. The final design is constructed and converted to an as-built set of edits that are reconciled and posted to the default version that exists at the time. There can be many such parallel editing processes at any time.

This is also a way for staff in remote offices to modify the geodatabase. Each office can take a copy of the geodatabase and update changes within its jurisdiction. These changes can be periodically posted and reconciled with changes from other offices to ensure alignment at jurisdictional edges. All editors are working on one instance of the ArcSDE geodatabase.

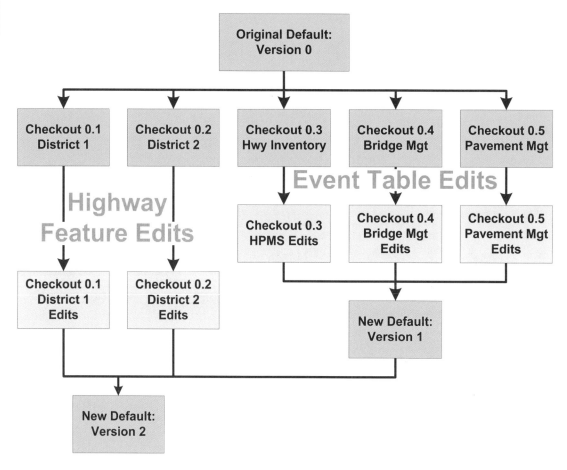

Figure 6.7 **Disconnected editing through versioning** In this example of disconnected editing, which is an extension of ArcGIS geodatabase versioning, various central office units are responsible for editing event tables in the geodatabase while the district offices maintain the road centerline features and selected event tables. The two sets of edits are periodically brought together to produce a revised default geodatabase containing feature and object class edits for redistribution throughout the agency and to outside users. (Derived from **Versioning Workflows**, an ESRI technical reference document, pp. 9-10.)

A more versatile form of multieditor geodatabase versioning is offered by disconnected editing that works with a check-in/checkout process. A multitier editing tree requires all editors to be connected to a single ArcSDE instance. This may be a problem for remote offices due to network latency issues. Disconnected editing allows a remote office to check out a copy of the default version, work on it in isolation, and then post and reconcile changes to the shared version at a future date. This editing approach is a form of data replication, which is discussed later in this chapter, and is especially useful when one group is editing

Continuous versioning

geometry and another is updating event tables. It is important that the workgroup responsible for creating new facility identifiers does so before others attempt to supply the related geometry and event data. As this requirement implies, disconnected editing requires a considerable amount of coordination and oversight by the ArcSDE administrator.

Continuous versioning

ArcSDE versioning and rollback logs supported by various RDBMS products offer a way to retain the editing history of the geodatabase. Eventually, though, you will need to compress the geodatabase and create a new default version in order to improve performance and commit edits to produce a new public geodatabase. At that time, the editing history will be lost. This section on continuous versioning will show a database design technique that will permanently preserve editing history.

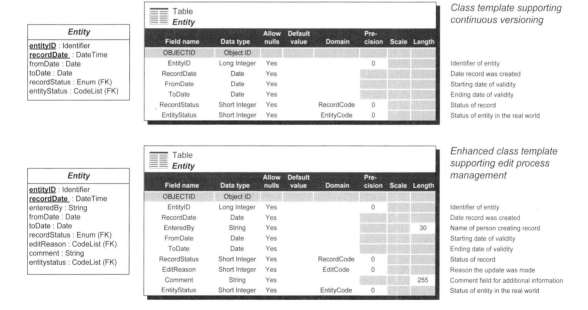

Figure 6.8 **Class templates** The Entity class template is an abstract superclass stereotype, or template, for developing all table and feature classes in a geodatabase for which you want to provide temporal support. This figure shows both logical and physical data model forms. In addition to full support of temporal aspects of the database, such an approach to database design can better manage the data editing process. An enhanced Entity template is also illustrated, which provides additional attributes to support editing process management.

123

The technique is based on adding a set of standard fields to all user-defined geodatabase classes. The *Entity* stereotype (template) shown in figure 6.8 illustrates the two choices you could follow, with the bottom one offering more functionality. Both templates retain the temporal data that apply to each class. The bottom template also supports enhanced editing-process management. We will examine each field to explain its role.

The most important field is the entity identifier, EntityID. Although ArcGIS supplies each class with an OBJECTID field that it uses to ensure unique identification of each member, the value of this field can change during certain operations, such as when you copy a geodatabase. To be certain you have a value that never changes for building foreign keys, you need to explicitly create a candidate primary key field.

The RecordDate field stores the data and time when the row was created. EntityID and RecordDate can work together to uniquely identify each row as a complex primary key that you can use with relational tables and files that exist outside the geodatabase. (ArcGIS presently supports only simple primary and foreign keys.) You can use RecordDate to chronologically order the various versions of a given entity in the geodatabase; i.e., rows with the same EntityID value but different RecordDate values.

The FromDate and ToDate fields store the period of validity for the information contained in the row. Every record should be given a FromDate value when it is created. A ToDate value will be added when the information in the row is no longer valid. FromDate is different from RecordDate because there is a latency period between the time a change occurs in the real world and the time it can be reflected in an update to the geodatabase.

The key to making the Entity class template support continuous versioning is the RecordStatus field. In its most basic implementation, the RecordStatus field can take one of three values: Active, for records that indicate the present state of the real-world entity; Retired, for records that show a previous valid state for the entity; and Replaced, for rows that were later found to be in error. To find the current version of the geodatabase, you select all the rows with a RecordStatus value of Active.

The RecordStatus field lets every class operate as if it had a permanent set of Adds and Deletes tables. Internally, this approach is fully compatible with all forms of ArcSDE versioning. Externally, it allows remote users to perform transaction updates rather than go through a process of completely replacing their version with a downloaded copy. This design feature is especially useful when the remote user mixes the downloaded source data with data it maintains locally. For example, a local government may periodically download information about state roads from the state DOT geodatabase and then integrate that data into a local geodatabase containing information on all roads within the local government's jurisdictional boundary. A more traditional drop-and-replace approach requires much more work to integrate updates.

The update process based on the history preservation design offered above involves three steps:
1. Read the existing record that may be changed by searching for the related entity identifier value (EntityID). If there is no such identifier value, then the desired action must be to simply add a new record to the table or feature class. (Skip to Step 3.)
2. Update the old record to reflect the edit action. Replace the original RecordDate with the current date and time. If the action is to retire the record as a result of changes in one or more attributes, then you will add a ToDate value for the last date the old information was correct and change the RecordStatus value to 7 (Retired). The ToDate value will indicate the last time the information in the record was valid. If the action is to replace the record, which means the historical data is incorrect, then you will change the FromDate value to null and change the RecordStatus value to 6 (Replaced). You may also want to ensure that the value of ToDate is null. The null values will indicate that the record has no validity for any period in time. You then write the modified record back to the database, which will update the old record.
3. Add all the new records with a RecordStatus value of 5 (Active) to the table or class. If the previous record was retired or replaced, then the active record will provide new information. If there was no previous record, then the active record presents information about a new entity of the type stored in the table or class being updated. Either way, the new active record should include an appropriate FromDate value. If the action is a replacement to correct an error, the new active record is likely to include the same FromDate value as the erroneous record it replaced. If the action is retirement, then the ToDate value of the old record is probably the same as the FromDate of the new record.

It would be useful to develop a little data entry routine that did all this for you whenever you selected a row for editing and changed a value. Such a tool would automatically modify the old row and create a new one based on the type of action.

The EntityStatus field offers a way to differentiate between the status of the row in the geodatabase and the status of the entity in the real world. For example, if the row represented a proposed facility, it would have a RecordStatus of Active and an EntityStatus of Proposed. Other typical EntityStatus field values could be Under Construction, Open to Traffic, Closed for Maintenance, and Abandoned. Such a field could be used to identify those facilities that need to be ignored by a pathfinding application because they are not open to traffic. To find the full list of all facilities that can be used today, you would select all the Active records with an EntityStatus value of Open to Traffic.

The longer version of the *Entity* template adds three fields that can facilitate editing process maintenance. EnteredBy stores the name or user ID of the person who made the edit. This allows a user to identify the person with whom they need to discuss the edit if they want to

Chapter 6: Data editing

Initial State 0

EntityID	RecordDate	FromDate	ToDate	RecordStatus	EntityStatus	...
1	08/31/2003	01/01/1978		5	6	
2	08/31/2003	01/01/1978		5	6	
3	08/31/2003	01/01/1978		5	6	
4	08/31/2003	01/01/1978		5	6	
5	08/31/2003	01/01/1978		5	6	
6	08/31/2003	01/01/1978		5	6	

How to use versioning To see the present state of the world, simply select all rows with a **RecordStatus** equal to 5 (active).To find a previous version, select all records with a **FromDate** value earlier than the historical date you seek to reproduce and either no **ToDate** value (active records that remain valid) or a value that is after the cutoff date you selected (now-retired records that were active at the time).

Updated State 1

EntityID	RecordDate	FromDate	ToDate	RecordStatus	EntityStatus	...
1	08/31/2003	01/01/1978		5	6	
2	08/31/2003	01/01/1978		5	6	
3	08/31/2003	01/01/1978		5	6	
4	08/31/2003	01/01/1978		5	6	
5	08/31/2003	01/01/1978	02/18/2007	7	6	
6	08/31/2003	01/01/1978		5	6	
7	02/23/2007	01/25/2007		5	6	
5	02/23/2007	02/19/2007		5	6	

Record retired because entity attribute(s) changed

New active record written with revised information

Updated State 2

EntityID	RecordDate	FromDate	ToDate	RecordStatus	EntityStatus	...
1	08/31/2003	01/01/1978		5	6	
2	08/31/2003	01/01/1978		5	6	
3	08/31/2003	01/01/1978		5	6	
4	08/31/2003	01/01/1978		5	6	
5	08/31/2003	01/01/1978	02/18/2007	7	6	
6	08/31/2003	01/01/1978		5	6	
5	02/23/2007	02/19/2007		5	6	
7	02/23/2007			**6**	6	
8	02/23/2007	01/19/2007		5	6	
9	02/23/2007	02/04/2007		5	6	
7	02/23/2007	01/01/1978		5	6	

The second updated version shows that the new row created for Entity 7 had an error in it and a corrected version must be written. Start by deleting the **FromDate** value for Entity 7 and changing its **RecordStatus** to 6 (replaced). Then write a new active record for Entity 7 with the correct information.

Record replaced because one or more attributes were in error

New active record written with revised information

Figure 6.9 **How continuous versioning works** The **RecordCode** attribute tracks all the various states of an entity's description in the table. This example starts with an initial state where all the records are active, representing the present state of the world. State 1, the first update, retires the original row for Entity 5 for which one or more attributes changed on 02/18/2007, and that becomes the new **ToDate** value. **RecordStatus** is also changed to 7 (retired) in the original record. A new active row is written with the revised value and a **FromDate** value of 02/19/2007. Because we use a complex primary key consisting of the **EntityID** and **RecordDate**, retaining the old **EntityID** value does not present a problem.

better understand why it was made. It also allows editors to take ownership of their work and for management to supervise staff. The EditReason and Comment fields provide space to explain edits. This information helps other users understand updates to the geodatabase and can be a reminder to the editor who made the change.

You can select those portions of the two *Entity* class templates that you want to use. You could even use different sets of the suggested fields for various classes. For example, event tables (object classes) could omit the EntityStatus field because the facility feature class or table that creates the facility identifier would contain this information. Or, you may choose to limit the editing process management fields to feature classes, where changes to one feature could inadvertently affect other features.

Coded value domain
RecordCode

Description *Record status code*
Field type *String*
Split policy *Default value*
Merge policy *Default value*

Code	Description
0	Work in progress
1	Proposed
2	Withdrawn
3	Rejected
4	Accepted
5	Active
6	Replaced
7	Retired

Coded value domain
EntityCode

Description *Entity status code*
Field type *String*
Split policy *Default value*
Merge policy *Default value*

Code	Description
1	Proposed
2	Under review
3	In design
4	Under construction
5	Substantial completion
6	Open to traffic
7	Accepted, in service
11	Damaged
12	Under repair
21	Jurisdiction transferred
31	Closed to traffic
32	Removed from service
33	Abandoned
34	Surplused
35	Stored
86	Demolished

Figure 6.10 **Status codes**
Two distinct status types are needed to support the temporal aspects of a geodatabase: the status of the record in the database and the status of the entity in the real world. Record status is part of workflow control and database management. Entity status is an attribute of the real-world entity the record describes. You can have an active record that describes a proposed transport facility, for example, or a proposed record that describes a facility presently open to traffic. Only **RecordCode** values of 5, 6, and 7 are needed to provide temporal support; the other values are for edit-process management.

It is also useful to increase the possible values for RecordStatus as a strategy to improve editing process management. The RecordCode domain shown in figure 6.10 lists four values that can be used during the editing process. Whether your agency uses direct editing of the default version, multitier version trees, or disconnected editing, you can employ these additional RecordStatus values as an internal management tool. The complete list of RecordStatus values and their intended meanings are:

0 *Work in Progress*, for identifying those edits a user has entered into the editing version of the geodatabase but which have not been submitted for validation. If an edit would result in a change to an existing row, the original row will remain unaffected by the update until the edit administrator accepts the new data. Ideally, the edit interface would present the editor's rows with a status value of 0 (Work in Progress) in a different color from that used to show official data. A single edit action may result in a cascade of other changes, so that any single edit may produce multiple new rows.

Coded value domain
EditCode

Description *Edit reason code*
Field type *String*
Split policy *Default value*
Merge policy *Default value*

Code	Description
0	Original record
1	Error discovered
2	New info received
3	Change in entity
4	Jurisdiction change
5	Construction started
6	Construction finished
7	Field revision
8	Maintenance
9	Jurisdiction transferred
10	Put out of service
99	Other

Figure 6.11 **Edit reason codes** The RecordCode and EntityCode domains provide most of what you will need for continuous versioning and full edit process management. However, it is often just as important to know why something changed in the database as what the change was. The **EditReason** and **Comment** attributes offer the record-level metadata needed to explain what resulted in a revised record being created. The domain suggested here is just a start for one that more closely matches your specific needs. You may even choose to develop separate reason code domains depending on the nature of the entity involved. For example, the edit reasons for conveyances might be substantially different than for the transport facilities upon which those conveyances operate.

1 *Proposed*, for identifying a row that has been formally proposed by an editor as an update to the database. This is not the same as saying the real-world entity itself is proposed, which would be in EntityStatus. Rows with a RecordStatus value of 0 (Work in Progress) should be changed to 1 (Proposed) when they are submitted for validation. At the time of proposal, the editor would select all rows with a RecordStatus value of 0 and change that value to 1. The edit administrator would then be able to identify the proposed edits using the RecordStatus value and, perhaps, the value of the EnteredBy field.

2 *Withdrawn*, for identifying a row that was originally part of a user's working update but was subsequently deleted from further processing, such as when a user corrects his or her own mistakes discovered during data validation. A withdrawn row has null beginning and ending validity dates. This value is required because the geodatabase operates under the general rule that no row is ever deleted; however, row deletes can occur to remove withdrawn rows and others that never reached the *RecordStatus* of 5 (Active) just before a new default version is created.

3   *Rejected*, for identifying a row that was rejected by an edit administrator. The geodatabase design may want to support editor feedback by including fields to indicate the reason for rejecting edits; e.g., EditorActionType and ActionReason. These fields would not be part of the published geodatabase. The applicable editor will be notified as to the need for further work, and must eventually change the RecordStatus value for rejected rows to 2 (Withdrawn) and generate a new proposed row for submission to the data validation process.

4   *Accepted*, for identifying a row that has been confirmed by the validation process and can be moved to the public geodatabase. Accepted rows are accumulated from all editors until the update cycle is completed, at which time the update is implemented and a new default version created. Only accepted rows can be given a RecordStatus value of 5 (Active). Rows with a RecordStatus value less than 5 should be purged from the geodatabase after the update is committed.

5   *Active*, for identifying a row that is an official statement regarding a transport facility or service. An active row has a validity period beginning date but not an ending date. The collection of all active rows represents current conditions.

6   *Replaced*, for identifying a row that previously had a status of 5 (Active) and was subsequently found to be in error. The RecordStatus value of such a row changes to 6 (Replaced) when a new row is entered with updated information. A replaced row has null beginning and ending validity dates. Note that it is not necessary for all replaced rows to actually correspond to an active row for the same facility and location since some entries may reflect events that did not happen, or which related to an entirely different facility or location. In other words, the error may have been such that it is no longer possible to relate the old row to the active row that replaced it.

7   *Retired*, for those rows that are no longer valid, such as when construction or another action physically alters a facility. A retired row has different beginning and ending validity dates, with the ending date matching the beginning date of the newer row with new information, if there is one. (Abandonments, for example, will not result in a new active row because the facility may no longer exist.) Only rows with a RecordStatus value of 5 (Active) and 7 (Retired) may be used to construct a valid historical sequence of states for the database.

Whether or not you utilize all these values for the RecordStatus field, managing the editing process is a necessary element of geodatabase maintenance within a decentralized organization. It is important for the geodatabase design to accommodate all aspects of the editing process and to provide tools to supply the best possible data.

Highway segment features

Simple feature class: HighwaySegment
Geometry: Polyline
Contains M values: No
Contains Z values: No

Field name	Data type	Allow nulls	Default value	Domain	Precision	Scale	Length	Description
OBJECTID	Object ID							
SegmentID	Long Integer	Yes			0			Identifier of highway segment
RecordDate	Date	Yes						Date record was created
EnteredBy	String	Yes					30	Name of person creating record
FromDate	Date	Yes						Starting date of validity
ToDate	Date	Yes						Ending date of validity
RecordStatus	Short Integer	Yes		RecordCode	0			Status of record
RouteNumber	String	Yes					32	Route number public key identifier
RoadStatus	Short Integer	Yes		EntityCode	0			Status of road in the real world
State	Short Integer	Yes			0			State in which roadway is located
CountyFIPS	String	Yes					255	FIPS code for county of roadway

Railroad tracks in transport geodatabase

Simple feature class: Track
Geometry: Polyline
Contains M values: Yes
Contains Z values: No

Field name	Data type	Allow nulls	Default value	Domain	Precision	Scale	Length	Description
OBJECTID	Object ID							
RailroadID	Long Integer	Yes			0			Identifier of railroad
RecordDate	Date	Yes						Date record was created
EnteredBy	String	Yes					30	Name of person creating record
FromDate	Date	Yes						Starting date of validity
ToDate	Date	Yes						Ending date of validity
RecordStatus	Short Integer	Yes		RecordCode	0			Status of record
EditReason	Short Integer	Yes			0			Reason the update was made
Comment	String	Yes					255	Comment field for additional information
TrackName	String	Yes					32	Track number public key identifier
TrackStatus	Short Integer	Yes		EntityCode	0			Status of track in the real world
TrackType	String	Yes		TrackCode	0			Type of service for track
Division	Short Integer	Yes		EntityCode	0			Identifier of operating division
Section	Short Integer	Yes			0			Identifier of maintenance division
TrackID	Short Integer	Yes			0			Identifier of track "route"
FromMeasure	Float	Yes			6	3		Linear event measure
ToMeasure	Float	Yes			6	3		Point event measure
RailWeight	Short Integer	Yes			0			Rail weight used
Owner	String	Yes					32	Name of owner

Figure 6.12 Using the *Entity* stereotypes These three example geodatabase classes illustrate how you can use the *Entity* abstract stereotype classes to create your physical data model. The Highway feature class represents a segmented structure used to construct a road map layer. It is based on the simpler stereotype without edit process management attributes. The Track feature class accommodates a linear referencing method plus some attributes that are typical of those found in railroad system databases. It is based on the enhanced stereotype. The Bridge table, which is also based on the enhanced version, shows how a point event class might be constructed to support dynamic segmentation to map the location of bridges. All of these are just examples to show how the concepts can be implemented. (Figure 6.12 continued on next page.)

Continuous versioning

Table Bridge

Field name	Data type	Allow nulls	Default value	Domain	Precision	Scale	Length	Bridges
OBJECTID	Object ID							
BridgeID	Long Integer	Yes			0			Identifier of bridge
RecordDate	Date	Yes						Date record was created
EnteredBy	String	Yes					30	Name of person creating record
FromDate	Date	Yes						Starting date of validity
ToDate	Date	Yes						Ending date of validity
RecordStatus	Short Integer	Yes		RecordCode	0			Status of record
BridgeStatus	Short Integer	Yes		EntityCode	0			Status of bridge in the real world
EditReason	Short Integer	Yes		EditCode	0			Reason the update was made
Comment	String	Yes					255	Comment field for additional information
NBINumber	String	Yes					32	National Bridge Inventory identifier
YearBuilt	Short Integer	Yes			0			Year bridge was built
Length	Short Integer	Yes			0			Bridge length, in feet
Width	Short Integer	Yes			0			Bridge width, in feet
StructureType	Short Integer	Yes		Structures	0			Type of structure
MaxVertClear	Short Integer	Yes			0			Maximum vertical clearance, in feet
MaxHorClear	Short Integer	Yes			0			Maximum horizontal clearance, in feet
MaintBy	String	Yes					32	Name of maintaining agency
LatCoord	Double	Yes			8	4		Latitude coordinate, in decimal degrees
LongCoord	Double	Yes			8	4		Longitude coordinate, in decimal degrees
RouteID	String	Yes					32	Identifier of locating route
AtMeasure	Single	Yes			6	3		Point event measure

Figure 6.12 (continued) Using the *Entity* stereotypes

Providing support for temporal aspects of a transportation geodatabase can offer a number of performance benefits to your users. It is not necessary for your agency to be a large, decentralized organization for you to provide temporal support in your geodatabase design. Even local governments may benefit from knowing how the transport system has evolved through such actions as annexation, reconstruction, and new construction.

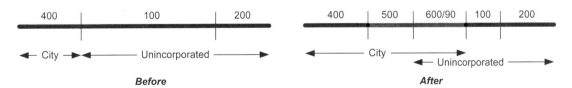

Figure 6.13 **Advantage of temporal geodatabases** In this illustration, a city has annexed all the adjacent portion of the county's Segment 100 and just one side of a middle portion; the jurisdictions will typically create a new Segment 500, subdivide part of Segment 100 by side of road, and leave a portion of Segment 100 unchanged. In the future, how will you be sure that a record pointing to Segment 100 relates to the new one? You may also want to illustrate how the city has grown or report the number of road miles annexed over a specified time period.

The design concepts presented in this section need not be restricted to geodatabases. They may be successfully employed in all types of databases whenever you may need to understand how changes occurred and to recover the historical state of the database at a given time. You need to compare conditions now to those of an earlier time. Normally, you would have to make snapshot copies of the database at various times. There is no way to recover the previous state of the database except to go to one of the historical copies. Of course, if the date for comparison does not correspond with the date of one of the available copies, it may not be possible to precisely perform the analysis. Continuous versioning allows for the recovery of any historical date during the life of the database and you will not overload your data storage devices with many old copies.

Preserving data dictionary history

Preserving the history of the geodatabase classes by using the *Entity* class templates may be inadequate for a large geodatabase to which many editors may add new classes or if you share the geodatabase with numerous outside users. In such a case, you may also want to support data dictionary history. This means telling users which fields are in which classes and what their previous structure may have been.

Just as rows evolve to reflect changes in the entities they describe, so, too, do the classes that contain the data. New fields may be added, others deleted, and domains changed. It does little good to preserve all the historical versions of the geodatabase if you no longer understand what the older data meant. To retain this information, you need to store data dictionary history.

The logical data model shown in figure 6.14 illustrates one approach to preserving data dictionary history so that you can understand all historical views of the geodatabase. This design is compatible with all editing methods.

Changes in transportation database designs are relatively common. For example, from time to time the Federal Highway Administration (FHWA) will change the domain of values used by a particular attribute to be submitted in the annual Highway Performance Mandatory System (HPMS) report. Another possibility is that an internal group, say the pavement management section, decides to change pavement condition from a quality rating using words (excellent, good, fair, poor) to one based on numeric values (a range of 1-10, with higher numbers representing better condition).

Such a change could be implemented in the geodatabase in a variety of ways. You could simply add the new values to the domain class, but that would not prevent editors from selecting an older value that is no longer in use. You could add a new field to offer both versions, but if you stop maintaining the older values, then they will become inconsistent

Preserving data dictionary history

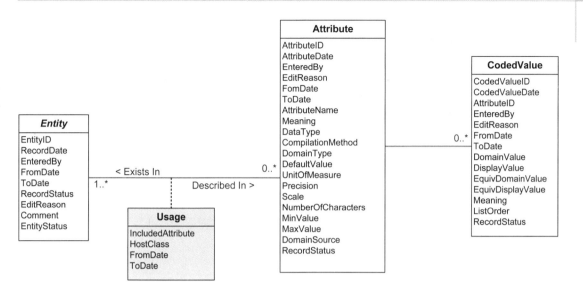

Figure 6.14 **The rest of the story** Restoring the state of the database at any historical point will provide you with the data as it was but not what it meant. This logical data model shows how to extend the temporal support system to retain the history of data meanings. Where *Entity* provides record-level metadata, these three additional classes supply column-level metadata. The Attribute class stores information about every user-defined field in the geodatabase. Those rows with a coded value domain are also included as a set of rows in the CodedValue class, one row for each value. The CodedValue class provides a history of the various versions of the related domain class, if any. The Usage attributed relationship manages the many-to-many relationship between *Entity* and Attribute.

with the new ones and no longer represent equivalent data. You could also convert all the old data to the new form, but sometimes that cannot be done unambiguously.

Another approach is to create a completely new class with the revised field definitions. All the rows would receive the same RecordDate value when you populate the new version. Any user seeking data for an earlier date would receive no rows in the return set, which would prompt them to look at the earlier version for their data. The new version of the class would need a slightly different name, as would domain classes, if their values had changed. For example, you could put the year of the change in the new class's name to indicate its vintage. For this approach to work, you will need to ensure that your users know there is an earlier version.

The best approach is to explicitly create a data dictionary that consists of tables in the geodatabase using the logical data model in figure 6.14. Figure 6.15 shows how the logical model could be implemented. This self-documenting approach will allow any user to understand how the geodatabase has changed over time and to always be able to interpret the information it contains. With this approach, your users will always have a full understanding of what those old coded values meant.

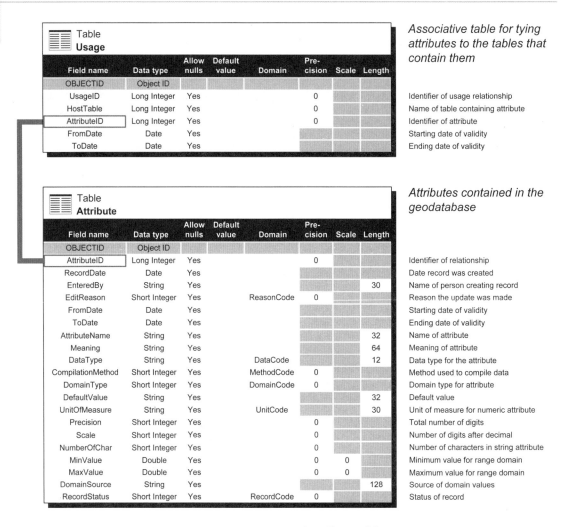

Figure 6.15 **Usage and attribute table**

Supplying data dictionary history also serves as a good geodatabase administration tool in that it fully describes each class. For example, it can find all the tables that utilize a particular domain class, so that if you are contemplating a change to the list of included values, you will be able to see which classes (and users) would be affected by the change. It would also supply information to support full dataset metadata of a type not supported by some formats in a comprehensive way, such as showing how the unit of measure has changed over time from feet to meters, or how the precision and scale specifications for a field were altered. This kind of change may not be obvious in a normal data dictionary view.

Preserving data dictionary history

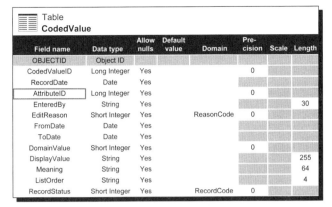

Figure 6.16 **Domain control**   The type of domain control offered by ArcGIS is more than adequate for static attribute types or those to which new values may only be added, not replaced or deleted. The CodedValue table offers a more complete solution for tracking changes in domains over time, as it retains past values that may no longer be in use.

Domains for several Attribute table fields can simplify creating and maintaining data dictionary history. Such a table can be used to ensure the consistent specification of fields across multiple classes, and to prevent the duplication of a field name with a different meaning.

Of course, one of the most useful applications for a data dictionary history is to tell others about the dataset when they incorporate it into their own. For instance, the proposed design for preserving data dictionary history can expand your understanding—and that

135

of your users—of various coded-value domains that may be included in the geodatabase. A CodedValue table would provide information about the derivation and history of each domain class member.

Separating the editing and published geodatabases

Several references have been made to completing edits and publishing a new version of the geodatabase, which are two separate things. An optimally designed editing database is structured to support the editing process and is managed accordingly. Such a geodatabase will be substantially normalized so as to reduce the number of actions required to implement a single change.

Normalization is generally not the design approach you want to use for an application dataset, where significant performance improvements can be realized through denormalization. Plus, you have to anticipate that different applications will use the same data in different forms, and a lot of data that users will need is derived from other data. For example, the number of lane miles is the product of length times the number of lanes, or the product of length times the roadway width divided by some standard lane width. It is much better to calculate such fields than to require them to be derived outside the geodatabase and manually entered. You will very likely need to produce several application-specific datasets from the editing geodatabase.

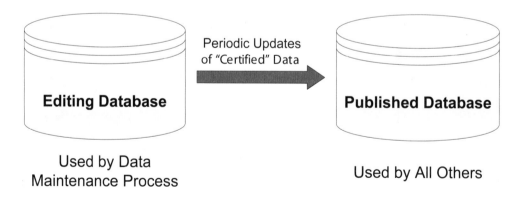

Figure 6.17 **Editing vs. published databases** The optimal design for a database that supports data maintenance is very different from that typically needed by other applications. For example, avoiding data redundancy and its resulting added editing headaches points an editing database toward a highly normalized design. In contrast, slight differences in data needs for multiple applications that may use the data require significant data redundancies, derived values, and different formats. If for no other reason than editing databases are "messy" and full of half-done updates, you will need separate editing and published databases.

Separating the two views of the geodatabase provides other benefits. First, it restricts the number of people initiating queries against the editing geodatabase. Nothing can ruin your day faster than the infamous "query from hell," which is initiated when a user accidentally or through ignorance consumes host RDBMS capacity by seeking to return every record it contains. Such a query can be applied directly to the RDBMS; it need not go through ArcGIS for you to suffer its performance consequences.

A second good reason is to allow the editing and published datasets to evolve independently as technology and user needs change. If you point them both to the same geodatabase, you may need to revise the editing environment every time a new version of the application is put into production. You would also be precluded from making changes to the geodatabase to improve editing performance if they would impact the user applications.

You also have to consider that a user who issues a query against a geodatabase while it is undergoing the editing process may get a different result each time because the geodatabase is constantly changing. All kinds of referential integrity issues could arise if someone runs an application while editing is under way. The alternative is to block everyone from using the geodatabase, which may not please users who demand an answer right now.

There are also users who absolutely do not want an updated version. They may need one that is stable during an entire fiscal year so that all analyses conducted during the year produce the same total of some quantity, like centerline miles by functional class. If you keep changing the geodatabase to make it "better" for some users by supplying the latest information, these once-a-year-update users will start making copies for themselves and you will soon have two different versions of the geodatabase in production application. It gets worse if they decide to "fix" problems they discover during the year.

For all these reasons and others, it is much better to have separate editing and published versions of the geodatabase. The proposed enhanced temporal editing version of the *Entity* class template and the RecordStatus domain are used to manage the process of publishing new public versions of the geodatabase. After the editing process has gone through its validation cycle, you publish the database by selecting all class rows with a RecordStatus of 4 (Accepted). These rows can form the next addition to the transaction database that remote users can access to find what has changed in the source geodatabase since their last update extract was created. This transaction database is like a permanent repository of Adds and Deletes table rows.

They are also applied to the editing database to make all the changes permanent. Records to be retired will get a new RecordStatus value of 7 and an ending period of validity date. Records to be replaced due to a discovered error in an earlier update will get a new RecordStatus value of 6 and have both beginning and ending validity period dates set to null. All rows with a value of 5 (Active) will be added to the geodatabase.

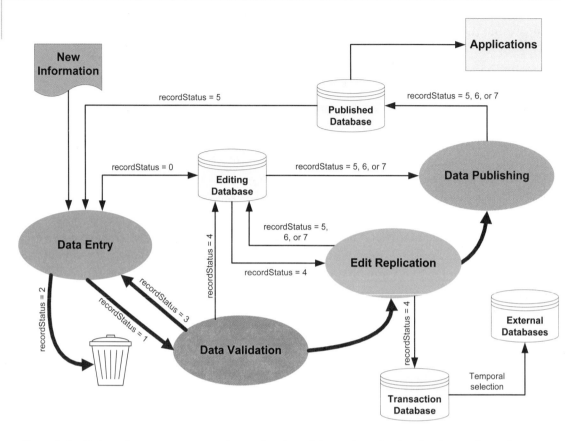

Figure 6.18 **Data maintenance process** Separating the editing and application databases allows the various application databases to evolve as their users may require, prevents users from using edited data until they have passed quality-assurance tests, and allows stronger edit-process management. The models in this book present geodatabase designs optimized for editing. The record status attribute helps manage the process.

Once the editing geodatabase is fully revised, you can then use it as input to the data publishing process. Here is where you would have the functions that calculate derived fields and generate the various alternative versions different users might need. Denormalization is the rule in published geodatabases, as it will help the user if each logical data grouping has all the information that particular application might need. For example, you can expect centerline miles and lane miles to be scattered throughout the published geodatabase, perhaps with various specifications for precision and scale. Some derived versions may be completely replaced; others may be supplemented with a new version. An example of the latter could be the official mileage report for a previous calendar year or the adopted work program for the prior fiscal year. More information on how to update the edit and publish geodatabases is included in chapters 14 and 15.

ArcGIS supports a replication process that facilitates data distribution in a split edit/publish structure. The data replication process works with both personal and enterprise geodatabases. It is based on the concept of a parent-child relationship between two datasets. In two-way replication, both the parent and the child can be a supplier and recipient of data updates. In one-way replication, the parent geodatabase supplies updates to the child geodatabase, which is a read-only version. Through disconnected database synchronization, the roles of parent and child can be changed in each update session.

The replication process has a feedback loop that allows the data sender to know that the recipient has successfully received the update. If the acknowledgment message is not transmitted by the child and received by the parent, then the data update will be transmitted again during the next replication session.

Geographic filters can be used to select areas of interest, such a county selecting a subset of edits from a state geodatabase. Geodatabase schema changes can be accommodated to a degree. The data dictionary preservation methods shown above provide a more robust way to manage schema changes.

Service-oriented architecture

One of the more recent advances effective in multieditor environments is the service-oriented architecture (SOA). An SOA goes beyond decentralized editing of a single geodatabase to create a decentralized geodatabase. Multiple sources can publish and register their data for discovery and use by any number of remote users. They can either download the entire dataset or use Web mapping services (WMS) to dynamically select and download just the portion required by the extent of their present map view.

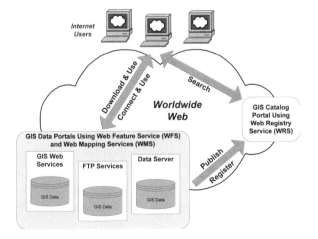

Figure 6.19 **Service-oriented architecture** This sample implementation uses Web-based services while other SOAs may use a more tightly coupled network, such as a LAN. Regardless of the underlying architecture, a user initiates a request for data through a search of catalog portals. Data suppliers register their published data with catalog portals. Once the desired data is found, the user can select the most appropriate source and either connect to it dynamically or download the dataset and use it locally.

Although ArcIMS and ArcGIS Server support Internet-based SOA structures, implementing an SOA does not require you to deploy Internet connections or publish your data on the World Wide Web. An SOA is equally useful within a local-area network (LAN), such as you may have within your agency. You could deploy SOA to create a consistent interface for internal users accessing both internal data stored in your geodatabase and external data supplied by other sources. You could also use an SOA to support field users employing mobile devices in a variety of data-automation processes.

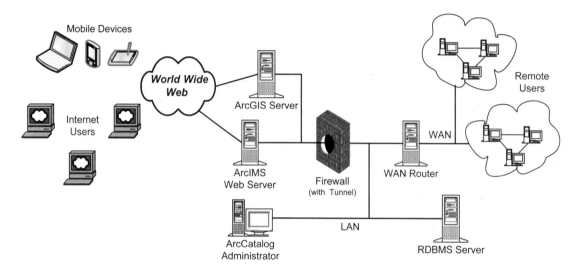

Figure 6.20 **ArcGIS Server and ArcIMS architectures** Service-oriented architectures can be supported by ArcIMS and ArcGIS Server, both of which offer out-of-the-box components that can immediately support WFS, WMS, and WCS. These services provide the means for remote users to have exceptional access to published data sources throughout the world.

SOA deployment will increasingly be a part of implementing multieditor and multiuser geodatabase designs. As the number of SOA service providers increases and new access tools are released, the richness of data beyond that maintained by your organization will grow, offering both an increase in the wealth of data available and a reduction in your editing workload.

SOA and the geodatabase design alternatives presented in this chapter set the foundation for data interoperability. The type of edit-and-publish process described above is called asynchronous editing and publishing. Such a process can be readily extended to support publish-and-subscribe data distribution that allows each user to seek the revised data at the times that are most appropriate for that user. Data sources would publish updates according to an agreed-upon cycle, perhaps with different datasets being revised along separate schedules.

The objective of an enterprise multimodal geodatabase is to measure once and cut twice. This is just a shorthand way to describe data interoperability. By extension, interoperability may go beyond a single enterprise to allow the data to be used by others and to allow the enterprise to accept data from other sources. An SOA can provide the mechanism to facilitate both internal and interagency data interoperability.

The best reason for deploying SOA is to provide a common form of governance for enterprise information resources. SOA is a way to organize the enterprise's resources to meet the needs of an evolving business more efficiently. The governance aspect of SOA is that it provides a shared platform for accessing resources across the enterprise, which allows the organization to apply those resources in ways not originally intended. SOA allows the entire information technology infrastructure to follow the mantra of measure once, cut twice. The resource governance aspect of SOA cannot be ignored, lest it become just another project creating its own functional silo. Yet, as with the agile methods discussed in earlier chapters, migration to the SOA needs to occur through one or more pilot projects to allow the organization to gain experience with the technology.

In order to meet these objectives, SOA must pierce the many functional silos within the organization. You cannot expect to get the entire enterprise to adopt SOA at one time. SOA and the other approaches suggested in this chapter are techniques that can be used for editing and compiling transportation geodatabases and other data themes. You do not need to implement all the ideas at one time. Although the original justification for the editing geodatabase design focus in this book was its ability to support multiple editors and users, you need not have multiple editors and users for the ideas presented to have merit. The techniques shown to preserve data dictionary history are equally useful to remind yourself (or your successor) what you did six months or six years ago.

Consolidating the editing task to a single group or geodatabase is also not a necessary near-term objective for the concepts of this chapter to be useful to you. Even if you don't believe that enterprise geodatabase operation will occur, it surely never will if you do not employ the multiparticipant design elements shown here.

chapter seven

Linear referencing methods

- The problem with large transportation datasets
- The advent of linear referencing systems
- Dynamic segmentation

Designing a geodatabase traditionally encourages the user to develop a set of map layers, each consisting of a set of discrete features with their attributes. This approach has been used to present various transportation geodatabase design alternatives. For example, the pavement management system of chapter 4 subdivides the transport system into segments defined by intersections and, when necessary, at midblock locations where a key attribute's value changes. These segments may be combined to create pavement management sections and indicate the impact of public-works projects.

Chapter 7: Linear referencing methods

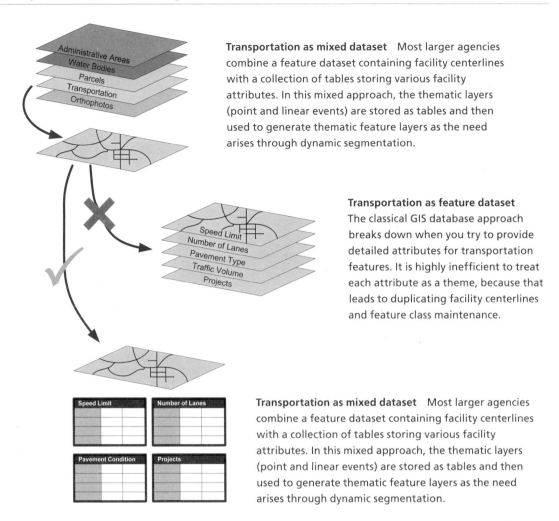

Transportation as mixed dataset Most larger agencies combine a feature dataset containing facility centerlines with a collection of tables storing various facility attributes. In this mixed approach, the thematic layers (point and linear events) are stored as tables and then used to generate thematic feature layers as the need arises through dynamic segmentation.

Transportation as feature dataset The classical GIS database approach breaks down when you try to provide detailed attributes for transportation features. It is highly inefficient to treat each attribute as a theme, because that leads to duplicating facility centerlines and feature class maintenance.

Transportation as mixed dataset Most larger agencies combine a feature dataset containing facility centerlines with a collection of tables storing various facility attributes. In this mixed approach, the thematic layers (point and linear events) are stored as tables and then used to generate thematic feature layers as the need arises through dynamic segmentation.

Figure 7.1 **Map metaphor does not apply** Most geodatabases appear as a set of feature classes forming map layers. Feature classes include discrete phenomena and their attributes. Each class is able to stand alone. Such is not the case for a rich set of transportation data, which is typically stored as a set of basic facility geometry and a number of attribute tables, called somewhat misleadingly by the name "event."

This design philosophy is consistent with a transportation theme approach, which works well if there is not a lot of data to maintain or if the segments created do not change very often. Theme design is especially difficult to implement at a state DOT, though, where segments defined by intersections may be several miles long. Because dozens of attributes state DOTs need will likely change at midblock locations, traditional geodatabase design will produce thousands of segments. Many of these segments will change monthly as projects, maintenance work, and annexations occur.

Adding to the editing process is the updating of linear referencing measures. For example, it is common for the location of an entity to be redefined based on recent field work using supposedly better methods. It is difficult to change a location reference, such as a milelog, because of the effect it may have on other attributes and application databases. This is because many transportation datasets define location using only the route identifier and milelog measure, making the location the foreign key. Changing that location, even if intended to make it better, will "move" the entity and break the foreign key relationship between that entity, defined only by its location, and its attributes. Common examples where these location foreign keys may be used are the definition of intersections by a route and milelog reference in a crash database and the location of a bridge in a structures inventory.

As suggested in the previous chapter, you will likely need to generate multiple views of the geodatabase for various applications. You should avoid maintaining multiple feature datasets for the transport system, one for each application. It is preferable to maintain one feature set describing the basic shape and position of all transport facilities, and then to modify this shared dataset to present the form each user community needs. Those communities may themselves maintain the attributes that concern them.

Such a design philosophy has been followed by state DOTs and other large transport organizations that use a mixed dataset of feature classes and attribute tables. Maintaining features is usually disconnected from, but loosely coordinated with, the process of updating attributes. This "divide and conquer" approach is necessary for efficiency and because of the organizational realities of a large agency. But it has a limited ability to supply an integrated dataset that serves the needs of all users.

The problem with large transportation datasets

The issues presented by a large transportation dataset are not limited to its size or to the way an agency maintains its data through a decentralized organizational structure. There are really four fundamental aspects of transportation datasets that present the primary design considerations.

The first and largest of these problems is transport facilities are often not discrete. Certainly, one can clearly identify a facility through its physical form. The problem is in defining the limits of that facility. If you work at a state DOT or a multijurisdictional agency, such as a metropolitan planning organization (MPO), then you will face the problem of defining the extent of linear facilities. For a city or a county that can use the short-segment design approach, this issue disappears to the point that it can be addressed with one of the multiple street name solutions presented in chapter 4. Such a solution breaks down when many jurisdictions are defining the extent of linear facilities.

Chapter 7: Linear referencing methods

Facilities overlap The traditional feature-based theme approach doesn't work for transportation facilities because they are often not distinct features. Sometimes they overlap, as in this example of two numbered routes traversing the same segment. Is the shared segment part of Rt. 27, Rt. 84, or both? The answer affects editing overhead.

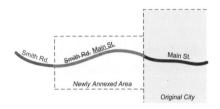

Facilities evolve Even when you don't have shared segments, as may be the case for a local government that segments roads by street name, segment attributes change. In this example, a city has annexed a portion of Smith Road, the name assigned by the county, which will now be called Main Street, the city's name. Is the new segment an entirely new feature, or is it an extension of the existing Main Street feature?

Attributes are continuously variable There is generally no correlation between the various attributes you may want to provide for transportation features. Each attribute is likely to define a different set of segments. As a result, you will have to either create very short segments and live with the resulting massive duplication of values, or incur the overhead of separate tables and dynamic segmentation.

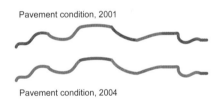

Attributes Change Over Time A common application of spatial data is to compare conditions as they change over time. For example, pavement condition will vary according to deterioration from traffic effects and rejuvenation from resurfacing. Using a segmentation scheme that is based on attribute values can produce different segments every time the database is updated, which can increase the difficulty of comparing time snapshots.

Figure 7.2 **The transportation database design problem** Here is a sampling of some of the issues presented by transportation data. Each of these issues and many more must be accommodated by any effective geodatabase design.

Linear facilities overlap, perhaps not physically, but certainly in the way we think about them. Take the case of a state road with its route number identifier. Each city and town it traverses will give it a local street name. There may even be overlapping state roads for portions of the route. As a result, it may be advantageous to separate the facility from its public identifier, which would become just another attribute. Of course, then you are left with having to develop some kind of segmentation schema in order to work with features of a manageable size and geographic extent. As a database designer, you will see that the relationship of facilities to routes is many to many. A facility may have many different routes traversing it, and a given route may include part of more than one facility.

Denormalized

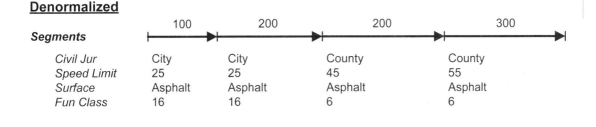

Normalized

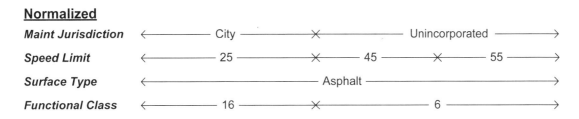

Figure 7.3 **Denormalized/normalized figure**

The second problem, one that was addressed in chapter 6, is that facilities and their attributes change over time. How you accommodate such changes in the database design is substantially determined by whether you have adopted a normalized design objective. A typical example may be presented by city annexation. In a normalized, multijurisdictional geodatabase, such a change will be implemented by altering the spatial extent of one or two attribute records. Even in a denormalized design with short segments, you may need to simply split a segment or change the attribute value for a couple of segments.

If, on the other hand, you are the recipient of several single-jurisdiction datasets and are trying to maintain your own dataset by accepting changes from your suppliers, then you have a much bigger editing problem. The city and/or the county may have modified the spatial extent of its features to reflect the fact that one lost mileage while the other gained it. The roads themselves may not have changed, but you could be faced with two very different views of the same facility as each supplies you with features covering the affected facilities.

This example points out the other two problems: attributes are continuously variable and their values change over time. The short-segment design approach works for local governments because they typically do not need a lot of attributes. A state DOT, on the other hand, has significant federal data-reporting requirements and other applications that require scores of attributes to be maintained on each facility. The values of these attributes typically do not change at intersections or other obvious points of segmentation.

Coupled with this phenomenon is the presence of many different segmentation schemas within a given organization. Block-by-block segmentation may work fine for some users, but it imposes unnecessary breaks for users who need longer segments, like pavement management and project planning staff. Although these users could combine segments to produce the features they need, the component short segments are constantly churning, which means the longer segments need to keep being constructed.

The problem goes in the other direction, too. Some users need to refer to a single point or short segment along a block. For instance, traffic-operations staff members may need to identify a no-parking zone or a traffic crash at a midblock location. If they only have intersections and blocks to which they can attach data, then they lack the resolution required. There are a number of local jurisdictions that attach all crashes to intersections for this very reason.

The spatial extent of a state DOT or MPO geodatabase can also be part of the problem. The value of some attribute is changing every day somewhere on the system. If the agency based its segmentation schema on subdividing features wherever an attribute value changed, then segments would also be changing daily. Not only does this create a large feature-maintenance workload, it presents an ever-changing set of features to the user communities.

The advent of linear referencing systems

As a solution to all these design issues, the railroad industry developed the concept of long routes and a linear referencing system (LRS) to identify positions on a rail line. Such a one-dimensional coordinate system uses cumulative distance from a point of origin to identify the location of a point on the facility. Each facility thus becomes its own datum for stating a location along its length. An LRS to describe locations along a linear facility has become a standard for many modes of travel and is the underlying structure of river reaches and other linear features of all types. Linear referencing offers the ultimate solution to restrict the number of feature classes as a strategy to reduce maintenance workload and improve geodatabase performance.

A linear referencing method (LRM) is the set of technical processes used to determine, specify, and recover a location within an LRS, which also includes business rules for managing the measurement process. For example, an LRS will specify how to define the origin and when to check the accuracy of distance measuring instruments (DMIs), which determine the offset distance form the origin. The one or more LRMs within the LRS will specify a unit of measure, a precision and scale for measured distances, and other parts of the measurement task. The distance from the origin to the point of interest is called a measure.

The traditional way to apply an LRM is called route-milelog. The route is the facility on which measures are determined and applied. The milelog is the measure stated in fractional miles, typically units of 0.01 or 0.001 mile (intervals equivalent to 52.8 feet or 5.28 feet, respectively). A route identifier was established for each road within the database. Although usually formed initially by following the path of a numbered route, some states elected to separate the route identifier from the route number, treating the latter as another attribute. Some states adopted a side-of-road structure as a way to better store and report data for divided highways.

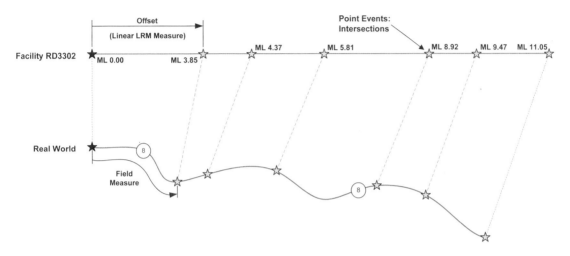

Figure 7.4 **Route-milelog** State DOTs are responsible for thousands of miles of highway where intersections are relatively far apart. They have used a route-milelog linear referencing method (LRM) that expresses location along a linear facility using an offset distance, called a measure. Such a 1D coordinate system allows easy location of positions in the field by using a vehicle's odometer.

State DOTs and the Federal Highway Administration (FHWA) adopted the route-milelog LRM when they created the Highway Performance Monitoring System (HPMS) in the early 1970s. As a result, almost every state DOT has a database with this form of LRM, and many local governments and MPOs that share data with the state need to accommodate this measurement system. The original LRM database designs predated the advent of GIS and CAD software, so they did not include any geometry. Many database management systems based on these old designs are still used by state and federal agencies.

Routes may be further subdivided at a jurisdictional boundary, such as a district or county line, to accommodate the organization's structure and resulting maintenance responsibilities. Some states refer to the atomic portion of a route as a section. The section becomes the route for defining each LRM datum.

Many states go so far as to place a mile marker at the approximate start of each whole mile interval, sometimes with the route or section identifier also indicated on the sign. A few states employ an explicit route marker to serve as the origin. Such a reference-marker

Chapter 7: Linear referencing methods

system bases measures on the location of the marker. For example, all interstate highways are required to be marked at one-mile intervals using a green sign with white numerals, with the point of origin being the state line. Many states use these milelog measures to number their interchanges, but most employ a completely different LRM for recording the location of events—including the interchanges—along these facilities. In recent years, some states have begun to place markers at shorter intervals so that motorists calling about an incident on a limited-access highway can describe their location with a nearby point of reference.

If you are unfamiliar with some of these terms, it is important to understand the way they are being used here. In ArcMap, a route is a polyline feature that has measure coordinates defined for its vertices. The discussion here is not so precise, where the meaning of a route is a numbered or name path through the transport system, whether it is a highway, railroad track, or navigable waterway—all of which use similar route-milelog LRMs.

A route event is anything referenced to the LRM; it is not just something that happens. Unless the context clearly demonstrates that the topic is a "real" event, then the term *event* refers to a generic phenomenon that describes an attribute or element of a linear route. The inherent cardinality for an LRM is one to many because one route has many route events.

You have probably been working with linear referencing and dynseg for many years but called it address geocoding. Point events are very much like addresses. They refer to a relative position along a linear facility, which is a route defined by a street name. The LRM is defined by the addressing system. Dynamic segmentation, abbreviated as "dynseg," is an enhanced way to locate a point along a linear feature that adds the capability to offset a point from the centerline (geocoding only puts the point on the centerline) and the capability to generate route segments by pairing two "addresses" to define starting and ending points.

Because street centerlines are often used for address geocoding, you may want to create address segment linear events to tie the addressing system to the route system. You could then use dynseg to create the block-length segments needed for traditional address geocoding and avoid having to maintain two sets of roadway centerline features, one for each application. You may also want to create address block linear events if you work at an MPO or DOT that receives data from local governments that use only the address-block segmentation schema.

The purpose of the linear referencing is to provide one-dimensional position statements for locations along a linear facility. These locations are called event points, and they are stated using two variables: the identifier of the facility and the linear measure. Most measures in the United States are stated in decimal fractions of a mile, such as 0.01 and 0.001 mile. Measures begin at an origin and increase linearly to the terminus of the facility. Over time, changes to the facility may alter the linear progression of measure values. Transport agencies have developed various methods for dealing with these situations.

A facility attribute may be a descriptive characteristic, an elemental component, or an action. Facility attributes are stored in route-event tables. Event points are used to express the location where the values of a facility attribute changes, with one important addition: the route measure describing the route event's location in conformance with the LRM. Some route events describe point-like phenomena, like signs, intersections, and light poles; point events have a single route measure position column. Other events describe linear phenomena, like speed limits, functional classification, number of lanes, construction projects, and pavement condition; linear events have two route measure position columns, one for the starting point and one for the ending point.

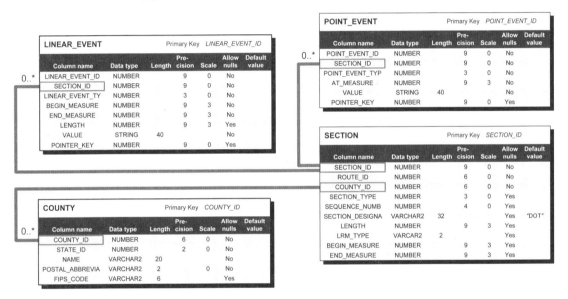

Figure 7.5 **Nongeodatabase tables**

The design shown in figure 7.5 is based on routes being subdivided into sections within each county. The attribute data is stored in route-event tables, often with different tables for point events (one locating measure) and linear events (two locating measures). Some states adopted a design where one table accommodates all route events. For example, a single linear-event table could be defined, with the different linear attributes—speed limit, functional class, pavement conditions, etc.—identified by an event type code. Point events may also be stored in this table by omitting a value for the second measure field. Other states adopted a variety of alternatives, from attribute-based segmentation to multiple attribute groupings.

Regardless of the database design, the origin of the section is generally assigned a measure of 0. The origin is typically the southern or western end of the section, with measure values increasing linearly along the route. Problems naturally arise when a route is realigned, resulting in a changed length. Some states revise the measure value of all downstream locations to reflect the results of realignments. Most states have adopted various LRS business rules that provide another means for reconciling measures to the new length of the route. Other LRS business rules were created to manage the result of shortening a facility by abandonment or jurisdictional changes, or lengthening a route by annexation and construction.

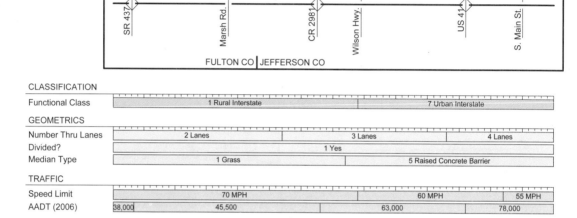

Figure 7.6 **Straight-line diagrams** State DOTs traditionally illustrate routes and their primary attributes using a straight-line diagram. The various attributes have been separated and grouped according to type in this example, but other forms are common. For instance, many states place symbols and annotation directly on the drawing. Curves are indicated by locating and labeling the point of curvature (PC) and point of tangency (PT), with the shape of the curve defined by the intersected angle (Δ) and degree of curvature (D). Bridges, intersections, and jurisdictional boundaries are also commonly included attributes.

The first form of geometry used to illustrate the data contained in a route-milelog dataset was the straight-line diagram (SLD). Such a graphical standard was a natural outgrowth of treating each facility as a one-dimensional object. Bridges, curves, intersections, and other major structural features are illustrated directly on the diagram using symbols and annotation. Most other attributes were displayed on bars parallel to the line that represented the facility. Since no SLD can accommodate all events, a state DOT may elect to include only a fixed subset or to offer multiple versions with various subsets.

SLDs were printed at regular intervals and used by office staff as an original data source. Smaller versions of the SLD set were commonly found in maintenance vehicles, or a copy of an SLD might be attached to a work order as a way of identifying the job site.

A few employees may have acquired the skills and permissions to use the agency's mainframe computer database through a remote terminal. Queries typically took days to get a response, as each required a computer program to be written to get the answer and produce a printed report.

CAD software and the large, expensive minicomputer platforms that ran it became widely available in the early 1980s. A number of state DOTs started maintaining both SLDs and more traditional 2D highway maps for the public using CAD. A few states even created automated programs within their CAD environment to produce SLDs. These innovations reduced the workload for SLD maintenance, but it did not really affect their effectiveness and versatility.

When personal computers became more widely available in the mid-1980s, it became possible for people to start to form and execute their own database queries against a copy of the mainframe database or through a portal supplied by the information systems management group. Enterprising staff members learned how to use these new desktop computers and began to write and execute their own queries, produce reports utilizing spreadsheet software, and otherwise increase the number of queries that could be performed by data users. Statistical reports that perhaps were previously done annually could now be generated more frequently. This demand for up-to-date information hastened the need to get revisions into the mainframe database with an equal frequency. Still, the SLD was the only graphical display mechanism available to most agency staff.

Dynamic segmentation

Paralleling changes in microcomputers and CAD software was development of the ArcGIS predecessors, which offered a 2D view of the world. However, unless a way could be found to transfer the information contained in route-event tables, using GIS was no different than CAD in terms of displaying the data. What was needed was a way to subdivide routes to match the spatial extent of selected event table records.

That solution is dynamic segmentation, or dynseg, which uses the equivalence between a feature and a route or section. Dynseg is the process of computing the map location of route events along a route feature so the events can be displayed, queried spatially, and analyzed in a GIS. Through straight-line interpolation, dynseg calculates proportions based on milelog measures to segment a centerline into pieces representing the extent of each linear event or to place a point event symbol in the corresponding location along the route feature.

Although somewhat foreign to local government transportation datasets, dynseg, an LRM, and route-event tables offer the opportunity for the various interest groups to separately identify the segments and points of interest along a common set of facility routes. This reduces the editing overhead of reconciling separate feature classes with varying segmentation schemas. Instead, one route is subdivided according to the linear data described by the LRM, with dynseg creating the features representing each segment when a map layer is needed.

A route suitable for dynseg is any linear feature with measure values assigned to its vertices. A route can be a road, river, rail line, sewer pipeline, or a member of any other polyline class that has a unique identifier for each facility and an LRS defining a 1D measurement datum. A measure is any position along the facility using a common unit of measure. A combination of route identifier and measure uniquely specifies a location within the system of all such facilities.

Route-event tables can include many different kinds of data and need not be part of the geodatabase. ArcGIS will accept text files, spreadsheets, and OLE-connectivity databases as inputs to the dynseg process.

Incidentally, the feature class that defines the geometry of a route and is itself an input to the dynseg process does not store the measure unit. Since dynseg works through proportionality, the measure coordinate is essentially dimensionless. There is no capability for you to tell ArcGIS that the unit of measure is miles, meters, or feet. You have to explicitly maintain this information in your event tables or in the data dictionary structure described in chapter 6. Be sure to check that your route-event tables all include the same kind of measures.

Dynseg requires that the source feature include a measure value for each vertex. You may have noticed in the data dictionary views presented thus far that there was a descriptive item in the upper right corner that said, "Contains M values," and a value of "No." That feature class property needs to be "Yes" to use linear referencing. The m coordinate is the measure value that can be assigned to a vertex. A complete discussion of how this can be done in ArcGIS is contained in the book, *ArcGIS 9: Linear Referencing in ArcGIS*, available from ESRI.

Dynseg produces a new feature class in memory; the original route features are unchanged. You can make the new feature class permanent using ArcGIS tools. To conduct a dynseg operation, you use the MakeRouteEventLayer tool. ArcGIS first identifies the proper linear feature, as defined by the route identifier field's value. ArcGIS then identifies the vertex with the m value that is less than or equal to the input measure. As necessary, dynseg will use straight-line interpolation to approximate the location along the line element that begins at that vertex. If the operation involves a point event, the ArcGIS dynseg function places the appropriate point symbol at the location. For a linear event, ArcGIS creates any required vertex for the from-measure and copies the original line until it reaches the ending point equal to the input to-measure. The ending point may also require the interpolation of a vertex.

Dynseg in ArcGIS creates and maintains a relationship between the original route geometry, the route-event table, and the dynseg output. This means that changes to the event table will be reflected in the dynseg output as they occur.

Figure 7.7 shows the basics of dynseg operation. Dynseg operates on feature classes, typically of the polyline type. A polyline is an ordered collection of paths, which need not be contiguous; i.e., the polyline can consist of multiple parts that are disjointed. In our example,

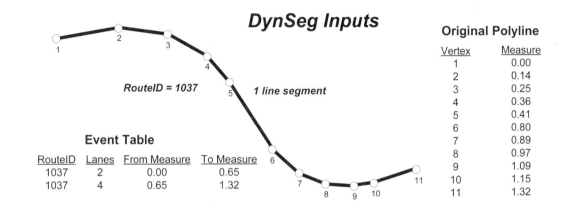

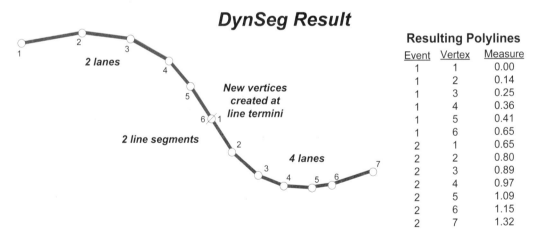

Figure 7.7 **Dynamic segmentation** A new spatial operation called dynamic segmentation (dynseg) was developed in the 1980s to help illustrate the location of linear facility attributes using the route-milelog LRM. A typical dynseg operation begins with a centerline feature class containing vertex measures and an event table containing facility attributes and measure extents. ArcGIS makes a copy of the selected centerline feature class, and then subdivides each route feature to match the extent of each linear event or the location of each point event. You can then use one or more attributes to assign symbology that illustrates each event's value.

there is one continuous path composed of 10 line segments defined by 11 vertices that collectively represents Route 1037. The table to the right shows the measure values stored for each vertex (the values for the other fields have been omitted). The bottom two shapes are the result of applying the linear-event table for number of lanes, which produced two event polylines. The two resulting features include the new, interpolated vertices needed to represent the end of the first event and the start of the second. The created features are part of a new feature layer; the original polyline is unchanged.

It is important to note that measure values need not increase linearly with distance, as in the example. Measure values can arbitrarily increase, decrease, or remain constant, as far as ArcGIS is concerned. However, your life will be much easier if they do increase linearly with direction and distance from the origin.

Measure values are independent of any geographic or other multidimensional coordinate system, which can impose its own problems for integrating data from multiple sources and coordinate systems. Measures can be in different units than what might be used in the coordinate system used to construct 2D and 3D features for mapping applications. Until a polyline element has stored measure values, each m value will be set to NaN, an ESRI abbreviation for "not a number." ArcGIS provides tools to assign measure values to features.

Although ArcGIS can accommodate simple and complex (self-intersecting) polyline classes, performance of dynseg operations will be enhanced by:
1. Removing zero-length segments and empty path records;
2. Ensuring adjacent segments are properly oriented end-to-end geometrically (ToPoint of one is connected to the FromPoint of the next);
3. Creating a new path for any discontinuous segments or segments with different attributes (each path is composed of only continuous segments); and
4. Correcting obvious spatial data gaps; i.e., where one of the two coincident endpoints of adjacent segments lacks a data value, it will be copied from the other endpoint.

You can utilize the dynseg function of ArcGIS to create geometric elements corresponding to event table rows, thereby creating a new feature layer with the event table's attributes. Multiple event-based feature layers may be created and then operated upon to conduct spatial analyses. For example, a feature layer constructed from a table of speed limits can be unioned with another feature layer constructed from a table of pavement types using a line-on-line overlay to produce a third feature layer containing the spatial join of both inputs. Alternatively, these tables could have been joined external to the ArcGIS environment then the result used to build a single event-based feature layer.

Dynseg allows a single set of geometric feature to be used over and over by various users to produce event-specific maps without requiring duplicative feature maintenance. The key is to establish a uniform set of route identifiers throughout the organization, and to create a centerline feature for each route's extent.

Dynamic segmentation

In the example shown in figure 7.8, we have chosen to segment the route system by county, such that a centerline represents that portion of a route—the segment—that is within a given county. Thus, there is a one-to-one cardinality between Centerline and Segment in our physical data model example. All the classes on the right side are tables. All the classes

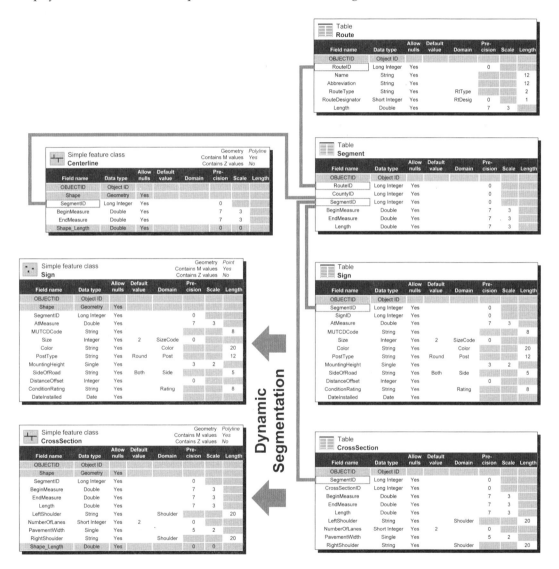

Figure 7.8 **Dynseg depends on LRM relationships** Dynseg allows you to produce and use geometry representing events without having to maintain it by utilizing the implicit relationship offered by the route identifier and LRM measures. In this example, segments are nonoverlapping portions of a route. SegmentID links the Centerline feature class containing polyline features matching the extent of each Segment table row to the Sign and CrossSection event tables.

157

on the left side contain features. Only the Centerline feature class is actually maintained by the traditional editing process. Different work groups within the agency may maintain event date represented here by the Sign and CrossSection tables.

Maintenance of Centerline feature geometry and event tables may occur in disconnected but loosely coupled ways according to a general schedule. When both update cycles are completed, ArcGIS can be used to generate the geometric representation of sign locations and cross-section segments by employing the dynseg function. The result is a collection of thematic layers that conform to the traditional multitheme view of GIS.

The next chapter will demonstrate the advanced dynseg functionality available in ArcGIS and demonstrate some transportation geodatabase design strategies that take advantage of that functionality.

Making routes

In most instances, you will already have routes defined in your agency. However, you may need to construct routes from block segments if your agency is migrating from one schema to another. Or, you may need to construct a new set of routes conforming to a different specification than the existing routes. In either case, it is rare to build a route system from scratch.

A common way to build routes from an existing dataset is to merge features based on a common identifier. For example, you could merge adjacent blocks with the same street name. If you want to keep routes as single-part features—always a good idea—then you will need to create a separate route feature for each disconnected portion of a multipart street or numbered route. This process can be readily automated.

A more hands-on way to get the job done is to create a new empty route feature class and load segments through manual selection. Of course, it's easiest to take the existing feature class without measures and toggle the HasM switch to Yes. You can then use ArcGIS tools to impose measure values on each vertex by declaring the value for the starting and ending vertices.

Routes need event tables. An automated method to create route tables uses existing thematically segmented feature classes. For example, if you have a set of segments for address geocoding and another used by pavement management, you can convert them to route-event tables using ArcGIS tools. You can even use the TransformEvents tool to change route-event table measures to conform to the new LRM.

As was noted earlier, a single facility may be included in more than one route, each of which has its own LRM. There are different ways to deal with this situation. One is to maintain the original geometry without measures, and then to use a calibration point feature class to apply measures from each route system to the geometry when you publish it for others to use. The calibration point feature class can have all the various measure values for the same points, with you deciding which to use for each route system, or you can have different calibration point features for each route system. The shared calibration point feature class offers the advantage of a Rosetta Stone (the ancient key to translating hieroglyphics) translation mechanism between the route systems. You can move route events from one LRM to another through spatial operators available in ArcGIS. Solutions presented in the next chapter show how to do the same thing in route-event tables through managing the multiple LRMs.

Another way to accommodate the multiple-LRM situation is to pick one as the master LRM and relate the others to it. This is implicitly what happens when you choose to create address block linear events in a route-milelog LRM: the address LRM is made subordinate to the route-milelog LRM. The organizational problem with this solution is that you have to get common agreement as to which LRM is the "real" one.

A third way to solve the problem is to separately maintain multiple route feature classes while using topology rules to save you the trouble of changing the geometry of each feature class separately. When you reshape one line, the other coincident lines can be brought into conformance through the topology rules. This solution cannot solve all the editing issues presented by multiple LRMs when the various route systems include different physical facilities. For example, if one part of the organization includes private roads and the other does not, then some facilities appear in only route system feature class. This solution also does nothing to reconcile differences in LRM measures for a given point along each route.

Traversals

Sometimes you need to create a hierarchy of routes to facilitate editing and usage operations. A common way to do this is to create traversals from routes. A traversal is a route composed of part or all of other routes. Traversals can be static or dynamic. A static route is one that is preserved in the geodatabase for multiple uses. For instance, if you had to create separate street route features due to discontinuities in the named street, you could construct a static multipart traversal that included all routes with the same street name. You could do the same thing with a state road that was originally subdivided at county lines as an aid to editing by combining the county-length segments to produce a state-length route.

A dynamic traversal is generated for a particular use and then discarded. An example would be the path determined to get from an origin to a destination. Even if you saved the traversal, it is only useful for that limited purpose and would probably be specific to one person or cargo trip.

Whether static or dynamic, you can impose a new LRM on the traversal. For an interstate highway, where interchanges are numbered to conform to the green milepost signs mounted on the side of the road, the traversal LRM would be derived from the milepost signs. Those signs may be located in the underlying route system according to a very different set of measures, perhaps values that reset at each county line or maintenance district boundary. A dynamic traversal might have an LRM derived from cumulative distance traveled or one that resets at each turn along the path. In both cases, the TransformEvents tool and related ArcGIS functions can be used to modify route events to serve as traversal events.

Traversals offer a way to accommodate multiple LRMs in accordance with the general practice of the second option shown above, which calls for selection of a base LRM for registration and derivation of the other LRMs. The route system LRM is the base version and the traversal LRM would be the derived version. If you need to permanently assign measure values within the traversal LRM so that the same value is assigned to the same point every time the traversal is generated, then you will need to store those measure values in the route-event tables or use an equivalency table that says a position on the base LRM is equivalent to a specific measure on the static traversal.

chapter eight

Advanced dynamic segmentation functions

- Offset events
- Other dynseg options
- Going beyond dynseg
- Adding element tables
- Intersections
- Accommodating multiple LRMs
- Creating traversals for pathfinding

ArcGIS offers special features within its dynseg toolkit. This chapter will present these advanced capabilities and show you how to take advantage of them when designing your geodatabase. Chapter 9 will present a complete geodatabase design for traffic monitoring using dynseg.

Chapter 8: Advanced dynamic segmentation functions

Offset events

Traditional dynseg segments a route-length centerline to create features that match the extent and attributes of an event table. But the end result is still a single linear facility, and this simplistic process does not well represent data arranged by side of road and other aspects of a transport facility. ArcGIS provides extended dynseg functions that can provide a more realistic mapping of event table data.

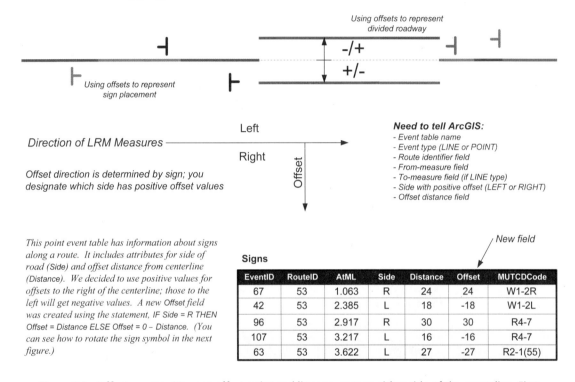

This point event table has information about signs along a route. It includes attributes for side of road (Side) and offset distance from centerline (Distance). We decided to use positive values for offsets to the right of the centerline; those to the left will get negative values. A new Offset field was created using the statement, IF Side = R THEN Offset = Distance ELSE Offset = 0 – Distance. (You can see how to rotate the sign symbol in the next figure.)

Signs

EventID	RouteID	AtML	Side	Distance	Offset	MUTCDCode
67	53	1.063	R	24	24	W1-2R
42	53	2.385	L	18	-18	W1-2L
96	53	2.917	R	30	30	R4-7
107	53	3.217	L	16	-16	R4-7
63	53	3.622	L	27	-27	R2-1(55)

Figure 8.1 **Offset events** You can offset point and linear events to either side of the centerline. First, convert your data in the offset distance field, with an offset having positive values to one side and an offset having negative values to the other. The magnitude of the value will determine the distance from the centerline. Right and left offsets are set by the direction of linear measures, not digitizing direction. Divided roads often include data stored by side of road, which will not display properly without using offsets to produce distinct carriageways for symbolization in a map layer.

One of the more commonly needed capabilities to properly map event-table data is to laterally offset the dynseg result feature from the facility centerline. This function can create two parallel centerlines to symbolize data by side of road, or to properly place sign symbols with the correct spacing from the facility centerline. You can also use the dynseg

Offset events

offset function to locate linear elements of the facility, such as highway guardrail or a drainage structure.

Right and left offsets can be accommodated by using negative and positive offset distances, with the sign indicating the direction of offset relative to the direction of increasing milelog measures. If your source data has offsets expressed in another way, such as by listing a direction and side, you can convert the data to the form required through simple field math tools in ArcGIS. You designate which sign relates to offsets in a particular direction. The example sign inventory point-event table has been structured such that a negative offset represents one to the left.

The only limitation is that the line so created will have the same starting and ending offset and will be parallel to the centerline. To overcome this limitation, the offset function can be further extended to support rotation of the dynseg-produced feature by specifying a rotation angle, which can either be normal or tangential, and can be either the measured angle or its complement. Normal and tangential are relative to the mapping plane with the point indicated by the measure serving as the center of rotation.

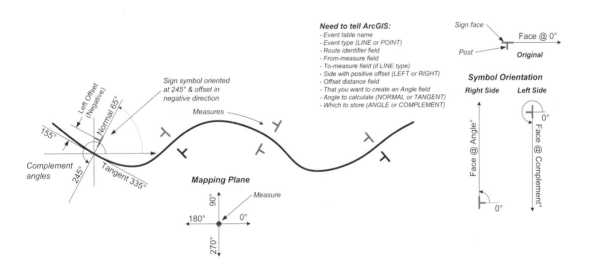

Figure 8.2 **Offset and rotate** Sign symbols in the previous figure were rotated so that the symbol represented the way the sign faced in the real world, visible to approaching traffic. This figure shows how to do the rotation. ArcGIS's dynseg function can calculate the "normal" (correct) angle, the tangent angle, or their complements based on the orientation of the centerline at the interpolated position for a given LRM measure. Angles are determined relative to the mapping plane. This figure also explains how to deal with signs that face the "wrong" way. (Figure 8.2 continued on next page.)

Chapter 8: Advanced dynamic segmentation functions

Signs

EventID	RouteID	AtML	Side	Distance	Offset	MUTCDCode	Traffic	Angle
67	53	3.490	L	24	-24	W1-2R	L	245
42	53	4.218	L	18	-18	W1-2L	L	127
96	53	4.290	R	30	30	R4-7	R	307
107	53	6.207	L	16	-16	R4-7	L	73
63	53	6.269	R	27	27	R2-1(55)	R	258
64	53	8.763	L	16	16	W14-3	L	134
65	53	8.763	R	18	18	R4-1	L	134

We created the T-shaped sign symbol by orienting it along the NORMAL angle axis for signs on the right side, so we went with the default NORMAL angle setting. Of course, this results in all the left-side signs being backwards, so we then used a simple program to calculate the complement angle for flipping the symbol to the right orientation: IF Side = "L" THEN Angle = Angle + 180; IF Angle > 360 THEN Angle = Angle − 360. We have to make a third pass through the table to flip all the signs that are oriented for reading by traffic on the other side of the road, such as R5-1a (Wrong Way) and W14-3 (No Passing Zone), by looking for rows where Side ≠ Traffic.

Figure 8.2 (continued) Offset and rotate

Figure 8.2 shows how to use the offset and rotate extension to the normal dynseg function to properly position and align sign symbols derived from a sign inventory stored in a point-event table. Signs are not always visible to drivers traveling on the side of the facility where the sign is located. Thus, in addition to using the angle of the road at the measure and offsetting the symbol by the appropriate distance, you will need to rotate the symbol to the correct viewing direction.

To support such an operation, you will need to have the appropriate information in the event table. The sign inventory point-event table from the previous figure has been expanded to include the additional attributes. The Traffic column indicates the direction of travel to which the sign face is presented for driver observation. The Angle field stores the results of the dynseg process.

You can implement this relatively complex dynseg process without having to create a special tool. Apply the offset-and-rotate function, ignoring the rotation required by signs that are on the "wrong" side of the road—oriented for drivers traveling in the opposite direction. If the value for side of the road were different from the value for the direction of reading, then you would calculate a complementary angle and reverse the symbol's rotation in a second step.

Other dynseg options

In addition to taking things apart, ArcGIS dynseg functions can put them together. ArcGIS can do line-on-line, line-on-point, and polygon-on-line overlays. A line-on-line overlay can combine the outputs of multiple dynseg operations employing linear event tables. The practical limit to the number of linear attributes you can display is three, with each variable controlling one aspect of line symbology: width, color, and pattern.

Other dynseg options

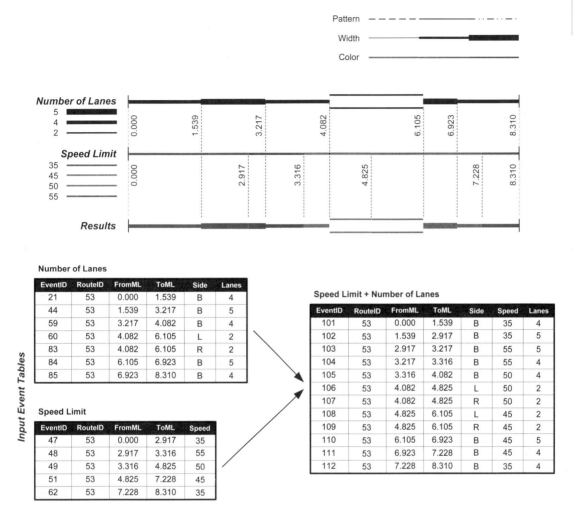

Figure 8.3 **Event overlays: line on line** You will often want to display multiple events on a single map. If your database is normalized, first join the different events into a single table. ArcGIS supports line-on-line overlays (joins) of linear event tables. Three is the practical limit of event tables you can overlay for mapping because that is how many line characteristics you can control: pattern, width, and color.

ArcGIS can also transfer line-event values onto point events, effectively increasing the number of attributes for the point event. For example, you could combine speed limit and pavement condition—both stored in linear event tables—onto crash data contained in a point-event table. As with line-event combinations, the practical limit is three attributes: symbol, size, and color. Of course, rotation is also available, if it provides useful information.

Chapter 8: Advanced dynamic segmentation functions

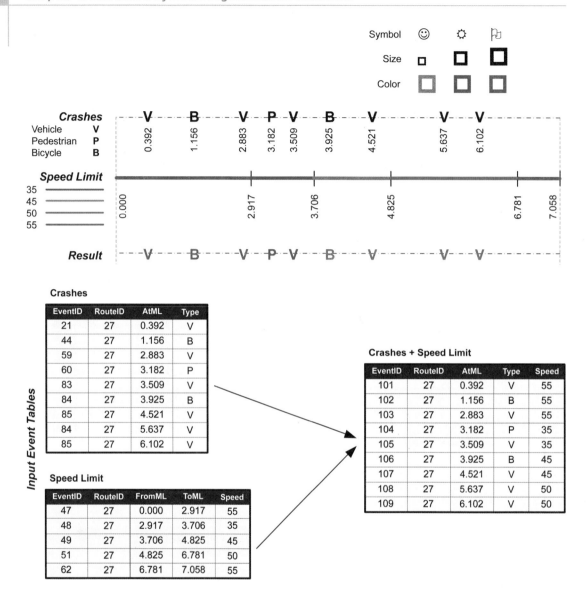

Figure 8.4 **Event overlays: line on point** ArcGIS supports line-on-point overlays (joins) of event tables. The result is a point event table with the characteristics of the linear event table. Again, the number of point symbol characteristics you can effectively use to display the results on a map has a practical limit of three: symbol, size, and color.

A typical use for linear events is to reflect the value of an area feature, such as which portion of a road is within a city. Rather than manually construct these event tables, you can use a polygon-on-line overlay in ArcGIS to generate linear events describing the extent of a road within each polygon.

Other dynseg options

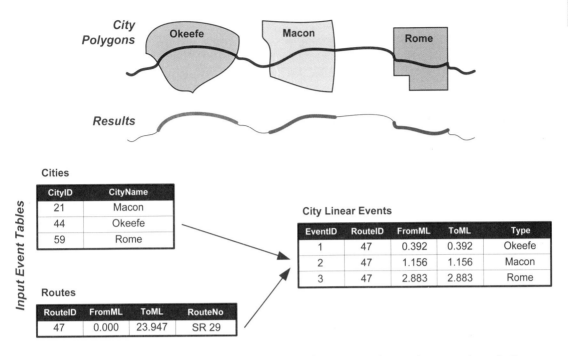

Figure 8.5 **Creating event tables** It is possible to create linear events from polygon overlays of a linear feature that includes LRM measures. A linear event is created for each segment of the centerline that is within a polygon. This figure shows how city polygons are overlaid on a road centerline.

One function available through ArcGIS linear referencing tools is compressing denormalized event-table rows to create a minimal number of rows with unique values. This process works in two ways. The dissolve option will merge overlapping and contiguous rows with the same attribute value into a single row. The concatenate option will only merge contiguous rows and will ignore overlapping rows.

This second option will extract a single event from a multievent table, as the compression is applied to a single field, which is the only nonposition value placed in the output table. If you have a multievent table containing event fields for speed limit, functional class, and number of lanes, you can tell ArcGIS to concatenate the data for speed limit. The output table will be a normalized list of atomic values for speed limit alone. Such an approach can help restructure a geodatabase received from a local government that uses the short-segment design approach.

Chapter 8: Advanced dynamic segmentation functions

Speed Limit + Number of Lanes

EventID	RouteID	FromML	ToML	Side	Speed	Lanes
101	53	0.000	1.539	B	35	4
102	53	1.539	2.917	B	35	5
103	53	2.917	3.217	B	55	5
104	53	3.217	3.316	B	55	4
105	53	3.316	4.082	B	50	4
106	53	4.082	4.825	L	50	2
107	53	4.082	4.825	R	50	2
108	53	4.825	6.105	L	45	2
109	53	4.825	6.105	R	45	2
110	53	6.105	6.923	B	45	5
111	53	6.923	7.228	B	45	4
112	53	7.228	8.310	B	35	4

Input Event Table

Need to tell ArcGIS:
- *Event table name*
- *Event type (LINE or POINT)*
- *Route identifier field*
- *From-measure field*
- *To-measure field (if LINE type)*
- *Output table parameters*
- *Operation (CONCATENATE or DISSOLVE)*

Number of Lanes

EventID	RouteID	FromML	ToML	Side	Lanes
21	53	0.000	1.539	B	4
44	53	1.539	3.217	B	5
59	53	3.217	4.082	B	4
60	53	4.082	6.105	L	2
83	53	4.082	6.105	R	2
84	53	6.105	6.923	B	5
85	53	6.923	8.310	B	4

Output Event Table

Figure 8.6 Concatenation Linear referencing functions in ArcGIS include the ability to concatenate denormalized records to produce a normalized event table for each attribute in a multievent input event table. This process will compress the event table by reducing the number of rows. Since the performance of dynseg is directly proportional to the number of event table rows, using concatenate on denormalized data can save you time in the long run. For this process to work, the **ToML** of each record has to match the **FromML** value of the next record.

For example, if you have received data from a geodatabase that was based on the short-segment structure and want to convert to a normalized design—essentially to select each attribute and merge adjacent segments with the same value—you can use the concatenation function in ArcGIS. This function can also be used if you store several attributes in an event table and want to pull out one attribute and compress the resulting set of rows to the minimal number of rows or to dissolve overlapping events.

Another potential application of these functions is quality assurance. Getting a different number of result rows for dissolve and concatenation actions on a given class will tell you that there are overlaps within the rows.

Number of Lanes

EventID	RouteID	FromML	ToML	Side	Lanes
21	53	0.000	1.539	B	4
44	53	1.539	3.217	B	5
59	53	3.217	4.082	B	4
60	53	4.082	6.105	L	2
83	53	4.082	6.105	R	2
84	53	6.105	6.519	B	5
85	53	6.923	8.310	B	4

Acceptable (Side is different) *Unacceptable (Side is same)*

Figure 8.7 Data scrubbing Linear referencing functions in ArcGIS can help you with data quality assurance at the end of the editing process in order to help ensure that the data you publish to the user community is the best it can be. If you are using an event table design similar to the one used in the examples, you will want to sort by **RouteID**, then **FromML** (or **AtML** for point events), and then by **Side**. Once you do, you are ready for data scrubbing. The text explains how you can identify erroneous gaps and overlaps using simple SQL statements and linear referencing functions.

You can also employ normal SQL statements in quality-assurance operations. To check a linear event table that should not include any gaps or overlaps, you can order the rows by route identifier and beginning measure. Next, use SQL to see if the ending measure for the previous row of the same route identifier matches the beginning measure of the next row. A smaller number is a gap; a larger number is an overlap. If you store data by side of road, then you can use SQL to ensure that row pairs have matching measures in addition to lacking gaps and overlaps.

Going beyond dynseg

Historically, the most common way to maintain data in a mixed feature/event table environment is to maintain the event data externally and periodically copy it to the geodatabase. Geometry is maintained in ArcGIS, which is combined with the imported event table data to perform spatial analyses. This practice is the result of legacy applications being used for event table maintenance, and may be facilitated by feature maintenance being assigned to a different workgroup than event table maintenance. While preserving the work practices of the past, it is based on data redundancy and can result in inconsistencies between the original data source external to the geodatabase and the data distributed through the geodatabase to the end users outside the source workgroup.

Preserving organizational practices, while logical, leaves a lot of built-in ArcGIS functionality on the table. You need to get more out of the software. ArcGIS was designed to support data editing without requiring a lot of user-developed functionality. Given the probability that the external software maintaining the data at the source was homegrown or, at least, a "vintage" off-the-shelf commercial product, it imposes its own costs on the organization just to keep it working and responsive to changing data requirements. One of these costs is increased risk to the workgroup and the organization from software failure, the labor required to deploy workarounds to accommodate changes unsupported by the software, or as a result of data-integrity issues. Some of these problems may lead users to start editing the geodatabase version in order to get what they need, creating its own data-integrity issue. You should at least consider migrating one or more non-GIS editing processes to the ArcGIS platform to prevent and solve these problems with a product that has continuing vendor support and widespread user commitment.

Chapter 8: Advanced dynamic segmentation functions

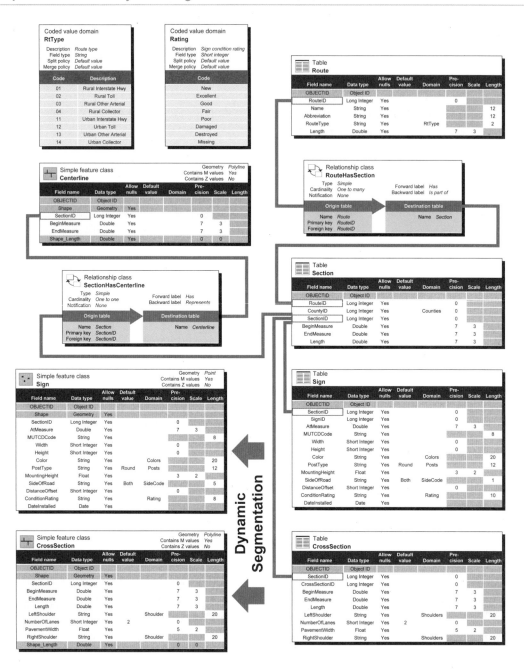

Figure 8.8 **Taking the next step** Limiting ArcGIS to making maps of transportation data is like buying an Italian sports car just to go to the grocery store. ArcGIS is made for data editing. Why not go ahead and use it to maintain the data? Establish explicit relationships and domains to keep bad data out, and run gap-and-overlap analyses to find consistency errors before they get out to the users. (Figure 8.8 continued on next page.)

Going beyond dynseg

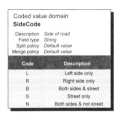

Figure 8.8 (continued) Taking the next step

The example database design from figure 7.8 in chapter 7 has been expanded in figure 8.8 to include relationship and domain classes that illustrate the types of simple enhancements useful in editing event tables. Workgroups that edit event-table data frequently employ ArcGIS for spatial analyses of such data so adopting the same platform for editing the data allows consolidation of software. This consolidation reduces software maintenance overhead within the organization, as fewer licenses and less user support are required. It can also enhance data quality when users can use ArcGIS functions to control data entry and check it after updates are completed. ArcGIS support for multieditor operations, which is not limited to feature editing, can be especially useful for event-table maintenance in a decentralized organization.

One strategy to migrate to ArcGIS as an editing environment for event tables is to develop data-maintenance routines in ArcGIS as various users identify new or changed data needs rather than revise the legacy software. Domain classes, valid-value tables, relationship classes, and referential integrity rules can be readily created in ArcGIS to control the data-entry process without developing any custom software.

The expanded design in figure 8.9 accommodates those facility inventories that treat all aspects of a facility as a route event. This is a common legacy database design, one that offered great benefits years ago when database management system performance was slow by today's standards Although more recent geodatabase designs separate route events into elements (facility components), aspects (facility attributes), and things that happen ("real" events), the more common legacy view is to treat them all the same. As a result, a common legacy database design that is often mirrored in the geodatabase is to create one or two event tables that contain all data. Although this design ignores the benefits flowing from an object-oriented solution where each class can be given data and behaviors specific to its needs, there is a considerable momentum and familiarity associated with the legacy design. Since this

171

Chapter 8: Advanced dynamic segmentation functions

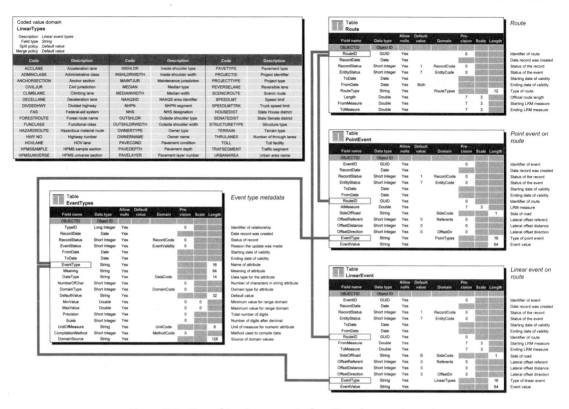

Figure 8.9 **Everything is an event**   (continued on next page)

duplication of legacy structure will likely continue for several years, figure 8.9 shows such a geodatabase with domain classes added. It also includes the enhanced metadata discussed in chapter 6, as a means of improving data quality, facilitating data-maintenance activities, and supplying important data to external users.

One design aspect of the event tables is that they include several fields that allow relatively precise position descriptions. In addition to the traditional route-milelog fields, there are opportunities to indicate side-of-road and offset locations. An offset referent is the element from which the offset distance is determined. Such elements could be a centerline, edgeline, face of curb, or back of sidewalk. The offset direction could be in or out, away or toward, or a compass direction, depending on what is normal for that element being located. For example, you could specify the location of a sidewalk as being five feet outside the curb face, or a sign as being 10 feet away from a bridge abutment.

The simplest version of such a design is to eliminate the point-event table and place all data within the linear event table. If your agency uses this design for its legacy systems, then you may want to include a field that indicates whether the event type requires one or two terminal measures. This way, you can check data as it is entered to ensure that the correct location description has been provided.

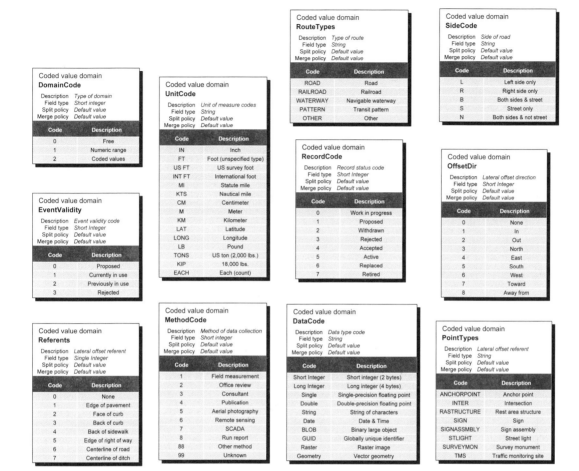

Figure 8.9 **(continued) Everything is an event** Some state DOTs have adopted a database design that treats everything as a point or linear event. This sample design uses extensive domain control and join (implied foreign key) relationships, and uses the temporal metadata design techniques discussed in chapter 2. The EventTypes table is a form of the valid values table design and includes temporal metadata. To view this figure in detail, go to http://www.esri.com/industries/transport/resources/data_model.html

Adding element tables

As we noted above, it is increasingly common in geodatabase design to separate events into classes for elements, aspects, and things that happen. This design offers a way to separate elements from other event types. An element is any physical part of the facility. For example, pavement, right of way, signs, guardrail, ditches, medians, and shoulders are all physical parts of a road facility. You can usually identify an element by your need to include multiple attributes for it.

Chapter 8: Advanced dynamic segmentation functions

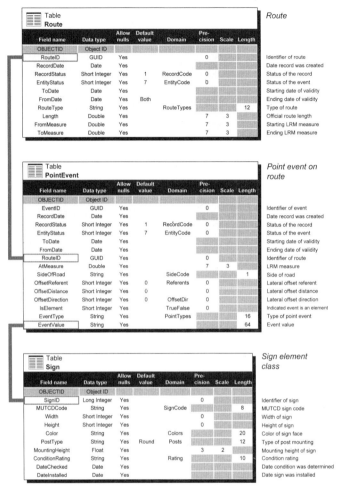
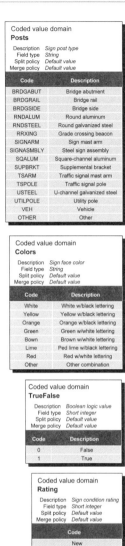
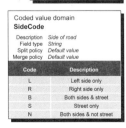

Figure 8.10 **Extending the model to accommodate elements.** The "everything is an event" data model is acceptable when there is only one value to store for each event type. But that restriction can seriously slow the ability to store information about elements and events that may require more than one attribute. In this example, the event data model is extended to use the event class as a locator for a physical element and a pointer to the record that contains descriptive information. When the value of the **IsElement** field is set to True, **EventType** points to the external data table and **EventValue** to the related row's identifier.

An element may be stored using a table or feature class. In many cases, the geometry for a feature class can be generated, as needed, using the dynseg function. Only if the geometry has some special characteristic, such as indicating the true path of a ditch line that is not oriented consistently relative to the facility, will you need to manually edit it.

For example, an outside shoulder is an element with multiple attributes, such as width and type. The common class structure for an event allows one attribute for event, such as functional class, speed limit, and pavement condition. This structure must be extended to accommodate physical elements with multiple descriptive attributes. You could use multiple events, such as one for outside shoulder type and another for outside shoulder width, but it enhances the editing process and data quality to have one place to enter both related attributes.

It is a simple matter to enhance the event-centric design to accommodate elements, each of which can be in its own class. Figure 8.10 illustrates how a geodatabase sign inventory could be constructed. The location of the sign is stored in a point-event table. The value of EventType tells you it is a sign, with EventValue serving as a foreign key to the Sign table.

If all you want to know is that a sign exists at a location, the PointEvent class row tells you that. If, however, you want to know which signs may need to be replaced, you can extract all the sign point events, use dynseg to create a point feature for each sign, and then do an on-the-fly join to add the Sign class attributes to the sign features. A simple SQL SELECT statement based on the value of ConditionRating will allow you to map those signs that need to be replaced.

Intersections

Putting the location data in a different class from the other descriptive data allows you to discover the existence of any element by searching a single table. It also lets you point to one element from multiple facilities without duplicating the data about the element. This design feature may not be useful for a sign inventory, but it can be very important in dealing with intersections. An intersection is, by definition, shared by multiple facilities. Each facility will define a position for the intersection based on that facility's LRM data, but you do not want to design your geodatabase to include two intersection elements.

This example design in figure 8.11 substitutes the Intersection simple junction feature class included in an earlier geometric network geodatabase design for the Sign table in the previous figure. The Intersection class could alternatively be a table or a simple point feature class; the class type does not matter.

One PointEvent row will exist for each involved route. The value of EventValue will point to a single, shared Intersection feature. Such a design avoids redundantly storing information

Chapter 8: Advanced dynamic segmentation functions

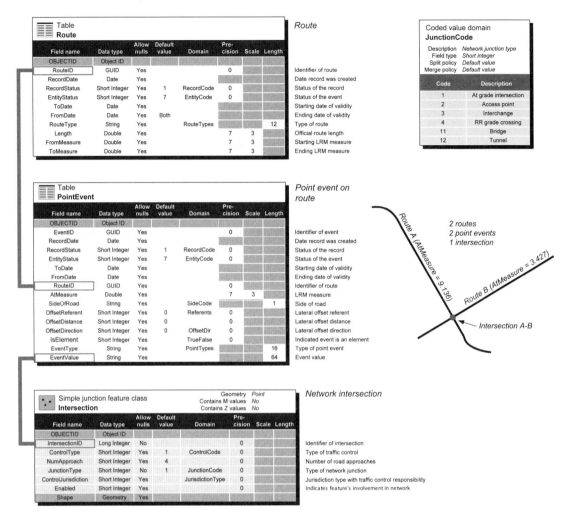

Figure 8.11 **Avoiding redundancy** Some system components can be claimed by more than one route. Such is the case with network junctions and the elements or facilities they represent. In this example, we have used the principle of separating location from other attributes in order to avoid redundantly creating intersection records. We still need two point events for an intersection of two routes, because each route's LRM imposes a different location description. Both point to the same Intersection feature. We do not have to explicitly store information about the involved routes because we find them by looking for the intersection identifier in the **EventValue** field. Junction types suggest a quality-control check by implying a numeric range for the number of involved routes: at least one for a bridge; at least two for at-grade intersections; exactly one for access points, railroad grade crossings, and tunnels; and exactly two for interchanges.

about the intersection for each route, and has the benefit of supplying an atomic intersection object for relating other data. For instance, you could use the Intersection class as a way to organize traffic-signal timing records, turning-movement counts, and crash data according to the intersection location where they occur. The values suggested for the JunctionCode domain class repeat those of the earlier example.

Accommodating multiple LRMs

The next step toward fully normalizing the geodatabase design separates the position description from the event table. The traditional approach is to treat the position description as an event attribute. The result of adding position fields to an event class is that there is only one LRM available. A serious side effect of this design practice is that a given event or element will be described by multiple rows from multiple classes, one for each LRM. Anyone wanting to work with multiple LRM references has to reconcile the two coordinate systems. Some methods to overcome this limitation were presented in chapter 7.

The design variation shown in figure 8.12 is based on the view that a location is separate from the various position descriptions that may be provided. Location is absolute while position is relative to a location reference. An intersection is in one location, but several position statements, such as one for each route-based LRM datum, may describe that location (the traditional route-milelog measure, an intersection node number for crash reporting, or as a street address).

This design example in figure 8.12 includes one class for locating point and linear events, with a point event having a null **ToMeasure** field value. Both event tables and element tables can point to a single LRMPosition table. You can easily employ this design to drive normal dynseg operations by selecting all the LRMPosition rows for a specified LRM type, and then doing an on-the-fly join of the extracted LRMPosition rows and any event table of interest.

All events and elements remain atomic entities in this design. Because the resulting relationship between Event and LRMPosition is one to many, multiple LRMPosition rows may carry the same LRMPositionID value. The true foreign key is complex in that it also includes LRMType. You will need to specify the **LRMType** to properly select the data in which the position is to be stated for a given application. If you wanted to do so, you could convert the relationship cardinality to many to many and add an attributed relationship class between Event and LRMPosition.

One of the other changes included in this geodatabase design is the accommodation of multiple route systems within a single dataset. Many route types may be overlaid on a single facility, with each designed to meet a particular application need. For instance, the route-milelog system used for HPMS sample sections may be different from that used by the maintenance division. These route systems may all describe the same set of physical facilities, but with different beginning and ending points, paths, and LRS rules. Nevertheless, each route still defines the LRM data for that LRM type.

Chapter 8: Advanced dynamic segmentation functions

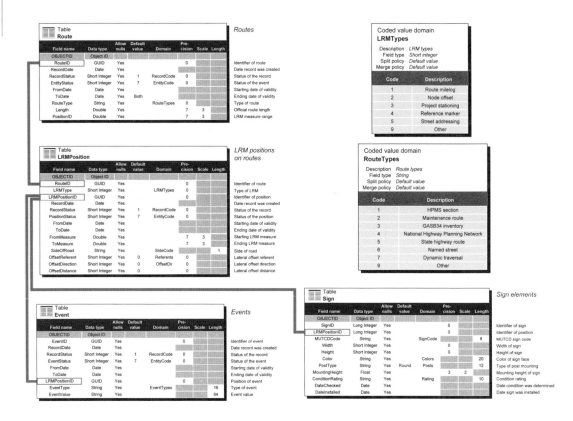

Figure 8.12 **Supporting multiple LRMs** When you have multiple versions of an attribute, as when there are multiple LRMs, you have to take the location information out of the class and treat it as a many-to-many relationship between the routes and its events and elements. Here, the relationship and its attributes are in a table to provide domain control. The route and measure attributes in point and linear event tables are replaced with the **PositionID** foreign key. An LRMPosition table row represents a combination of **RouteID** and **LRMType** for a given location. Each event is now in the database one time, but can be combined with the appropriate **RouteID** and measure values to produce a version for use by a given constituency. There no longer needs to be separate point and linear event classes; a point event has only a **FromMeasure** value in the LRMPosition table; a linear event also has a **ToMeasure** value. This polymorphic behavior means a single event can be represented as both a point and linear event, depending on the LRM type assigned. To view this figure in detail, go to http://www.esri.com/industries/transport/resources/data_model.html

This design choice has the added advantage of keeping all location data in one table, resulting in smaller event tables. It also allows you to check data based on location more directly, such as mapping each element with various LRM types to see if they produce coincident points. Chapter 12 covers designing tables to preserve the equivalency between positions stated in different LRMs. This design relies on equivalencies being discovered through line-on-line overlays following dynseg operations applied to each route system.

The design suggested in figure 8.12 can be applied to feature classes by establishing an LRM-LRM equivalency to the LRM type used to determine the measure values attached to polyline vertices. A dynseg operation undertaken with a different LRM type than the one applied to the feature class will require conversion of linear measures to the base type. Chapter 14 describes an alternative approach that allows multiple measure values to be assigned to each vertex.

Creating traversals for pathfinding

One of the increasingly common uses of dynseg operations is to prepare data for pathfinding applications. An earlier geodatabase design example indicated that pathfinding applications need to know about restrictions to traffic flow, whether imposed by regulation (no left turn from 4:00 to 6:00 p.m.) or physical structures (weight-bearing capacity of a bridge).

A traversal is a path through the transport system constructed as an ordered sequence of route segments. Some traversals may be constructed using simple features to represent higher order facilities, such as combining all the segments with a given route number. These static traversals are defined by one or more common attributes and may be routinely created whenever a geodatabase is published. It is simple to select, say, the route number events and concatenate them to produce static traversals.

Other traversals illustrate how one could travel from an origin to a destination. Such dynamic traversals are determined as much by the nature of the vehicle as by the facilities it may move along. Assuming that you have already defined network features, the traditional event data must be prepared through a four-step process to support pathfinding applications.

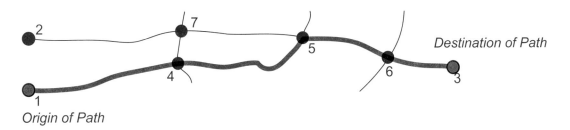

Figure 8.13 **Traversals** In this example, the traversal extends from Point 1 to Point 3 and combines Segments 1-4, 4-5, 5-6, and 6-3. A static traversal is one that is persisted within the dataset, while a dynamic traversal is one compiled for a single use and then discarded. A typical static traversal would be the route features created by joining all the street segments with the same name. A typical dynamic traversal would be the permitted path of an overweight vehicle through a highway network.

In figure 8.14, the first step involves creating a polyline feature class that has all the attributes necessary for the application. Such attributes are likely to include number of lanes, speed limit, and traffic conditions. You would use the ArcGIS dynseg function to produce a simple feature class with the attributed centerline, and then perform a line-on-line overlay of the dynseg result and the network elements you constructed earlier. The output of this step will be a set of network edge features with linear event attributes.

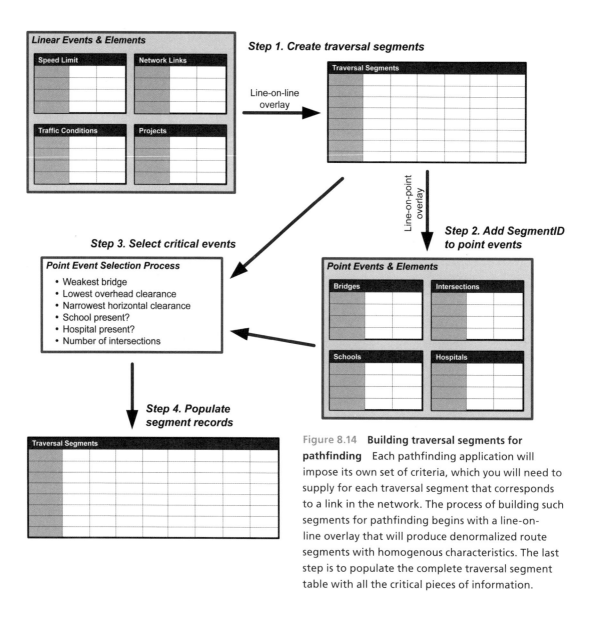

Figure 8.14 **Building traversal segments for pathfinding** Each pathfinding application will impose its own set of criteria, which you will need to supply for each traversal segment that corresponds to a link in the network. The process of building such segments for pathfinding begins with a line-on-line overlay that will produce denormalized route segments with homogenous characteristics. The last step is to populate the complete traversal segment table with all the critical pieces of information.

The second step allows you to prepare point events that will also be useful by attaching edge identifiers to the appropriate point events using a line-on-point overlay. Such event types include bridges and intersections. If your geodatabase also includes information on schools and hospitals, such data will be useful in routing hazardous materials.

The third step involves selecting the point events on each segment that may represent a limiting element, such as bridges and intersections. You will use the segment identifiers to group the point events by segment, and then select the ones that present the lowest value for an attribute of interest, such as the weakest bridge, the narrowest horizontal clearance, and the lowest overhead clearance.

In the fourth step, the selected point events become edge attributes. You now have fully populated edge elements for pathfinding applications. A vehicle that is too wide or heavy to traverse the most limiting element of the segment will cause that segment not to be included in the traversal generated by the pathfinding application.

The approach to building network features treats linear features as a special type of element. Separating the process of maintaining network feature extents from their attributes allows you to meet the objective to measure once and cut many times by employing ArcGIS dynseg functions.

chapter nine

Traffic monitoring systems

- Traffic monitoring sites
- Traffic monitoring system
- The data model
- TMS event tables
- Seasonal factor groups
- Equipment inventory
- Traffic monitoring site maintenance
- Traffic counts

Most larger transport agencies maintain a set of event tables that involves traffic monitoring. Using the geodatabase to store and process the data generated by such a system is less common. This chapter will show how to design such a geodatabase and explain the advantages of using ArcGIS in this way. It also covers methods for working with event tables and data stored using a linear referencing method (LRM).

The traffic monitoring system discovers traffic patterns for facility design and operational management. For example, a road with high truck traffic will wear pavement quicker than an equivalent road with less truck traffic. If pavement must last 10 years, it is necessary to forecast the amount of truck wear on that pavement for that lifespan. There are also operational considerations, such as predicting the number of traffic lanes needed to reduce congestion.

Since this book is about data editing in transportation organizations, this chapter will focus on traffic data creation, not data use. Transport agencies of all sizes conduct traffic counts. The steps in this chapter are common in the data creation and editing process leading to data publication, although smaller agencies may not follow all the steps.

A set of traffic characteristics for each traffic section is a typical end-of-the-year output. How this process is conducted is likely to be the biggest difference between smaller and larger organizations. Nevertheless, almost all traffic monitoring systems that use count data have traffic statistics as their end product. The most common such statistical indicators are annual average daily traffic (AADT), percentage of trucks (*t*-factor), percentage of traffic occurring during the highest hour of the day (*k*-factor), and directional bias in traffic during the highest hour (*d*-factor). Additional indicators can be generated, such as the peak-hour factor that describes the proportion of the highest hour volume that occurs during the peak 15 minutes. A statistic called the volume-to-capacity ratio (V/C) describes the proportion of the total facility capacity that is being used during the peak hour. Traffic operations and pavement management staff may also use estimated design-hour volumes.

> A traffic pattern describes the nature of a traffic flow that must be accommodated by a new design or reflects the current demands being placed on a facility to anticipate maintenance requirements. One of the most obvious elements of a traffic pattern is the cycle of traffic volumes. The cycle typically is determined by the nature of the travel destinations and origins supported by the facility. Arranging hourly traffic volumes from highest to lowest shows the basic patterns in figure 9.1.
>
> A recreation area access road would show the greatest volatility in traffic volumes. Weekend travel will likely be higher than weekday travel with large seasonal variance. In contrast, urban-center roads that serve workday travel will show a fairly consistent volume throughout the year.
>
> There are times when the traffic pattern is altered by capacity limitations. Congested conditions represent times when demand exceeds capacity. As a result of congestion, it is not possible to observe free-flow traffic patterns, and the highest volume hours essentially form a flat line when charted. Deviation from expected patterns, given the nature of traffic traversing the monitoring site, can be used to identify congestion and the functional capacity of the facility.
>
> One application for traffic pattern data is to produce forecasts of design-hour volumes for increases to the system's planning capacity. Commonly chosen values are those of the 30th and 100th highest

volume hour of the year. The implicit assumption is that it is inefficient to design a facility for the highest volume hour that, by definition, will only occur once a year. For urban roads, the 30th highest volume hour typically reflects the seasonal high value for the afternoon peak, while the 100th highest hour represents an average afternoon peak condition.

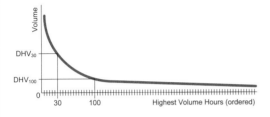

Recreation access road A road that provides access to a recreational facility, with its inherent special event peaks and low off-peak volumes, exhibits a pattern where the highest volume hours are many times greater than average.

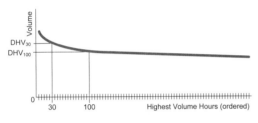

Urban center road Roads in an urban activity center typically have fairly repeatable patterns that exhibit consistent morning and afternoon peaks. Shopping centers will exhibit higher seasonal variance, such as one with a Christmas peak. Office centers will have almost no seasonal variance, but weekend volumes will be substantially below those of weekdays.

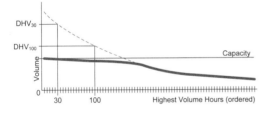

Capacity-constrained road When traffic demand exceeds road capacity, peak hours will be almost exactly the same volume. What will differ is the duration of the congestion produced by inadequate capacity.

Figure 9.1 **Traffic patterns** One output of the traffic monitoring system is a graph of highest volume hours, which can be used to guide roadway design and traffic signal programming. Several traffic characteristics are based on a selected design hour, often the 30th or 100th highest volume hour. Such factors include the percentage of daily traffic occurring during the highest volume hour (k-factor), percentage of traffic going in the dominant direction (d-factor), and the percentage of trucks in the traffic stream (t-factor).

Traffic monitoring sites

An automated traffic recorder (ATR) monitors traffic conditions at a designated site. ATRs come in three basic versions based on the capabilities of the unit. The most limited version can only count vehicles using pneumatic road tubes. Each two actuations form one vehicle. Vehicles with more than two axles, such as trailer trucks, produce overcounts. When used to count more than one lane of traffic, simultaneous actuations can produce undercounts. These ATRs reach saturation at a relatively low volume due to the cycle time needed to repressurize the pneumatic hose before the next actuation can be detected. Traffic counts may be conducted for two-way traffic, by direction of travel, or by lane with the proper arrangement of pneumatic hoses. Subtotals may be reported by 15-minute interval with hour totals or some other combination of time intervals. Each subtotal is called a bin. A single bin holds the volume counted during the interval for that sensor, which could be traffic in one lane, one direction, or two directions.

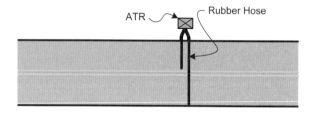

Simple count site The most common count site consists of a portable accumulation unit placed on the side of the road and one or two rubber hoses detect passing tires on the road. Every two compressions count as one vehicle, which usually results in an overcount.

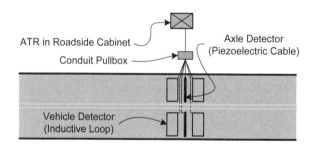

Vehicle classification site Vehicle classification is based on the number and spacing of axles. Two inductive loops sense the presence of a vehicle, with the difference in detection being used to calculate its speed. Axle detectors sense tires passing on the pavement. Vehicle speed and the time between axles can determine axle spacing.

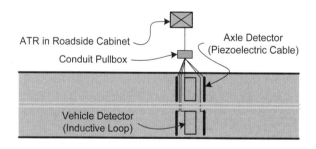

WIM site A vehicle traveling down the road at 70 mph is difficult to weigh. An ATR capable of conducting a WIM survey is expensive and must be carefully calibrated. The design shown here is only one possible configuration. Another common configuration employs bending-plate sensors and multiple inductive loops.

Figure 9.2 **Traffic monitoring sites** Traffic monitoring sites come in three basic types: simple count, vehicle classification, and weigh in motion (WIM). Permanent sites are more likely to have classification or WIM capabilities, although portable equipment can be used. Vehicle classification counts describe the types of vehicles in the traffic stream and derive the t-factor for a count site. WIM data is used to develop pavement designs, which are based on the accumulated wear caused by heavy vehicles.

Vehicle classification counts supply a subtotal for each vehicle class for each time interval, usually 15 minutes or one hour. Due to detector requirements, classification counts are conducted by lane of travel. This means that the data stream can include many bins for a single time interval. For example, a count conducted in 15-minute intervals using the 13-class schema will generate 1,248 bins of data in a 24-hour period. Weigh-in-motion (WIM) counts add information about axle and vehicle loads, which are an important input to pavement design and performance management.

> Traffic counters range from simple pneumatic impulse detectors, often seen on the side of the road connected to one or more rubber hoses stretched across the pavement, to complex units coupled to multiple sensors that can detect the number, direction, class, and weight of passing vehicles. In terms of raw numbers, the volume counter is the most numerous. It likely has one rubber hose, called a pneumatic sensor, attached to it. Every two impulses, each created by the passage of a single vehicle axle, are counted as one vehicle. Some machines may use two hoses to identify the direction of travel, determined by which hose is contacted first, in order to provide directional counts.
>
> A more versatile ATR type is the classification counter. In addition to counting vehicles accurately, it can classify vehicles by type. The most commonly used classification schema is the 13-class version developed by the Federal Highway Administration (FHWA) for vehicles based on axle number and spacing. Vehicle sensors, typically electromagnetic resonance antennas (aka loops) embedded in the pavement, detect the presence of a vehicle. Axle detectors, typically formed by piezoelectric cables, are placed between the vehicle detectors. A piezoelectric cable converts vibrations caused by tire impacts into an electrical signal. Vehicle speed is determined by the time required to traverse the distance between the loop detectors. Axle spacing is determined by the time between axle detections and vehicle speed.
>
> Weigh in motion (WIM), the most complicated ATR type, is capable of weighing vehicles as they pass at normal highway speeds. Originally, these ATRs used a bending plate to weigh each wheel. One plate is required in each wheel path. A more recently deployed version has replaced the bending plates with piezoelectric cables. The magnitude of the electrical signal generated by a piezoelectric cable as a result of a tire impact is roughly proportional to the weight of the vehicle. The "roughly proportional" qualifier is necessary because other factors other than weight, such as tire pressure and composition, can impact signal magnitude. As a result, WIM sites normally use two piezoelectric cables in each wheel path and calculate an average signal magnitude. Only a single vehicle detector is required, because the passage time between piezoelectric cables is used to determine vehicle speed.
>
> WIM data is typically generated as a wheel path subtotal for each axle of a vehicle, with a total weight for all axles and a vehicle classification indicated. WIM data may actually consist of a record for each vehicle observed, although some WIM ATRs ignore lighter vehicles except to provide a total volume. It is common for this data to be summarized daily to produce a data stream that looks similar to a classification count with equivalent single-axle load (ESAL) values.

> Pavement design is based in large part on the forecast number of ESALs to be experienced over the desired years of serviceable life. The wear-causing nature of traffic can be characterized as a single dimensionless number, the ESAL, based on standard 18,000-pound axles with dual wheels on each end (18 kip loads). ESALs increase geometrically with weight. A vehicle that has a weight of 2 ESALs has four times the pavement wear impact as one weighing 1 ESAL. A tandem trailer truck can have the wear impact of 10,000 passenger cars.
>
> Classification and WIM ATRs normally have the capability to perform less rigorous counts, such as simple volume totals by time interval. They also can do speed studies, where counts are reported in 5- or 10-mph subtotals by time interval, if there are two vehicle sensors for each lane of travel. Speed is determined by the time required to traverse the distance between the sensors, if the distance between sensors is known.

Traffic monitoring sites may perform a variety of functions beyond those of the traffic monitoring program. For example, vehicle detectors may be placed on major highways to supply traffic flow data to a central facility that manages changeable message signs to report traffic conditions ahead. Overweight trucks can greatly accelerate pavement wear, which is why all states operate weight-enforcement programs and position truck weigh stations along major highways. WIM ATRs may be placed in the ramp leading to the weigh station as a means of identifying those vehicles that may need to be weighed on static scales. Others will be allowed to bypass the static scales and will not need to stop. Such enforcement programs may also use portable WIM equipment for spot checks on weigh station bypass routes.

Traffic monitoring system

A traffic monitoring system develops traffic pattern information and compiles other statistical data for widespread applications. The system is composed of two primary components. Permanently installed, continuously operated ATRs at a few traffic monitoring sites feed data to a central location through periodic downloads via telemetry connections. Because of their continuous monitoring of traffic flows, these sites can supply very detailed data for the road segments they monitor. A sufficient number of such sites must be distributed geographically to fully reflect all traffic patterns that exist in the jurisdiction.

The other component consists of portable ATRs that are temporarily placed at many traffic monitoring sites to conduct short-term monitoring sessions lasting from two days to a week. Traffic characteristics compiled at the continuously monitored sites can be used to adjust the short-term counts for pattern effects, such as those reflecting differences by day of week and time of year.

Data from both continuous and short-term sites is checked against data-quality standards. Checks may compare the recently observed volume to that observed on the same day in earlier years, or examine the traffic volume distribution across all lanes of travel. Some differences may be due to equipment failure, such as the loss of a vehicle detector. Such data must be discarded. Other differences may be due to unusual conditions, such as one lane of traffic being closed for maintenance work, or a special event that greatly increases traffic flow compared to other days. Data produced by these short-term aberrations in the normal flow must be retained, as it truly reflects traffic conditions experienced during the year.

A traffic monitoring site is assigned to a traffic section, which is a road segment with homogeneous traffic characteristics. FHWA requires states to monitor each traffic section at least once every three years, with data for sections in nonobserved years extrapolated from growth rates at similar observed sites.

At the end of each year, data is collected from all traffic monitoring sites. Data compiled at continuously monitored sites will go through additional quality checks to ensure that each site supplied enough days in each seasonal period. Any site that fails to provide statistically complete data must be removed from further processing. Such problems can arise as a result of equipment failures or construction that takes a site out of service for an extended period.

The accepted data then goes through an end-of-year process that extracts useful information. Sites are grouped according to seasonal traffic patterns. It is important to fully understand the nature of traffic passing each monitoring site in order to place it in the correct group. Some states will have many more seasonal groups than others. For example, a state with significant recreational facilities, such as the ski resorts in Colorado or Vermont, must construct recreational traffic groups, perhaps one for summer and another for winter activities. A primarily rural state in the northern United States may have seasonal patterns affecting the ability to travel based on road conditions or activities, such as harvest season generating more loaded truck travel than at other times of the year.

The resulting group-based seasonal patterns exhibited by continuously monitored sites are used to produce seasonal volume and axle-adjustment factors for short-term count sites. These factors compensate for sample errors at sites monitored for a short time. An axle-adjustment factor derived from classification counts at similar sites must be applied to adjust the raw volume reported by an ATR using pneumatic hose detectors to produce a closer representation of the actual number of vehicles observed.

All short-term count data is then adjusted for day of the week and time of year. Each day of the short-term count may be separately adjusted to maximize the number of days when data can be collected. It is typical for the adjustment factors to be based on month or week of year, although some states use a different way to define seasons to more closely reflect the patterns they experience. Since the objective is to produce data that describes traffic conditions over the entire year, counts should be distributed across seasons and throughout the week.

Chapter 9: Traffic monitoring systems

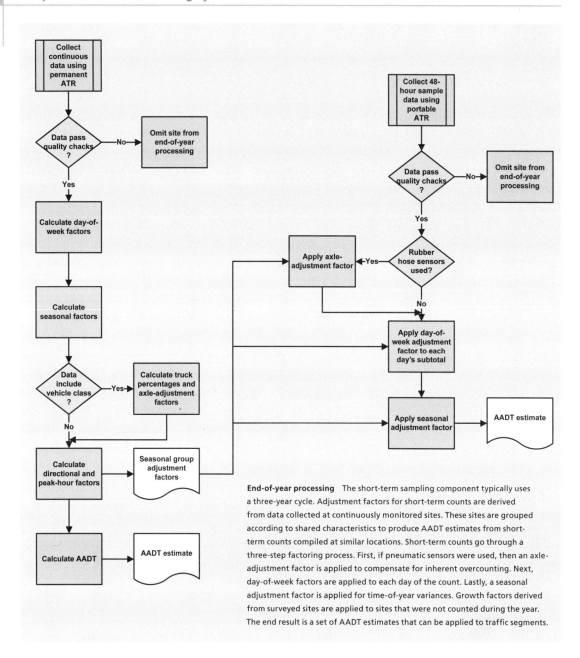

End-of-year processing The short-term sampling component typically uses a three-year cycle. Adjustment factors for short-term counts are derived from data collected at continuously monitored sites. These sites are grouped according to shared characteristics to produce AADT estimates from short-term counts compiled at similar locations. Short-term counts go through a three-step factoring process. First, if pneumatic sensors were used, then an axle-adjustment factor is applied to compensate for inherent overcounting. Next, day-of-week factors are applied to each day of the count. Lastly, a seasonal adjustment factor is applied for time-of-year variances. Growth factors derived from surveyed sites are applied to sites that were not counted during the year. The end result is a set of AADT estimates that can be applied to traffic segments.

Figure 9.3 **Developing traffic characteristics** Traffic characteristics are generated at the end of the count year. It is a fairly straightforward process to derive traffic characteristics at continuously monitored sites. Short-term sites, where two days of data may be collected only once every three years, require more work. Seasonal, day-of-week, and other factors compiled at continuously monitored sites can compensate for the short duration of portable ATR data collection sessions.

The one exception to the general rule requiring extensive samples is the WIM program. It is possible to use a few WIM sites to produce data that can be applied to all traffic sections with similar traffic patterns. Once the WIM sites determine the ESAL profile for the various truck-related vehicle classes, classification counts at other sites can supply information on the number of vehicles, by class, present in the traffic stream. This allows a WIM profile to be established at each vehicle-classification count site.

The data model

A traffic monitoring system geodatabase must contain classes to store counts, each of which is composed of any number of bin counts. The site where they are collected provides a means to organize counts. Each such traffic monitoring site consists of an ATR and one or more sensors, and may be located as a point event on a route using an LRM position. Each site is assigned to at least one traffic section, which may be stored as a linear event. Traffic monitoring sites generate and traffic sections receive annual statistical summaries; both are assigned to seasonal groups.

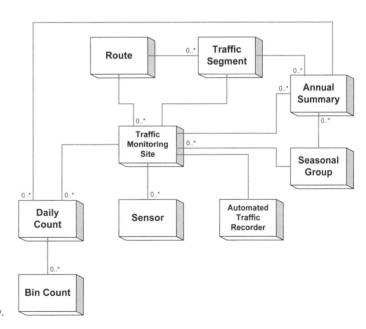

Figure 9.4 **Conceptual database design.** A traffic monitoring system consists of a collection of sites where various types of counts are taken. Each site contains an automated traffic recorder (ATR) and one or more sensors, and is assigned to a traffic section with homogeneous traffic characteristics. Each site is also assigned to a seasonal group by which statistical data may be derived. An ATR may be used at multiple sites for short-term counts, as part of a coverage count component, or it may be permanently installed at one site with data collected through telemetry.

Now that the general requirements are known for a geodatabase supporting a traffic monitoring system, it is possible to propose a conceptual data model. One or more 24-hour traffic counts may be conducted at a given traffic monitoring site, with each count consisting or one

191

or more bin counts. The daily counts are inputs to the production of annual traffic statistics, which are organized by seasonal group. A traffic monitoring site must include an ATR and can include one or more sensors. In all cases, a multiplicity of 0 indicates that data for this entity may not be present in the database.

A one-to-one relationship is shown for traffic monitoring site and traffic segment, although different cardinalities may be reflected in various agencies' practices. For example, continuously monitored sites may be used solely to produce adjustment factors and other summary statistics, without being assigned to a specific traffic section. It may also be necessary to assign one traffic monitoring site to multiple traffic sections, such as when multiple sections must be created between limited-access highway interchanges because of intervening political boundaries. It is not physically possible for traffic conditions to change along the path between interchanges. If the two traffic sections on either side of a boundary line had separate traffic monitoring sites, it is likely that slight, although illogical, differences would exist in the resulting traffic statistics. Some states are known to follow the opposite practice, sometimes assigning multiple count sites to a single traffic section in order to reflect the composite of traffic patterns present.

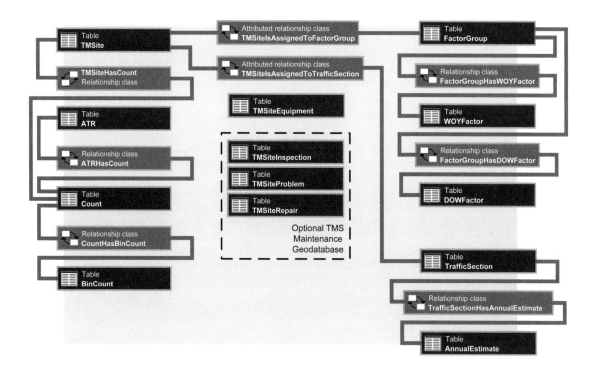

Figure 9.5 **Geodatabase structure**

The data model

This logical data model presents the various classes needed to implement the conceptual data model in ArcGIS. The design includes an optional traffic monitoring system maintenance component that will be described later in this chapter. There are no feature classes because this design is intended for an agency that uses event tables and dynseg to illustrate the location of traffic monitoring sites and traffic sections. There are almost as many relationship classes as there are tables. The classes are structured around four logical units: traffic monitoring sites, traffic counts, factor groups, and traffic sections.

Traffic data requirements Traffic counts are conducted according to various criteria, such as by time interval, direction of travel, and vehicle class. These criteria determine the number of data bins to be created for a given monitoring session. For example, a bin may be created for each vehicle class and lane of travel. The total volume in each bin is recorded at the end of the selected time interval. The type of traffic data that can be compiled at any given site depends on the ATR and sensors used. Common sensors include pneumatic road tubes, piezoelectric cables, inductive loops, magnetometers, video cameras, and bending plates. The Count table acts as an attributed relationship class since the relationship between sites and ATRs is many to many.

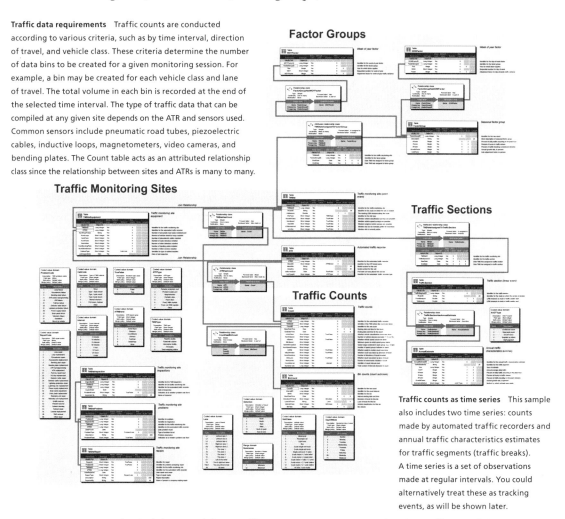

Traffic counts as time series This sample also includes two time series: counts made by automated traffic recorders and annual traffic characteristics estimates for traffic segments (traffic breaks). A time series is a set of observations made at regular intervals. You could alternatively treat these as tracking events, as will be shown later.

Figure 9.6 **TMS physical data model** A traffic monitoring system geodatabase offers useful tools to manage traffic data processing. This model can be extended to help select traffic monitoring sites and identify traffic sections. Although you could use implicit (join) relationships supported by foreign keys, using explicit relationship classes allows ArcGIS to preserve referential integrity. To view this figure in detail, go to http://www.esri.com/industries/transport/resources/data_model.html

Each unit contains tables and relationship classes. Three units also include domain classes to help manage data entry. In practice, the traffic count unit may be replaced by proprietary systems supplied by ATR manufacturers and only summary data reported in the geodatabase through a restructured Count table. Each class will be addressed in more detail in the following sections.

TMS event tables

Traffic monitoring sites are stored in a TMSite point event table with a single LRM position description (RouteID and AtMeasure). Traffic sections are described in a linear event table, also with a single LRM position description (RouteID, FromMeasure, and ToMeasure). It would be a straightforward change to adapt the model to an alternative LRM position design, such as those shown in chapter 8.

The TMSite class includes fields to describe the attributes of a site. SiteType, controlled by the TMSType coded-value domain, has options to indicate whether the site is permanent and whether a cabinet is provided to house the equipment. The sensor entity of the conceptual data model has been converted in the physical model into a set of Boolean indicators. They indirectly describe which attributes are present by indicating the types of traffic monitoring that can be conducted at the location. A final Boolean field indicates whether the site is still in use.

The TrafficSection class defines only its physical location. Fields to store the latest traffic statistics could be easily added although this design uses a separate class.

The conceptual data model showed a one-to-one relationship between a traffic monitoring site and the traffic section to which it may be assigned. The physical data model, however, reflects a many-to-many relationship. This is because the assignment may change over time because traffic monitoring sites may be abandoned or relocated. Traffic sections may also be redefined or eliminated. Thus, the attributed relationship that expresses the assignment must include fields for the beginning and ending dates of the assignment period.

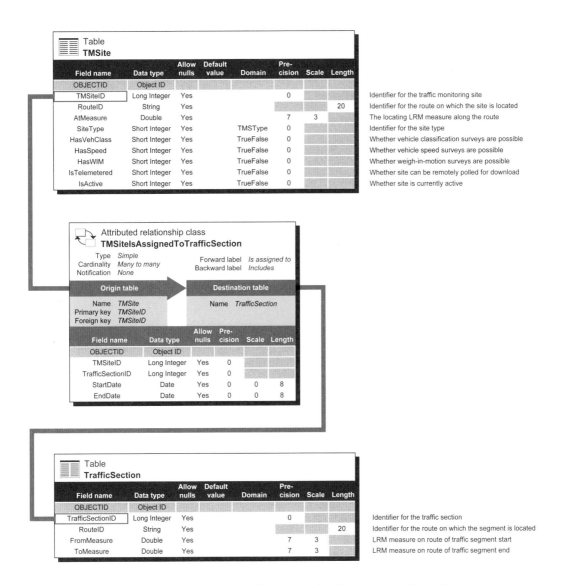

Figure 9.7 **Event tables** Although this geodatabase design does not include any feature classes, you can still map the data contained here. TMSite is a point event table that describes the location of each traffic monitoring site using a linear referencing method (route-milelog). TrafficSegment is a linear event table that similarly describes its location using beginning and ending locations. Both can be mapped using dynamic segmentation and a road centerline feature class with the required route-milelog structure. Spatial operators can be used to identify which segments lack a traffic monitoring site or do other quality control checks. Data contained in other tables can be mapped through their relationships with the event tables. For example, you could readily show which sites were surveyed in a given year, which ones were continuously monitored, or which were in a particular seasonal factor group.

Seasonal factor groups

Factor groups provide axle and seasonal adjustment factors tied to each discernable traffic pattern. The example design is for an agency that uses day-of-week and week-of-year adjustment factors, which reflects a relatively high volatility in traffic patterns. Other periods may be used without affecting the design. Day-of-week and week-of-year factors are arranged by factor group. Each traffic monitoring site is assigned to a factor group, either as a source for factor calculation for continuously monitored sites, or as a recipient of the calculated factors for short-term count sites. All adjustment factors may vary by year. Each factor group will have seven rows in the DOWFactor table and 52 rows in the WOYFactor table for each year of data.

Although it is not common practice to manage traffic data within a geodatabase, such a design offers practical benefits. Not the least of these is the ability to better understand the types of traffic that may be present at a traffic monitoring site. Census, land use, and traffic demand model data may all be viewed within a unifying platform by using ArcGIS. For example, the location of a traffic monitoring site near a college campus suggests that the traffic pattern it observes will be subject to seasonality tied to the school calendar. The ebbs and flows of an agricultural economy may similarly affect a site in a rural area, with ESAL peaks exhibited during harvest season. Shopping centers normally experience peak travel during the Christmas holiday season. Polygon overlays using census and land use data can generate linear events that may serve as the input to a traffic section identification and seasonal group assignment process.

Seasonal factor groups

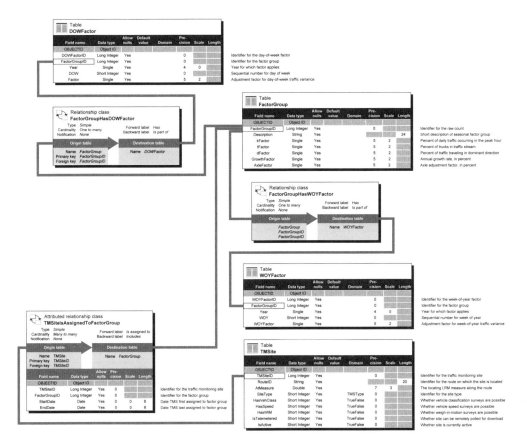

Figure 9.8 **An unusual solution** Although most agencies don't put their traffic data in a geodatabase, there are many advantages to doing so. In addition to being able to map the data for such applications as count cycle planning and quality control, a geodatabase approach can also help develop factor groups and growth factors for traffic segments where counts are not made during the year. For instance, you can use census and land-use data as polygon overlays to help suggest the appropriate seasonal factor group for a given traffic monitoring site or traffic section. **Traffic factor groups** Traffic counts are conducted at a wide variety of locations, each with its own set of descriptive characteristics and seasonal patterns. For example, resort areas exhibit very different seasonal patterns compared to, say, big-city downtowns. Similar traffic monitoring sites are often grouped together to form traffic factor groups for which more statistically accurate characteristics may be determined through the traditional traffic sampling method. A given traffic monitoring site may be assigned to multiple factor groups, and the mix of member traffic monitoring sites may change over time for any given group. The resulting many-to-many relationship is accommodated by an attributed relationship class, TMSIsAssignedToFactor-Group. **Raw-count adjustment factors** Short-term counts taken with portable ATRs generally need to be adjusted to reflect day of week and week of year (or some similar seasonal pattern). When road tube sensors are used, the raw counts also need an axle-adjustment factor. Day-of-week adjustments may be made for individual days, weekdays vs. weekends, or other regular patterns. This figure models week-of-year seasonal patterns, but other, less specific approaches may be used, such as month of year. To view this figure in detail, go to http://www.esri.com/industries/transport/resources/data_model.html

Chapter 9: Traffic monitoring systems

Equipment inventory

If an agency desires to manage the inventory of counting equipment, it can add tables for ATRs and other kinds of equipment used at traffic monitoring sites. Such tables can be the foundation for the optional maintenance record discussed in the next section.

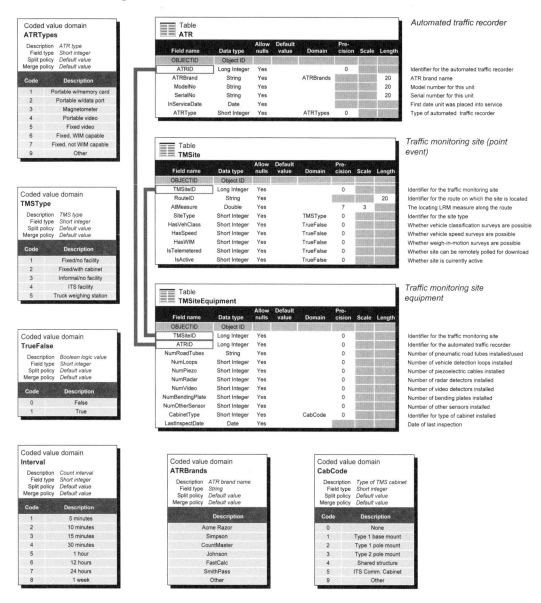

Figure 9.9 **Equipment inventory physical data model**

In the geodatabase design shown in figure 9.9, the equipment at a traffic monitoring site is listed in a TMSEquipment table. One field in this table points to an ATR table. An alternative design would be to provide an attributed relationship class to store a many-to-many relationship with validity dates, if ATRs are frequently exchanged between sites. However, ATR assignment to a traffic monitoring site is really only meaningful here for continuously monitored sites where ATRs are permanently installed. An ATR need not be assigned to any traffic monitoring site, so there could be many more rows in the ATR table than in the TMSite table.

A CabinetType field describes the fixture, if any, in which the ATR, sensor receivers, and telemetry equipment are housed. Cabinets may be mounted on a ground pad or on a post. Some ATRs are completely self-contained and may be used without any external housing. An ATRType domain provides choices that describe the capabilities of the ATR itself. ATR capabilities may not be the same as that of the traffic monitoring site due to a lack of suitable sensors or an ATR's inability to receive the output of some sensor types. For example, a WIM-capable ATR may be installed at a site without the sensors required for WIM operation in order to conduct vehicle classification counts.

Sensor data fields have been expanded from the simple on/off domain of the TMSite table to indicate the number of each sensor type present.

Traffic monitoring site maintenance

A logical extension of the equipment inventory is to help retain maintenance data regarding that equipment. A LastInspectionDate field in the TMSEquipment table allows retention

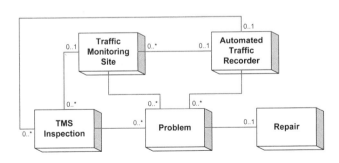

Figure 9.10 **Adding traffic monitoring system maintenance** The basic geodatabase model can be extended to support traffic monitoring system maintenance activities. Besides providing a more complete set of information about the system and the conditions under which the data was collected, this extension will also support spatial analyses regarding problems and solutions. Such information can be used to better allocate maintenance staff and suggest repair part stockpile quantities. For example, you may decide to make more frequent inspections of problem locations, or move troublesome ATRs to less critical locations.

of the date and time when the site was last reviewed for proper operation. A maintenance inventory needs to include more detailed information about such inspections, the problems found, and corrective action taken.

The simple conceptual data model shown in figure 9.10 includes entities for inspections, problems, and repairs. A traffic monitoring site may be inspected. A problem identified during an inspection must relate to the traffic monitoring site and may relate to an ATR. Repairs always relate to a problem. This model indicates the possibly optimistic view that a single repair will correct any problem. You may choose to be more realistic in your implementation.

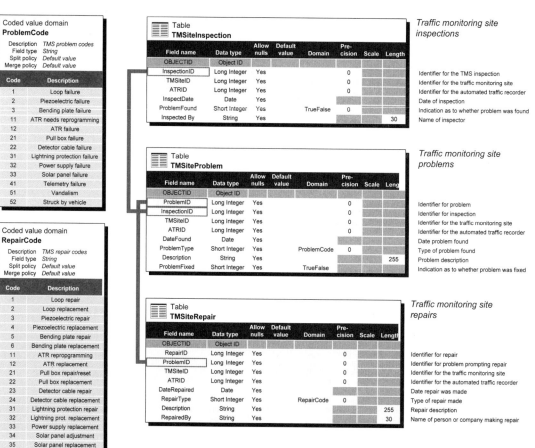

Figure 9.11 **TMS maintenance geodatabase** This figure shows how you could implement the TMS maintenance data model in the geodatabase. An inspection identifies problems that are subsequently repaired. The codes adopted for repairs follow the same numerical arrangement as those for problems. Both domains can be revised to match the problems and solutions appropriate for your agency.

Problem and repair code domains are suggested in figure 9.11, a physical geodatabase data model implementing the conceptual design. Using domain control allows similar problems to be classified uniformly. Potential applications include identifying problems associated with particular ATR models or sensor types.

As a result of using ProblemID as a foreign key, it is possible for this implementation to support a one-to-many relationship between problems and the repairs intended to correct them. If you choose to use that option, you can order repairs by date to see the sequence of actions taken to solve a problem.

Traffic counts

The heart of any traffic monitoring database is the traffic count data. The design reflects the business rule that a count is conducted by an ATR but is "owned" by the traffic monitoring site where the ATR was located at the time. The Count table includes foreign keys linking it to both the TMSite and ATR tables. The Count table is the destination class in both relationships, as the quality-assurance process may reject individual counts. A count is a time series of observations for a traffic monitoring site, which is a stationary temporal object.

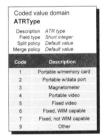

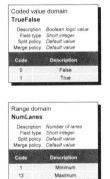

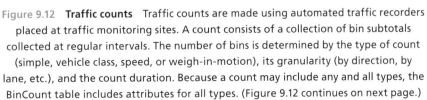

Figure 9.12 **Traffic counts** Traffic counts are made using automated traffic recorders placed at traffic monitoring sites. A count consists of a collection of bin subtotals collected at regular intervals. The number of bins is determined by the type of count (simple, vehicle class, speed, or weigh-in-motion), its granularity (by direction, by lane, etc.), and the count duration. Because a count may include any and all types, the BinCount table includes attributes for all types. (Figure 9.12 continues on next page.)

Chapter 9: Traffic monitoring systems

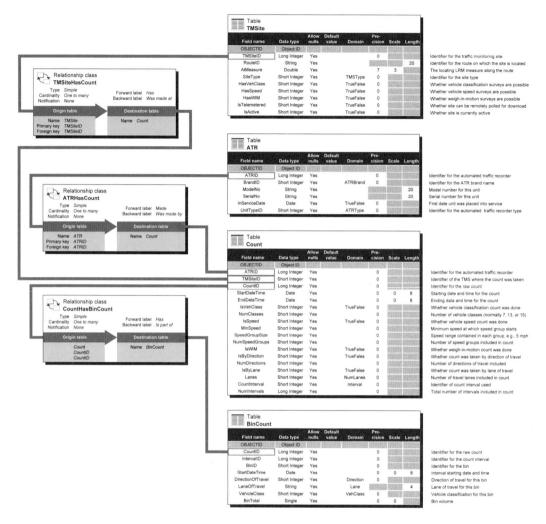

Figure 9.12 **(continued) Traffic counts** To view this figure in detail, go to http://www.esri.com/industries/transport/resources/data_model.html

The BinCount table is a proposed repository for all count bins. The class includes fields for indicating all possible count types (simple volume, volume by direction of travel, volume by lane, or volume by vehicle class). WIM data is summarized by time interval and vehicle class by the time they are placed in this table. As noted earlier, proprietary data structures and processing software may be substituted for the BinCount class. If that is done, a substitute table to store hourly count subtotals, percent of trucks, and similar data should be provided to supply the count data with sufficient detail for calculating summary statistics in the end-of-year process.

Traffic counts

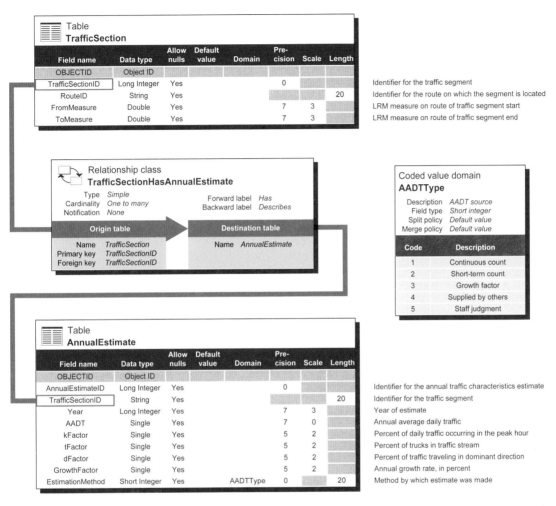

Figure 9.13 **Traffic characteristics** Traffic monitoring systems generate traffic characteristics on an annual cycle. Since conditions change from year to year, and change analysis is a typical application served by traffic data, the AADT estimates and other characteristics are in a time series table, AnnualEstimate. A one-to-many relationship links TrafficSection to AnnualEstimate.

An annual statistical summary of traffic characteristics is prepared for each traffic section during the end-of-year process. At a minimum, the annual estimate table should include fields for the applicable year, the estimated AADT; k-, d-, and t-factors; an annual growth rate; and the manner in which the AADT estimate was made. The data may be derived from continuous or short-term counts, or by inference. In the latter case, data compiled at similar sites during the year may be applied to estimate values for sites not observed during the count year, and an AADT estimate will be provided using the previous value multiplied by a growth factor.

The annual statistics table is a time series, and could be viewed as a temporal observation table for a stationary tracking object, the traffic section. Since the seasonal group to which a traffic section is assigned may change due to conditions affecting the traffic flow it experiences, the factor group identifier is part of the AnnualEstimate table, not the TrafficSection table.

chapter ten

Classic transportation data models

- The ArcInfo route system
- Linear datums
- Providing a linear datum in ArcGIS
- NCHRP 20-27(2)
- FGDC data exchange standard
- The GDF format
- NCHRP 20-27(3)
- Other transportation data models

Over the past 20 years, comprehensive data models were designed for the special needs of transportation datasets. This chapter will present the most influential versions for transport agencies. They include the ArcInfo route system structure supported by the coverage data format, National Cooperative Highway Research Program (NCHRP) Project 20-27, the Transportation Feature Identification Standard proposed by the Federal Geographic Data Committee (FGDC), and the Unified Network for Transportation (UNETRANS) developed at the University of California at Santa Barbara through ESRI sponsorship.

The ArcInfo route system

An Arc Data File contains the geometric shape elements that form lines. An Arc Attribute Table stores references to the geometric elements for use by other files. The coverage data structure was extended almost 20 years ago to accommodate linear measurement systems through a route system option that added two files to the normal ArcInfo database. Multiple route systems can be defined for a single set of arcs.

A route consists of one or more sections. Each section includes all or part of an arc. Each route defines a linear referencing method (LRM) domain. Route measures are stored in the Section file for each section, along with the matching geometric position. A route file has the extension of .rat, short for Route Attribute Table. The section file has a file name extension of .sec. The Arc Macro Language (AML) supported basic dynseg functions.

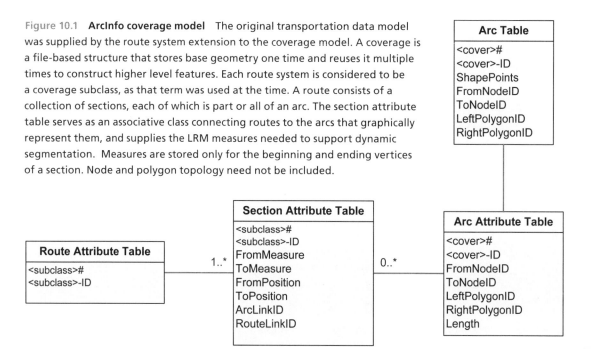

Figure 10.1 **ArcInfo coverage model** The original transportation data model was supplied by the route system extension to the coverage model. A coverage is a file-based structure that stores base geometry one time and reuses it multiple times to construct higher level features. Each route system is considered to be a coverage subclass, as that term was used at the time. A route consists of a collection of sections, each of which is part or all of an arc. The section attribute table serves as an associative class connecting routes to the arcs that graphically represent them, and supplies the LRM measures needed to support dynamic segmentation. Measures are stored only for the beginning and ending vertices of a section. Node and polygon topology need not be included.

The general coverage model allows the geometric representation of a route to be stored as a single set of line segments. Since a section could not be constructed from more than one arc, and a route was a whole number of sections, an awareness of route extents had to be part of the original digitizing process.

The ArcInfo route system

One shortcoming of the coverage route system structure compared to that of the geodatabase is that measures are only stored for section termini. Positions can be more closely approximated through dynseg when measures are assigned to each vertex—as is done for geodatabase feature classes and shapefiles—since the straight-line interpolation works with a single line segment rather than an entire polyline path.

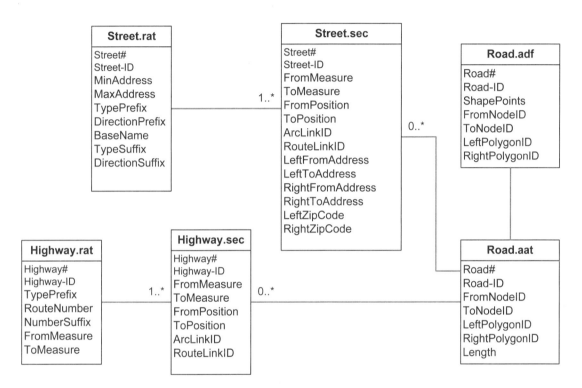

Figure 10.2 **Coverage example** This example shows a Road coverage with two defined route systems. The Street route system supports address geocoding and includes the additional user-defined attributes required for that application. The Highway route system similarly has the user-defined attributes needed to support an LRM. Both route systems are tied to the same set of arcs and serve as Road coverage subclasses.

Figure 10.2 shows how the coverage file structure can be employed to implement two route systems, one for streets and another for highways. Both route systems are constructed from the set of arcs contained in a Road coverage. The expectation is that numbered highways will often overlap named streets. Linear measures in the form of a route based LRM are provided for highways. Streets have both a route-milelog and an address-based LRM. The Street.rat file stores the street name and the upper and lower bounds of the address range. The Highway.rat similarly stores the route number and the upper and lower bounds of the

207

Chapter 10: Classic transportation data models

milelog measures for each route. Measure fields are supplied in the Street.sec file, but these fields will likely only be populated for sections that overlap those defined for highways. Limited-access highways that do not have addresses will not be included in the Street route system.

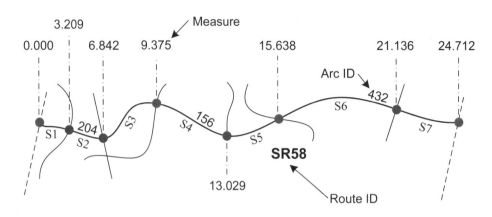

Highway.rat

Highway#	Highway-ID	TypePrefix	RouteNumber	NumberSuffix	FromMeasure	ToMeasure
...	...	...	...	...	...	...
102	SR58	SR	58		0.000	24.712
...	...	...	...	...	...	...

Highway.sec

Highway#	Highway-ID	FromMeasure	ToMeasure	FromPosition	ToPosition	ArcLinkID	RouteLinkID
101	S1	0.000	3.209	x1,y1	x2,y2	204	102
102	S2	3.209	6.842	x2,y2	x3,y3	204	102
103	S3	6.842	9.375	x3,y3	x4,y4	204	102
104	S4	9.375	13.029	x4,y4	x5,y5	156	102
105	S5	13.029	15.638	x5,y5	x6,y6	156	102
106	S6	15.638	21.136	x6,y6	x7,y7	432	102
107	S7	21.136	24.712	x7,y7	x8,y8	432	102

Figure 10.3 **Route system implementation** This figure illustrates the coverage data structure using an example route (SR58), which is composed of seven sections (S1-S7) that define all or part of three arcs (204, 156, and 432). The function of the section table (Highway.sec) to resolve a many-to-many relationship is made clear by the values in each field. **RouteLinkID** in the section table points to the Highway# field in the route table, with the **FromMeasure** and **ToMeasure** fields supplying the "where" component. Similarly, **ArcLinkID** points to the primary key in the associated arc attribute table, with **FromPosition** and **ToPosition** identifying the portion of the arc used by each section.

Figure 10.3 illustrates how the Highway route system could be implemented for State Road 58. Three arcs are used to form seven sections to geometrically describe the route. The termini of sections are established by road intersections so that absolute measures can be applied to these locations. This is a strategy for ensuring that events are mapped on the correct side of an intersection.

Linear datums

As transport agencies increasingly employed GIS to map their facility inventories, they needed improved spatial accuracy. Data quality had not kept pace with improvements in map accuracy. Field data collection methods were still tied to the same distance measurement instruments that had always been used. Measurement accuracies of plus or minus 250 feet were commonly acceptable over the distances traversed by most routes.

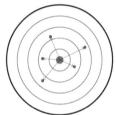

Accuracy Accuracy can only be estimated, never known, because there is no absolute reference for comparison; everything is a sample from which an estimate of the true value is made. Accuracy standards generally apply to well-defined phenomena and are stated by the percentage of values that are allowed to fall outside a given tolerance.

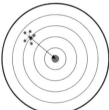

Precision Even if all the estimated values are quite different from the expected "true" value, high precision allows a correction factor to be applied that can greatly increase overall accuracy. The effect is akin to adjusting the sights on a rifle. However, if precision is low, then there is no simple way to improve accuracy.

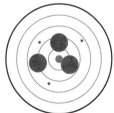

Resolution If you specify a unit of 0.01 mile for linear measures, then two locations have to be at least 26.4 feet apart in order for them to be assigned different measure values. The situation is similar to needing better aim with a rifle than a howitzer. Bigger map symbols and greater uncertainty of phenomena classification can point to lower resolution requirements.

Figure 10.4 **Accuracy, precision, and resolution** Perhaps nothing is more fundamental and misunderstood than the three principal qualities of spatial data: accuracy, precision, and resolution. These characteristics are almost exclusively determined by the manner of data collection and are specified by the application's needs. For example, if you are going to display your data at a scale of 1:100,000, then it is just a waste of money to set accuracy to +/- 1 foot. The three characteristics usually interact quite extensively and must be evaluated together.

The concept of a formal linear datum was devised to add measurement rules to the LRS concept.[1] Any such set of rules must address three aspects of a measurement system to be considered complete: accuracy, precision, and resolution. Accuracy is a statement regarding the average error in measurement for a set of samples. Every measurement is a sample; the true measure can never be completely known. An accuracy standard can only be applied to well-defined phenomena, such as the location of a physical element.

Precision represents the degree of repeatability in making a measurement. High precision will result in values that are numerically similar. As long as precision is high, system-level adjustments can be made to increase accuracy. Such adjustments are not effective when precision is low, because the impact is effectively random.

Resolution reflects the ability to differentiate between two positions that are close together. Resolution in an LRM is determined, in large part, by the unit of measure used. The closest two locations that can be distinguished by different measure values is a distance equal to one-half the measure unit. If the measure unit is 0.001 mile, then the resolution is one-half of 5.28 feet, or 2.64 feet.

Each route for which an LRM is created sets its own datum. At the very least, the datum consists of a defined path for the route, a starting point (the origin) and a way of making measurements (the LRM). Some agencies have gone a step further and created formal datum specifications for their LRM. Such a linear datum can be established to increase the accuracy of field measures. Taking the logical form of a link-node structure, a linear datum uses anchor points as nodes and anchor sections as links.

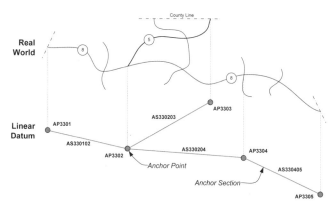

Figure 10.5 **Linear datum** Each route or link establishes a linear datum that defines the domain for valid offset distances and/or measure values, which generally must be within the numeric range implied by beginning and ending measures. In the 1990s, the concept of an explicit linear datum was put forth as a way to increase the spatial accuracy of transportation data by extending domain control to the field. Anchor points are established at identifiable locations in the field and anchor sections are defined between them to provide an "official length" attribute. All field measurements must be consistent with the stated length. Linear datum procedures and other rules are added to the LRM to form a linear referencing system (LRS).

Anchor points have a unique identifier, a location description, and some type of Earth-coordinate location as their attributes. Anchor sections have attributes of a unique identifier, designation of terminal anchor points, and length. The terminal anchor points may be supplemented by an intermediate anchor point to uniquely describe one of several possible paths between the two termini. Anchor sections must also specify a direction for increasing measure values, which can be indicated through ordering of the terminal anchor points, an explicit direction attribute, or specification of an origin node.

Anchor-section length serves as a quality control check for the accuracy of LRM field measurements. The ability to detect and correct measurement errors varies directly with the number of anchor points and inversely with the average anchor section length. Section length has the biggest quality-control effect because measurement error is cumulative. The degree of positional accuracy desired determines the density of anchor points; the greater the desired accuracy, the greater the number of anchor points needed. This is the same as making anchor sections shorter.

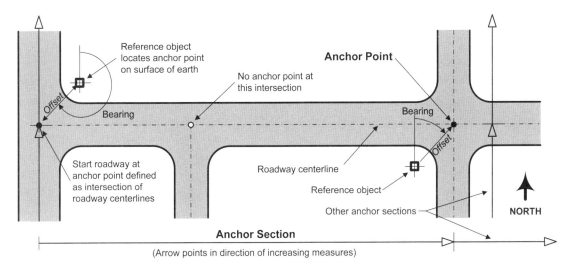

Figure 10.6 **How it works** Anchor points, like road intersections and jurisdictional boundaries, are conceptually simple but difficult to unambiguously locate. A precise anchor section length cannot be derived if the endpoints are not known with the same degree of accuracy. Reference objects can serve as monuments to specify the anchor point location. Given sufficient specification of anchor point locations, it is possible to register data compiled at different times and for different purposes. For example, project stations (high accuracy) can be related to the roadway inventory (medium accuracy) if they both contain anchor point references.

Anchor points may be difficult to precisely identify and capture in the field, as they are usually points only within the spatial abstraction. (Everything in the real world is an area.) Many common locations used to form anchor points are hard to unambiguously locate.

Where, for instance, is the exact center of an intersection or how do you really know where the county line is? To answer such a question, you can base the anchor point position on a reference object. Reference objects can be anything that is not readily movable, such as a mark chiseled into a curb, a bridge end, a traffic-signal pole, or a survey marker. The anchor-point position is stated as a distance and bearing from the reference object.

Anchor-point accuracy requirements should be determined by looking at both the users' needs and the limitations of available field procedures. A linear datum must meet the highest accuracy, precision, and resolution requirements of any application it is intended to support. Applications requiring lower levels of spatial and temporal accuracy and precision can use resolutions expressed with less accuracy. For example, if the production of county highway maps is the desired "highest use" of the data, then the largest scale at which roadway features may be mapped is perhaps one inch to one-half mile and working positional accuracy is, say, plus or minus 66 feet.[2]

Providing a linear datum in ArcGIS

ArcGIS offers some advantages for developing a linear datum with anchor points and anchor sections. There are only three entities required for a basic implementation. ArcGIS can also use attributed and other relationships and domain classes to aid the process of building and using the linear datum content.

The linear datum physical data model is based on three feature classes. By including geometry, these classes support the reconciliation of the linear datum that exists in the field with its expression that exists in the geodatabase. ArcGIS can determine the length of the geometry using the LRM scale's unit of measure, and then compare this value to the field-measured length established for the anchor section to create a metric of relative conformance. An anchor point will represent a point of registration between the field-determined measure values and those of the geometry.

The relationship between ReferencePoint and AnchorPoint is many to many, as one reference point may be used to locate more than one anchor point, and a given anchor point may be most accurately located through triangulation using two or more reference points. Thus, an attributed relationship class is required to tie reference points to the anchor points for which they supply an unambiguous position description. The attributes of that relationship include the information required to locate the anchor point relative to the reference point; i.e., the bearing angle and distance.

AnchorPoint and AnchorSection have two relationships between them, one for each terminal node. It will be necessary to subdivide any anchor section that has the same two anchor points, as may be the case when anchor points are placed at the two ends of a divided

Providing a linear datum in ArcGIS

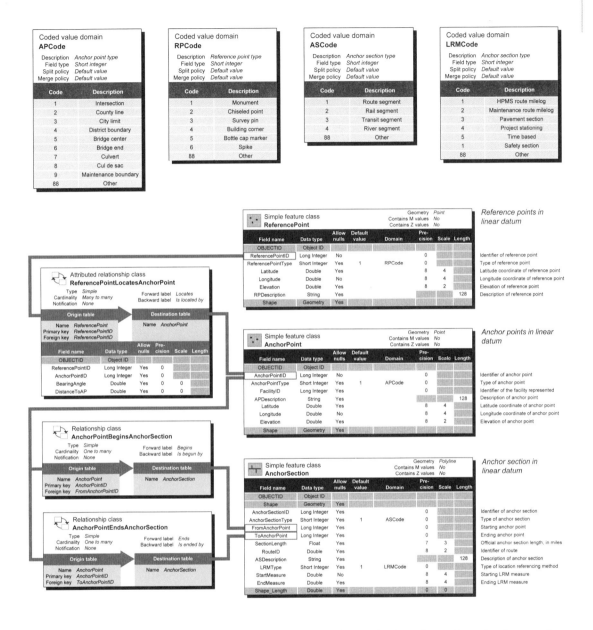

Figure 10.7 **Implementing the linear datum in ArcGIS** A simple geodatabase linear datum implementation requires two point feature classes and a polyline feature class. Since the relationship between reference points and the anchor points they may unambiguously locate is many-to-many, an attributed relationship is required to connect them. Two other relationship classes are needed to manage the beginning and ending anchor points for an anchor section. One of the assumptions of this design is the expectation that the linear datum supports at least one location referencing method.

Chapter 10: Classic transportation data models

highway segment. An intermediate anchor point will uniquely identify each directional path. Note that the same location can serve as both the beginning and ending anchor point for an anchor section that forms a loop.

Domain classes and their suggested values are supplied in figure 10.7. As the ASCode class shows, it is possible to define linear datums for any kind of linear facility.

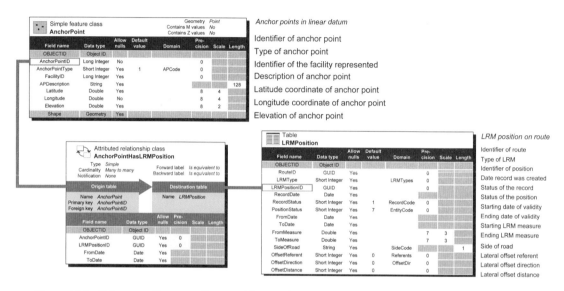

Figure 10.8 **Tying coordinates to LRM positions** The growing use of GPS data collection devices and condition-rating vehicles that produce coordinate data streams makes relating geographic coordinates to LRM positions important. The geographic coordinates for an anchor point can be translated as one or more LRM positions through the use of the anchor point to form the terminus of multiple anchor sections. Rather than indirectly relate anchor points to their LRM positions through the AnchorSection feature class, you could construct a direct connection through an attributed relationship class. Besides the two foreign keys required, this example includes **FromDate** and **ToDate** to indicate the period of validity, as establishing equivalencies may occur long after the involved coordinate or LRM position rows are created. New data may also eliminate former equivalencies at either end of the relationship, making the former equivalency obsolete. In addition to relating anchor points to LRM positions, you could tie reference markers or any other coordinate-based position to its equivalent LRM positions.

Since an anchor point is defined at a point on the Earth and may be at the junction of two or more anchor sections, it is not possible to store the LRM position description within the AnchorPoint feature class. It is necessary to use an attributed relationship class to supply an LRM position description for each anchor point from the perspective of each intersecting route. The physical data model shown in figure 10.8 assumes the use of LRMPosition class from chapter 8. The relationship class includes attributes that define the period of validity for any such relationship, as it is should be anticipated that refinements in the position description will occur over time.

NCHRP 20-27(2)

After the NCHRP 20-27 project published its survey of state DOT practices in the then-new field of applying GIS to transportation datasets, it was extended in a second phase to develop a uniform model for GIS deployment. This phase was motivated by the discovery that each agency was developing its own database design for dealing with the problem, resulting in much costly duplication of effort and complicating integration of these datasets at the national level.

It was difficult to develop a uniform, enterprise data model for 50 state DOTs. The principal organizational structure used by such agencies is function based. Organizational units are aligned with specific work tasks, often defined for each mode of travel. Agencies typically have project planning and programming, design, environmental assessment, construction, and maintenance divisions. Each mode of travel has its own physical operating environment and unique manner of describing its facilities, both in the field and in representational spatial databases.

There was also the matter of the coordinate systems used to describe positions on linear facilities. GIS databases typically use a 2D coordinate system based on Earth geometry. Transportation datasets use a 1D system defined by distance along a facility. It was difficult to place a 1D world within a 2D platform. Worse, there were likely multiple 1D datasets containing duplicative but dissimilar route systems. For example, the pavement management workgroup might use a different LRM than the maintenance forces, for a variety of application-specific reasons. The maintenance division has to manage things that are not on the road (e.g., roadside mowing and drainage structures), while the pavement management bureau may need to consistently evaluate the same pieces of pavement during each inspection cycle to correctly observe changing conditions. Transit services use stops, time points, and transit segments to say where things are located. The planning and programming division works at the relatively small scale of 1:24,000 compared to the design division, where a scale of 1:240 is common.

NCHRP 20-27(2) established four functional requirements for any LRM serving the needs of a transportation agency: locate, position, place, and transform.[3] "Locate" is the act of defining where one object may be found relative to another; i.e., determining the measure value for a position along a linear transportation facility. "Position" is a description of the location in a database. "Place" means being able to convert a database position description to a location in the real world. "Transform" is the ability to translate between location referencing methods, such as from a situs address to a route-milelog value. These functional requirements apply equally to any process that will use transportation data where position is defined by an LRM.

Chapter 10: Classic transportation data models

The 20-27(2) data model was among the first published approaches to developing an LRM datum using anchor points and anchor sections. In addition to the model, a major product was agreement on the other terms to use for transportation elements, such as traversals and events, and the distinction between the LRM and the LRS of which it was a component. The data model was based on the link-node data structure discussed in chapter 4 and was developed at the time when the coverage data structure was the dominant form.

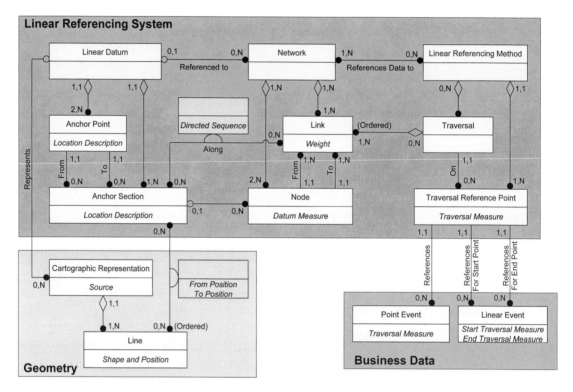

Figure 10.9 **NCHRP 20-27(2)** The first national transportation data model to accommodate linear referencing methods was created by a group of business area experts working under the sponsorship of the National Cooperative Highway Research Program's Project 20-27(2). The data model combined the concepts of a topological-vector structure, a linear datum, and the separation of geometry from business data. Many of the database component terms in use today—traversal, event, linear referencing method, measure, anchor point, and anchor section—originated with or came into common use as a result of NCHRP 20-27(2).

The data model includes three primary components: (1) the LRS and its linear datum; (2) business data; and (3) cartographic representation. The LRS is composed of three parts represented by the three higher-level classes along the top of the figure. The first is the linear datum, which is the framework within which measurements are made. Anchor sections are the foundation of the datum and establish the official length of highway segments.

The second part of the LRS component is the logical network of links (edges) that can be combined to construct the anchor sections. The position of a node is specified in terms of its location on an anchor section, thus providing a geometric sequence to the links. Nodes (junctions), which include an attribute for datum measure, were used as registration points between the anchor section and LRM positions along its length. The datum measure is a distance offset, not an LRM measure.

The third part is the LRM. The atomic object in this part is the traversal, which is defined as an ordered collection of links. LRM measures are assigned at the traversal level, with traversals—like anchor sections—being composed of one or more whole links. Overlapping traversals were permitted, as a link could be part of more than one traversal. Multiple LRM could also be used to define positions on a given traversal, but a location on a traversal could be stated only within the context of a single LRM. The expectation was that a traversal would be constructed using a sequence of anchor sections that supplied the linear datum for LRM positions along the traversal. A traversal could begin or end in the middle of an anchor section by stating an offset distance from one of the terminal anchor points, but a node had to be defined for that location.

Business data consisting of point and linear events is tied to traversals (routes) by reference to the route identifier and LRM measures are the second model component. The model's cardinality for the event classes to Traversal Reference Point is one to many, with a linear event having two such relationships. The cardinality for the relationship of Linear Referencing Method to Traversal Reference Point is also one to many. Thus, point and linear events may be related to only one LRM. The use of multiple LRM—or multiple measure descriptions for a common location—would require duplicative storage of the business data.

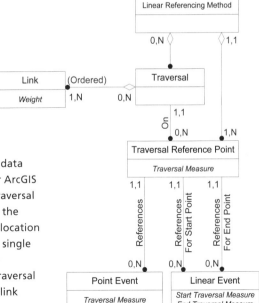

Figure 10.10 **NCHRP 20-27(2) traversals** The 20-27(2) data model's traversals are analogous to the routes of earlier ArcGIS geodatabase designs. The model separates positions (Traversal Reference Point) from the events they locate. Although the model allows a given traversal to be served by multiple location referencing methods, it restricts each LRM position to a single method. The model requires an included network to be reconciled to traversal extents; i.e., no link can cross a traversal boundary. However, there is no explicit way to locate a link along the traversal using LRM measures.

The geometry component supplies the cartographic representation of linear transport facilities. The model does not include point geometry. The only way to express business data via dynseg operations on the geometry is to go through the intervening network and linear datum.

Because the NCHRP 20-27(2) data model has been central to transportation spatial database design for many years, it is appropriate to look at it more closely. The traversal portion of the model defines routes and positions along those routes stated within the context of a single LRM. Although the relationship between Link and Traversal is many to many, there is no accommodation of traversals that may begin or end at other than the end of a link. This requirement could limit the ability of the model to support pathfinding applications, which often involve origins and destinations that are located along an edge (link).

Note also that a traversal must consist of at least one link, but a link need not be part of any traversal. This means that the network structure is a mandatory component of the dataset, a requirement reinforced by the cardinality of the relationship between Network and Linear Referencing Method. Indeed, the model explicitly states that the LRM references data to the network. Thus, the minimal dataset that conforms to the model is composed of a network constructed from links and nodes.

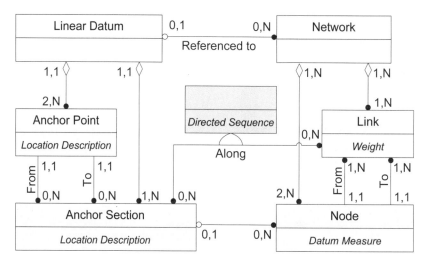

Figure 10.11 **NCHRP 20-27(2) linear datum** The researchers who led the NCHRP 20-27(2) project were also among the originators of the linear datum concept, so it should be no surprise that the 20-27(2) model included linear datum entities. What may be surprising is the relationship of the linear datum to the network, not the LRM. Note also the many-to-many relationship between Anchor Section and Link, where an anchor section may be a directed sequence of links, and the location of nodes using measures referenced to the anchor section's origin.

The model raises the implementation issue of how to define the location of a link within the context of the traversals of which it may be a part. No relationship class is shown for the association between Link and Traversal, as was provided for the association between Link and Anchor Section.

There is also the issue of placing one attribute, Traversal Measure, in two classes. For example, the location of a point event is defined both by a traversal measure in Point Event and another in Traversal Reference Point. This duplication of data could be a potential problem for referential integrity and, at the very least, imposes additional data maintenance overhead.

The linear datum is a strong part of the data model, and serves as the intermediary between the network and its representative geometry. An attributed relationship class is shown to manage the many-to-many relationship between Anchor Section and Link so as to supply the ordering required to place links in the correct sequence. The ordering may also be inferred from the datum

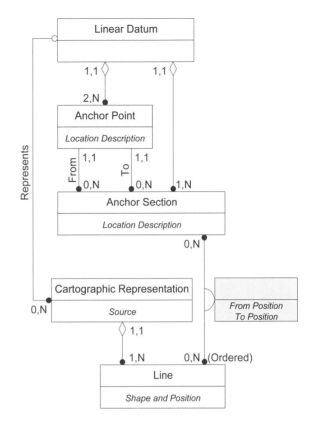

Figure 10.12 **NCHRP 20-27(2) geometry** The association of geometry to the linear datum suggests that the purpose of the datum is to make the geometry better match the real world linear features it represents. LRM measures applied to the line geometry are those of the linear datum. A line can cross anchor section boundaries. The relationship of Anchor Section to Link, and then to Traversal will allow you to identify the traversal of which a line is a representative part. Events can be displayed via dynseg only when the datum LRM and that of the parent traversal are the same.

measure for the terminal nodes of each link. A weight may be assigned to a link to represent impedance; i.e., the relative ease or difficulty of movement along the link.

Geometry features are tied to the linear datum and, through the datum, to the network and then to traversals and the business data. Although reasonably founded on the principle that the linear datum represents the registration of the geometry to the real world, it creates a design requirement for both the linear datum and the network to exist in order to map business data using a route-milelog LRM.

The structure of anchor sections and the lines that represent them looks very much like that of routes and sections in the coverage model. A line may describe all or part of an anchor section and is defined by endpoint positions. One significant difference from the coverage model is the absence of measures in the NCHRP 20-27(2) model. This model does not include measures for line segments, so it cannot be directly implemented to support dynseg operations. The model only provides geometry for all or part of an anchor section. Thus, although the total model includes an LRM as part of a traversal, it actually supports the short segment database design, where each link defines an atomic portion of the transport system, not the route-milelog database design, where the equivalency would be between traversals and geometry.

Of course, it must be remembered that the 20-27(2) model is, at most, a logical data model. It could be implemented in a way that combines the aspects of the linear datum, geometry, and network connectivity into a single geometric network. Such an implementation was developed in the UNETRANS data model. But before going to that model, which is described in chapter 11, we need to review some other transportation data models that were derived substantially from the 20-27(2) model.

FGDC data exchange standard

By the mid-1990s, the Spatial Data Transfer Standard (SDTS) had been completed and a transportation network profile was in development. These two data models, however, were not especially supportive of the business data that comprises the primary interest for data sharing among transport agencies. The SDTS had almost no provision for the exchange of rich attribute datasets, and the transportation network profile was concerned only with expressing network topology attributes. Shortly after the NCHRP 20-27(2) effort was concluded, the FGDC initiated an effort to develop a data exchange standard for transportation data based on the 20-27(2) model.

It was quickly recognized that a central component of any transportation data exchange standard required unambiguous facility descriptions relating to the data. Since multiple actors can create numerous datasets describing overlapping and redundant facilities, the standard had to be able to create equivalencies between the various facility identifiers.

Taking the concept of anchor points and anchor sections from the NCHRP 20-27(2) data model, the draft FGDC standard was based on the primary entities of Framework Transportation Segment (FTSeg) and Framework Transportation Reference Point (FTRP). Including the term "Framework" was a show of support for the National Spatial Data Infrastructure (NSDI), for which an implementation framework consisting of seven data themes had been proposed. One of those themes was transportation.

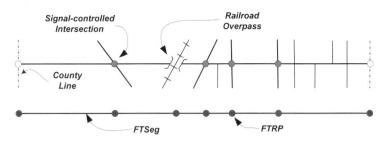

Figure 10.13 **FGDC transportation feature identification standard** The Ground Transportation Subcommittee of the Federal Geographic Data Committee (FGDC) in the late 1990s began to create a national standard for transportation data exchange. This standard was based on the NCHRP 20-27(2) data model, and was part of an effort to implement the National Spatial Data Infrastructure (NSDI). The standard was founded on an approach that combined a data supplier (authority) identifier and a sequence number unique to that supplier to create a globally unique feature identifier. A national registry was required to manage authority identifiers. The draft standard was published in 2001. Although never formally adopted, it did suggest a valid approach for data sharing.

The basic concept was that a unique identifier would be assigned to each data supplier, called an authority. Within each authority, each facility would be given a unique identifier. The combination of authority and facility identifiers would be globally unique. The expectation was that, over time, the specification of FTSeg and FTRP locations would improve, with general locations becoming more specific and accurate as measurement methods were improved. The draft standard was published for public comment in May 2001 but was never formally adopted as a result of a new data modeling effort being undertaken by the Geospatial One-Stop initiative. The subsequent Geospatial One-Stop data model is described below.

The GDF format

Not all transportation data exchange standard work was being done in the United States. A European effort supporting intelligent transportation system (ITS) applications, particularly in-vehicle navigation system development, resulted in the Geographic Data File (GDF) data exchange standard. The scope of GDF eventually went beyond a spatial data model and exchange format to describing how data was compiled and relations between facilities were defined. Information on travel origins and destinations, in addition to transport facilities, are covered by the standard.

GDF is available as an international standard, ISO IS 14825, and is managed by ISO Technical Committee 204. At the time of this writing, GDF was at Version 4.0 and included variations for major data suppliers. The successor version, an XML specification, was in development.

Support for point, line, and polygon features is provided through a multilevel data structure. Basic geometry building blocks are included in Level-0. These basic components form the topological structure of the transport network using nodes, edges, and faces. Level-1 contains the points, lines, and polygons that graphically express the shape and location of transport network facilities. A given Level-1 facility is described by both Level-1 geometry and Level-0 topology. Level-2 features are complex and represent a collection of simple features.

GDF features are described in feature catalogs, while attributes for those features are included in attribute catalogs. Topology information is contained in its own catalog. All three catalogs form a set of metadata records that describe the structure of a GDF dataset.

Although GDF may be a widely deployed standard for data exchange, it is not suitable for data editing. For example, GDF, being a navigable dataset, is a geometry-based structure that requires network topology to be supplied for every geometric feature. This is similar to the NCHRP 20-27(2) data model, which requires network connectivity to construct any transport dataset. Of course, many transportation datasets do not require network topology, and even geometric representations may be little more than an afterthought. GDF is also highly denormalized as a result of the need for each edge to contain all the data that might affect its being selected for a particular path. There is also a different version for each database vendor. Still, it meets a need in the spatial data community for the exchange of navigable databases as part of a complete ITS solution. A GDF-compliant dataset can be part of a published transport geodatabase.

NCHRP 20-27(3)

Once the issue of an LRM ontology and its representation had been addressed in 20-27(2), some users voiced concerns about existing data models and software being able to accommodate temporal aspects of transportation systems. For example, network components change over time as roads are built, closed, and reconstructed. Crashes that occurred on a two-lane road need to be viewed in that historical context after the road is widened to four lanes. Transportation planning analyses often need to look at system changes over time, particularly as it may look under various future scenarios.

The 20-27(2) data model had no clear way to explicitly accommodate evolving transport systems so as to recover historical states. To address this issue, the NCHRP 20-27 project was extended a second time to become Project 20-27(3). Accommodating temporal aspects and providing greater positional accuracies led to a decision to develop an enterprise, multimodal data model that could serve a broad range of GIS user needs. The project's stated objective was to construct a model that would facilitate "interoperability," which was defined as support for data portability and cooperative process control.

A workshop in December 1998 sought user input on the needs of four application areas: (1) transportation planning, highway construction, and asset management—the traditional heart of GIS applications for transportation; (2) highway safety and incident management; (3) traffic management and highway operations; and (4) transit facilities and operation, commercial vehicles, and fleet management, a group of roadway users not previously included in the work of Project 20-27(2).

Ten core functions were proposed for a comprehensive "multidimensional" LRM data model as a result of the workshop's findings.[4] Most of these functions dealt with accommodating

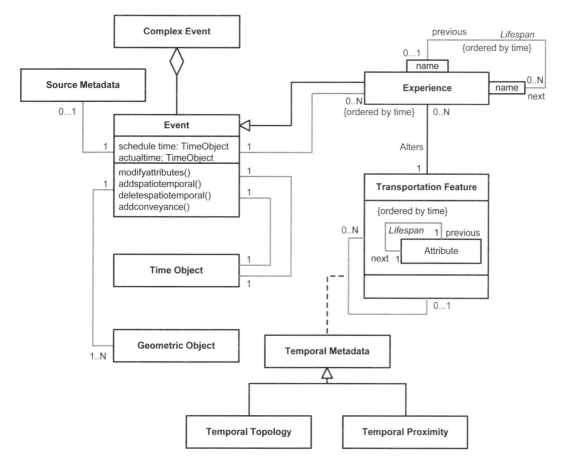

Figure 10.14 **NCHRP 20-27(3)** A continuation of NCHRP 20-27(2) attempted to address the need for temporal data support and move to a full object-oriented structure. The resulting data model, presented here in its original UML form, additionally attempted to change some of the terminology of the previous model. For example, "event" was changed to mean only something that happened—something that the facility might "experience." Although not directly implemented, the effort did educate the GIS-T community about object-based structures and design techniques.

multiple versions of the data to reflect changes over time, a variety of geometric and topological representations, and the transformation of data between linear datums.

The proposed conceptual data model contained four principal components labeled "what," "where," "when," and "how." The "what" component included transportation features and their attributes. The "where" component covered linear and other spatial datums. The "when" component addressed the temporal elements of the dataset. The "how" component represented events and their impacts on affected transportation features. ("Event," as defined at the workshop, was restricted to mean things that happen. It nevertheless remains more common in transportation database design to define "event" as including all attributes, subordinate elements, and things that happen.)

The core portion of the resulting 20-27(3) data model addressed the primary needs of the four components. It showed the basic objects for transportation features and the events that occur on them. In this context, event is something like a parade or a crash. In 20-27(3), elements were defined as other transportation features, resulting in the recursive relationship shown for Transportation Feature; i.e., a transportation feature could be complex. Characteristics (attributes) were placed within the Transportation Feature object as members of an Attribute class.

The 20-27(3) model also introduced new entities to the 20-27(2) model, such as conveyances, that move along transportation features and experiences, like construction projects, that affect transportation features. Temporal aspects are expressed in various temporal object classes. Events and experiences may have a specified duration. Combination "spatial-temporal" object classes were constructed to address the various spatial configurations of transportation features that result from the evolution forced by experiences. A Spatial Referencing System was proposed as a successor to the Linear Referencing System of the 20-27(2) model. The new system was placed within the broader context of all datums that may be used in a multifeature GIS, such as where transportation might be combined with property (cadastral) and environmental data.

Many elementary components can be combined into composite classes. For example, an Event instance may be part of a Complex Event, such as a multicollision crash. It is possible to use the model to construct a dataset that represents time in many ways for a single event. An example was the event occurrence time for a car crash being linked to the time a call was made by an observer of the event, the time the call was received, the time a dispatcher contacted a responding unit, the time the unit left its station to go to the scene, the time of arrival at the scene of the crash, and other subsequent components of the event.

Other transportation data models

Although the NCHRP 20-27(3) data model is multimodal in nature, it has not spawned multimodal database implementations. The restriction of most transportation data models to specific modes has continued to the present day in the new federally sponsored data standards being pursued to implement the NSDI. These data standards were developed under the auspices of the Geospatial One-Stop in 2003, although the project is now being led by the FGDC.

As with the original NSDI framework, seven spatial data themes were identified. Each theme went through a user-driven data modeling process. The NSDI transportation theme was defined as a collection of five modal standards plus a presentation of basic concepts, called the Base Model. An LRM mechanism is a major part of the Base Model and is applied in conceptually similar ways to road, public transit, railroad, and inland waterway modal models, and to the model developed for the hydrography theme. The aviation data model, being restricted to landside facilities, does not deal with the movement of airplanes.

The Base Model is structured like a geometric network composed of directed chains defined in the SDTS. Providing the mandatory endpoint junctions will present a hardship to the majority of transport data suppliers who do not include these features in their products. The standard does not show how to relate a conforming dataset of one part to another part, such as to use street centerlines in a road database as a framework for building transit routes.

Other researchers have suggested elements of a multimodal database design. For example, a paper from Portland State University proposed an enterprise database design for transportation.[5] A subsequent paper showed how transit agencies could combine parts of a highway data model with one for transit operations.[6] Researchers working at the University of Wisconsin at Madison proposed requirements for a data model that accommodated more complex multimodal databases, and noted that the network information required to actually apply such a design to multimodal problems was not widely available.[7]

As stated earlier, one of the primary uses of transportation databases is to manage street addresses. The National Emergency Number Association (NENA) provides a standard source of guidance for such databases in its two primary address standards.[8] NENA has also issued guidelines to help users implement the standards.[9] The basic NENA standard-compliant address database will include the following fields: (1) House Number, 10 char.; (2) House Number Suffix, 4 char.; (3) Prefix Directional, 2 char., domain N, E, S, W, NE, SE, SW, NW[10]; (4) Street Name, 60 char.; (5) Street Type Suffix, 4 char., codelist AVE, ST, RD, HWY, etc.; and (6) Post Directional, 2 char. These fields provide support for the information required from local exchange carriers (telephone companies), but are insufficient for a number of other applications.

The NENA standard is just one way to go with the design of a street address database. Complete address databases must include mailing addresses, too. An effort by a task force formed by the Urban and Regional Information Systems Association under the auspices of the FGDC was developing a more comprehensive address model for data exchange. A draft model provided numerous standard ways to convey address information and suggests a number of field definitions that may considered in designing any geodatabase intended to support address geocoding.

Notes

[1] Vonderohe, A., & T. Hepworth. 1998. "A methodology for design of measurement systems for linear referencing," *Journal of the Urban and Regional Information Systems Association.* Vol. 10, No. 1, pp. 48-56.

[2] This figure is calculated using a map scale of 1:31,680 (number of inches in half a mile) and the national map accuracy standard of error being no greater than 1 in 40: 31,680 ÷ 40 = 792 inches; 792 ÷ 12 = 66 feet.

[3] Adams, Teresa M., Nicholas Koncz, and Alan P. Vonderohe. June 2000. "Functional requirements for a comprehensive location referencing system." Madison, Wisconsin: Transportation Research Board, *Proceedings of the North American Travel Monitoring Exhibition and Conference.*

[4] Koncz, Nicholas, and Teresa M. Adams. 2002. A data model for multidimensional transportation location referencing systems. *Journal of the Urban and Regional Information Systems Association.* Vol. 14, No. 2, pp. 27-41.

[5] Dueker, Kenneth J., and J. Allison Butler. GIS-T enterprise data model with suggested implementation choices. *URISA Journal,* Vol. 10 No. 1.

[6] Peng, Zhong-Ren, J.N. Groff, and K.J. Dueker. 1998. An enterprise GIS database design for agency-wide transit applications. *URISA Journal,* Vol. 11, No. 1.

[7] Walter, Clyde. 1998. *Multimodal investment analysis methodology phase I—part III: The application of GIS to multimodal investment analysis.* Ames, Iowa: Center for Transportation Research and Education, Iowa State Univ.

[8] *NENA Data Standards for Local Exchange Carriers, ALI Service Providers & 9-1-1 Jurisdictions* (NENA-02-011, Nov. 9, 2004; and *NENA Recommended Formats & Protocols for ALI Data, Exchange, ALI Response & GIS Mapping* (NENA-02-010), Nov. 9, 2004. (ALI means address location identification.)

[9] See, for example, Ozanich, Beth, *E 9-1-1 data base guide, 2nd Ed.,* Coshocton, Ohio: NENA, 1996; and Ozanich, Beth, et al., *Enhanced 9-1-1 data base: Instruction manual,* Coshocton, Ohio: NENA, May 1994.

[10] A useful guide for the domain of street types is supplied by Appendix C1—Street Suffix Abbreviations in *U.S. Postal Service Publication 28*, http://www.nena.org/9-1-1TechStandards/Standards_PDF/USPSPub28.pdf.

chapter eleven

The original UNETRANS data model

- Network features
- Routes and location referencing
- Other key structures in the UNETRANS model

ESRI has long supported developing industry-specific data models and provides additional guidance on designing physical geodatabase data models and deploying applications using ArcGIS. The University of California at Santa Barbara created one such model for the transportation industry with funding support from ESRI. This model is called by the acronym of UNETRANS (from Unified Network for Transportation). The UNETRANS model uses a geometric network as its underlying structure.

Chapter 11: The original UNETRANS data model

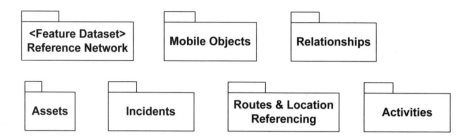

Figure 11.1 **UNETRANS** UNETRANS includes seven packages, or logical groups of feature and object classes. Although never completed, the model introduced design aspects that are still valid for any transportation geodatabase, and several implementation examples have been deployed.

UNETRANS is divided into seven object packages, or thematic groups:
- Reference network
- Assets
- Incidents
- Activities
- Routes and location referencing
- Mobile objects
- Relationships

Each package includes various feature and object classes organized around the geometric network consisting of complex edge features and simple junction features in the Reference Network Package. The UNETRANS data model was never completed, but it does offer some useful ideas for the design of geodatabases containing transportation data.

Network features

Since the geometric network is the core of the UNETRANS data model, it is a good place to start. The network consists of complex edge features and simple junction features. The derivation of the creatable geodatabase classes offers insight into how the Reference Network Package was to be used for pathfinding applications.

Based on the ComplexEdgeFeature class, the abstract *TransportEdge* class adds attributes that describe the basic characteristics of a linear transport facility. Such attributes include the date the facility was constructed and its current lifecycle status. The *RoutableEdge* class adds attributes useful to pathfinding applications, such as minimum horizontal and vertical clearances and link impedances. The implication is that UNETRANS could accommodate other subclasses of *TransportEdge* that were not defined.

Network features

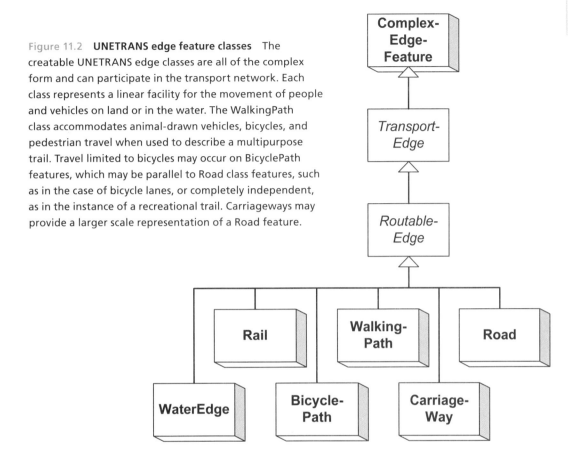

Figure 11.2 **UNETRANS edge feature classes** The creatable UNETRANS edge classes are all of the complex form and can participate in the transport network. Each class represents a linear facility for the movement of people and vehicles on land or in the water. The WalkingPath class accommodates animal-drawn vehicles, bicycles, and pedestrian travel when used to describe a multipurpose trail. Travel limited to bicycles may occur on BicyclePath features, which may be parallel to Road class features, such as in the case of bicycle lanes, or completely independent, as in the instance of a recreational trail. Carriageways may provide a larger scale representation of a Road feature.

The creatable classes included in the package to represent linear facilities are the various forms typically found in a transport system (roads, railroad tracks, bike paths, and navigable waterways). These are really just examples; any number of such classes could be defined for various modes of travel. The useful element of this part of the overall design is that the geodatabase allows feature classes to more closely approximate the entities they represent than do other available data models that include only a generic Route or Road entity.

The set of creatable simple junction feature classes in figure 11.3 also begins by adding common transport facility attributes to the original geodatabase class to produce an abstract *TransportJunction* class that serves as the general template for all junction types. Five kinds of intersection features are supported: at-grade, four-way stop, rotary (traffic circle), cloverleaf (limited-access highway interchange), and brunnel (a composite representing bridges and tunnels). Three kinds of station features were also included: train station, water port, and bus station. Again, these creatable classes may be viewed as merely examples of what could be done.

Chapter 11: The original UNETRANS data model

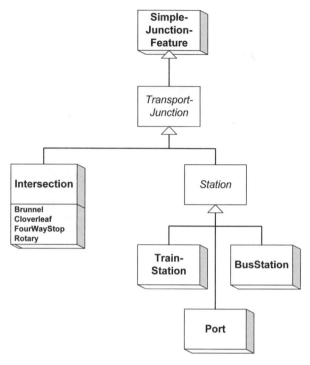

Figure 11.3 **UNETRANS junction feature classes** The creatable UNETRANS junction classes may be classified as either intersections (or their planar graph equivalents) and terminal destinations. Four *Intersection* subclasses and three *Station* subclasses were provided. The classes were determined by logical attribute groupings and their potential roles in the transport network that underlies it all. Each of the creatable classes can have multiple subtypes.

Routes and location referencing

The next most important package in the UNETRANS data model offers routes and elements. These classes are all geodatabase tables derived from the ObjectClass class. Among these tables, the one to which most route attributes and elements will be related is TransportRoute, which is the basic route that defines an LRM datum. A transport route would be related to one or more transport edges, and would supply the route identifier needed to relate elements to the facilities of which they are a part.

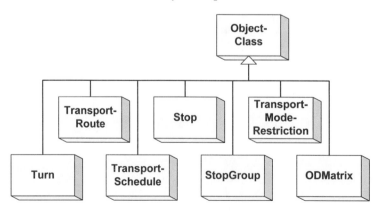

Figure 11.4 **UNETRANS object classes** The UNETRANS data model recognizes that nonfeature classes are useful to provide the full range of data structures required for a complete transportation geodatabase. This model, which can only display inheritance relationships, merely hints at the supported associative relationships.

Routes and location referencing

A notable aspect of the UNETRANS data model is that characteristics are part of each edge or junction feature, meaning that there was no need for a linear referencing structure to express the location of these attributes. The expectation was that an element, such as a bus stop, would be related to a junction and supply information about that junction within the context of a bus transit dataset. There was also support for complex features, such as the construction of a stop group from a collection of stops to represent the various termini of bus routes within a bus station.

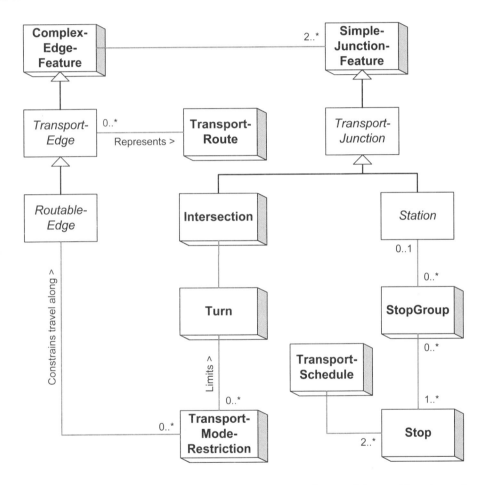

Figure 11.5 **UNETRANS association relationships** The UNETRANS data model uses object classes for various functions within the geometric network. This figure shows some of the primary examples. A station can serve as a stop-group location, and a route can be represented by one or more transport edges as it participates in multiple networks. A more complex example is using transport mode restrictions to prevent a particular vehicle from making a turn or traveling down a route segment. The UNETRANS data model does not include explicit association relationships; only generalization (inheritance).

Chapter 11: The original UNETRANS data model

Figure 11.5 demonstrates the types of association relationships anticipated by the UNETRANS data model. One or more transport edges represent a transport route within the geometric network. A turn describes the various paths away from a junction, while a transport mode restriction may supply constraints on those choices, such as to prohibit a left turn during the afternoon peak traffic period. A transport mode restriction may also be applied to a transport edge to represent travel restrictions imposed on a particular vehicle class, or to show that only one-way travel is permitted at all times.

A station is represented as a stop group that consists of one or more stops. A stop group may also represent the situation where stops are close enough to each other to provide a walking connection between them so as to support pathfinding across nonintersecting transit routes. The ordered sequence of stops in a transit pattern is formed by a transport schedule.

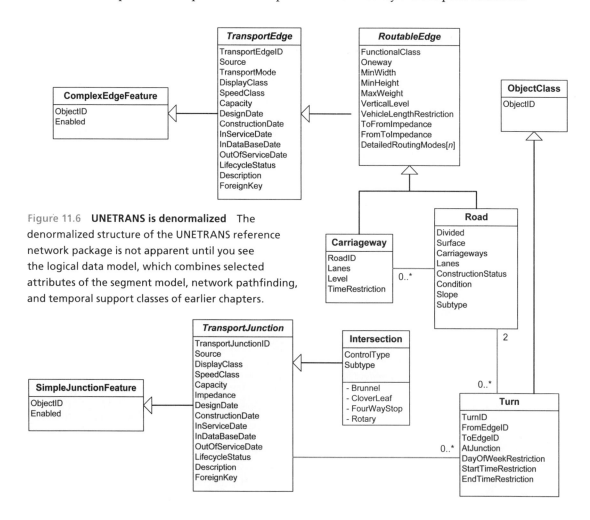

Figure 11.6 **UNETRANS is denormalized** The denormalized structure of the UNETRANS reference network package is not apparent until you see the logical data model, which combines selected attributes of the segment model, network pathfinding, and temporal support classes of earlier chapters.

Routes and location referencing

One of the issues presented by the original UNETRANS model is its denormalized design, which follows the short-segment design philosophy that treats small pieces of longer facilities as atomic elements of the transport system. The basic entity relationships that would be expected are provided, such as for a turn to be located at an intersection and to represent the connection between two roads, or for a road to include carriageways. However, the denormalized structure produces a great deal of data redundancy and limited support for the traditional event table data structure already in use at state DOTs and other larger transport agencies.

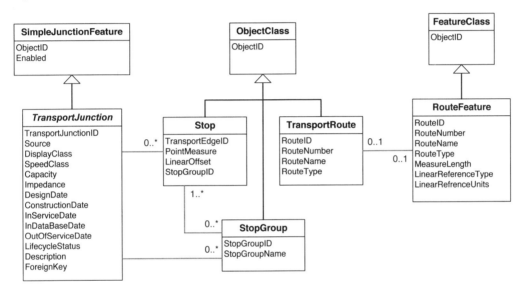

Figure 11.7 **Mix and match** This figure illustrates mixing classes from two packages: Reference Network (edge and junction classes) and Routes and Location Referencing (object and feature classes). As this model shows, UNETRANS offers both "plain" features and geometric network features for display of LRM routes and related events. Some association relationships are implied by foreign keys.

One of the more effective aspects of the UNETRANS design was its recognition that feature classes other than those offered by the geometric network may be needed. For example, a route feature may be used to map the path of a transport route without requiring construction of a geometric network. This form of geometry can support dynseg operations and event tables, which are supplied by the Assets, Activities, and Incidents packages. No event tables were supplied for route characteristics or elements not included in a specified class.

One of the possible implementations of the UNETRANS model could be to use the geometric network for a transit system, where pathfinding is a common need, and simple feature classes for the highway inventory, where event tables could be added to the data model to support dynseg.

Other key structures in the UNETRANS model

In the original UNETRANS data model, an asset is an identifiable, discrete element of a transport facility with or without its own geometric representation. As such, either a point or linear event table may describe an asset's location along a route feature using an LRM position. However, there is no explicit requirement for an asset to be tied to a route feature; it could be a standalone facility.

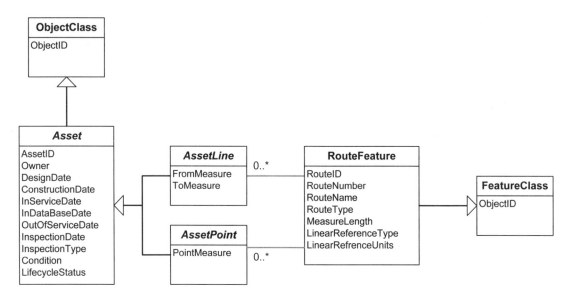

Figure 11.8 **Assets** Although UNETRANS is denormalized in its treatment of facility characteristics, it is somewhat normalized when accommodating physical elements of the transport system—what UNETRANS calls assets. The two Asset subclasses include LRM measure properties that are composites of multiple attributes; e.g., route ID, linear reference type, linear reference units, and LRM measure.

The Activities and Incidents packages present tables and feature classes to store information about things that happen. As with other packages, the classes supplied are examples offered to show what is possible to include in a complete geodatabase design. In addition to the event table structure, which can produce point and line features through dynseg, the Incident Package also offers a polygon feature class to describe the area extent of a capital-improvement project or other similar incidents with a broader spatial extent, such as a material spill that spreads to surrounding area and off the transport facility.

Other key structures in the UNETRANS model

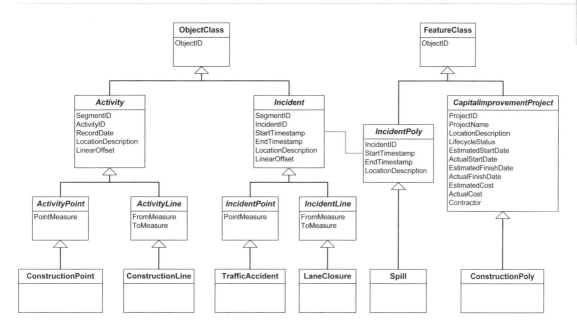

Figure 11.9 **Events happen** The Activities and Incident packages offer creatable classes for things that happen, planned and otherwise, such as construction projects, traffic crashes, lane closures, and material spills. The *Activity* class includes a **SegmentID** property that is found only in the *Incident* class; its purpose is not described in the available documentation. These two classes also share a common **LinearOffset** property, which suggests that they support lateral offsets from a facility centerline in dynseg operations. The model includes point, line, and polygon varieties for creatable classes, but only the polygon form is a feature class. This may be due to the expectation that point and linear geometry could be provided through the included LRM position descriptions and dynseg. Lastly, although *Activity* and *CapitalImprovementProject* are both in the Activities Package, there is no foreign key by which one could illustrate the geographic extent of an activity and using a construction project polygon.

Using ArcGIS and the capabilities of the geodatabase to more closely match class behaviors to those of the real-world entities they represent is one of the major innovations offered by the UNETRANS model. For example, it is possible to link crashes to the intersections at which they occur by dealing directly with Crash and Intersection classes. It is also possible to link projects and the facilities they may affect through an attributed relationship class.

Another area of innovation is a set of classes that had not appeared in any earlier transportation data model: the Mobile Objects Package. UNETRANS was the first widely published transportation data model to include system-user entities, such as vehicles, pedestrians, and cargo containers. Vehicles can be part of a fleet, or temporarily included as part of a mobile object group, such as a train or convoy. It would be a simple matter to treat a mobile object as a tracking event for display of the object's location through the functionality offered by the ArcGIS Tracking Analyst extension.

Chapter 11: The original UNETRANS data model

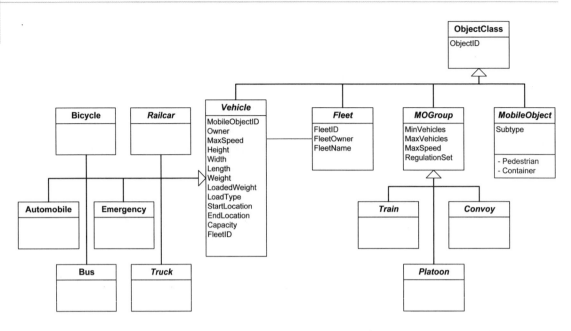

Figure 11.10 **Mobile Objects** UNETRANS was the first data model to include conveyances and other users of transport systems. This part of the model was included in the Mobile Objects package, which supplies only object classes. Although the model's documentation discusses the aggregation of individual vehicles into groups, no foreign key relationship is provided to support such an action. Similarly, the documentation discusses the use of a vehicle identification number (VIN) to uniquely identify each vehicle, but no such property was included in the model diagrams from which this figure is taken. Still, the inclusion of mobile objects in the data model was a significant step forward.

Some implementations of the basic UNETRANS model have occurred in larger agencies. For example, highway inventories maintained by the New York and Virginia DOTs in the United States and the Canadian highway network are all based on designs derived from the basic UNETRANS data model.

In the next chapter, revisions to the original UNETRANS data model are offered to increase its performance and ease of adoption by large and small transport agencies.

PART III *Enterprise-level solutions and modal data models*

chapter twelve

Improving the UNETRANS data model

- Editing support
- Describing LRM positions
- Aspects and elements
- Mode-specific Inventory Package classes
- Modal class groups
- Association relationships
- Intersections
- Bridges
- Events
- Pavement management example
- Mobile Objects
- Using all three packages

Chapter 12: Improving the UNETRANS data model

Experience with UNETRANS and the evolution of ArcGIS technology suggested ways to improve the original model. UNETRANS is a design philosophy more than prescriptive geodatabase design because there are so many components of the transportation "industry." The revised UNETRANS data model offers numerous enhancements that will benefit all transport industry sectors—whether facility provider or facility user—and works equally well for short-segment or fully normalized geodatabase structure.

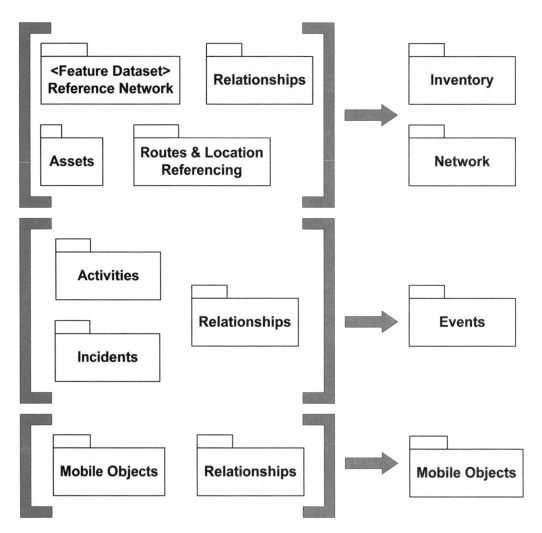

Figure 12.1 **Revising the class packages** The first thing we did to improve UNETRANS was to restructure the class packages. The end result was the aggregation of the original seven packages into four: Inventory, Network, Events, and Users. The former Relationship Package is now integrated into each of the others, while the Reference Network Package was split to produce separate facility inventory and network packages.

One of the proposed changes is to regroup the feature classes, tables, and relationship classes into fewer packages structured by application area. The original model included seven packages while four packages form the revised model: Inventory, Network, Events, and Users. A related change is integrating the former Relationship Package into the four packages and explicitly including domain classes.

The Inventory package includes support for transportation facilities of all types, including their characteristics and elements. The Network Package takes advantage of the new transportation-specific network data model used by ArcGIS Network Analyst extension, which replaces the geometric network model employed by the original UNETRANS. The Events Package combines the concepts of the former Activities and Incidents packages into a comprehensive group of entities representing things that happen on and to transport facilities. The fourth package involves users of the transport system and is an expanded version of the original Mobile Objects Package. This chapter presents an overview of these packages and details for their implementation except for the Network Package, which is the subject of chapter 13.

Editing support

The new UNETRANS model implicitly includes support for all the editing improvements offered by ArcGIS, working with ArcSDE to support the variety of versioning methods described in chapter 6. The complete set of standard temporal fields included in the two versions of the abstract *Entity* class template is part of this support. However, it is not required to adopt these class templates.

An agency implementing the UNETRANS data model can use any of the three basic forms of these classes, or some other combination of fields for temporal and editing support. The simplest version is to include only a facility identifier and type field. The identifier allows you to support on-the-fly joins, dynseg, and explicit relationship classes in your database design, in addition to supplying you with a stable candidate primary key. The type field allows you to create or to otherwise distinguish among various facility subtypes.

The next step up in functionality is the version that includes temporal data fields. This version adds a field for the date the row was added to the class, two fields for a validity period, one field for the status of the row in the table (active, replaced, or retired), and a field to reflect the status of the represented real-world entity (proposed, under construction, open to traffic, etc.). The minimum fields required to support some form of temporal operation, such as the ability to recover a previous state of the geodatabase, are RecordDate and RecordStatus.

Chapter 12: Improving the UNETRANS data model

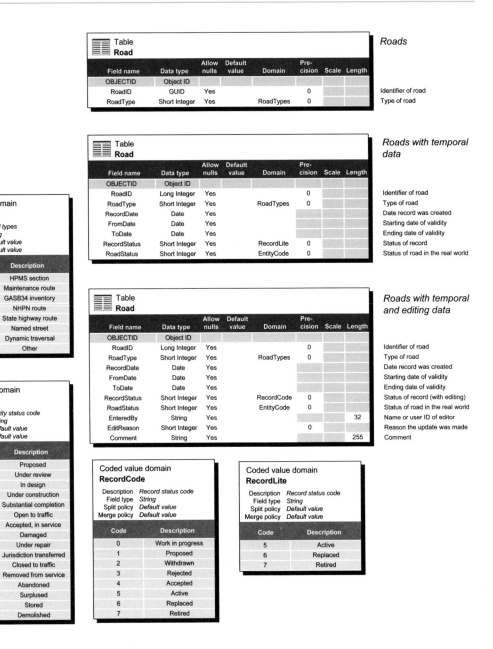

Figure 12.2 **Temporal and editing support** The UNETRANS data model remains neutral on the issue of whether to provide the class properties required to support temporal aspects of an evolving transportation geodatabase or to manage the data-editing process. This figure shows the three basic versions of a UNETRANS Road class. The safest course to take is to include temporal and editing support in the database design, even if you will not initially populate all those "extra" fields.

The highest level of functionality, which supports both temporal aspects of an evolving database and management of the editing process, adds three more fields. One field, EnteredBy, records who made the edit. The other two fields describe the reason for the change and supply details about what was done. The EditReason field offers a controlled domain that allows managers to search classes to develop descriptive analyses regarding the nature of edits to the database. Such information can supply metrics indicating the amount of effort required by each type of change and provide guidance on staff resource allocations based on the anticipated volume of work by edit type. The free-text Comment field offers space to store information that provides additional details about the edit or the entity represented. For example, information about the source of the change or its cause could be placed in this field.

One or more of the various alternatives may be employed for different classes. Feature classes may include the full set of fields, while tables include only the temporal fields due to the volume of classes and updates.

Whether you select one of these complete prototype versions or develop your own, it is important to follow a common class structure and field-naming convention. Use of the full data dictionary history database extension described in chapter 6 can also provide solid documentation on the contents of the geodatabase and make its management in a large organization a more achievable objective.

Describing LRM positions

The new UNETRANS data model endorses the practice of separating the position data from the other entity attributes so as to allow multiple datums to be accommodated, for both LRM and geographic positions. The LRMPosition table is repeated from chapter 8. The GeoPosition table is new to this chapter. Equivalences between LRM positions and geographic positions are supported, as is a recursive relationship for LRM positions.

An attributed relationship class handles the LRM position to geographic position equivalencies, as this is a many-to-many relationship. The class could be extended to store the valid period of equivalency.

The recursive equivalency relationship for LRMPosition requires a little more work to accommodate it. Since ArcGIS does not directly support recursive relationships within the geodatabase, an explicit associative table class is required to manage LRM position equivalencies among the various LRM datums that may be in the dataset. A period of validity was included for this class because they are expected to change as measurements are refined during update cycles. Chapter 14 shows how to do the same thing with a calibration point feature class.

241

Chapter 12: Improving the UNETRANS data model

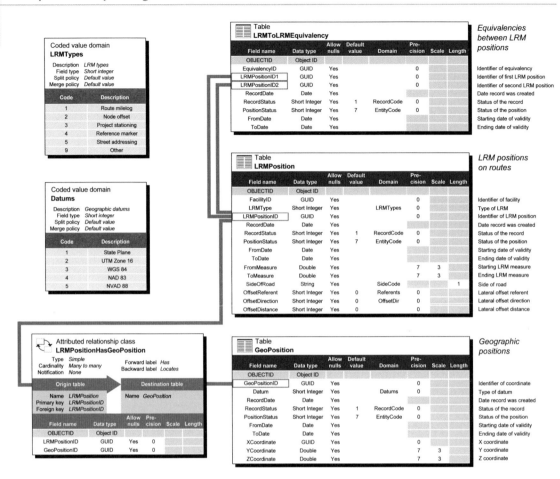

Figure 12.3 **Position classes** The modified UNETRANS data model uses the previously discussed approach of separating the position description from the entity it locates. The Inventory Package includes two basic kinds of position descriptions: LRM positions and geographic coordinates. Equivalencies between the two position types can be handled through an attributed relationship, but equivalencies between two LRM positions must be accommodated through an object class since ArcGIS does not support recursive relationships.

All three classes are shown with the temporal support fields included. Separation of the LRM position data from other attributes is recommended even if your agency has only a single LRM datum. For example, it allows a bridge to be represented as a point element at smaller scales and a linear element at larger scales. It also allows multiple LRM references for a shared facility, thereby avoiding duplication of data for that facility, which could be a bridge or an intersection. Refinements in position descriptions can be accommodated using a separate LRMPosition table without having to modify the characteristic and element rows they serve.

Three new domain classes are also included in this part of the revised UNETRANS geodatabase model.

Aspects and elements

One of the recognized shortcomings of the original UNETRANS model was its lack of explicit support for event tables that represented facility characteristics. As shown in chapter 7, the short-segment approach to transportation database design—one adopted by the original UNETRANS model—does not accommodate attributes that change at midblock locations. The denormalized database design required by this data structure also imposes significant editing overhead compared to a more normalized design provided by full use of an LRM.

The revised UNETRANS model extends support for LRM positions to allow a normalized design for all descriptive aspects of a linear facility. In the traditional legacy route-event

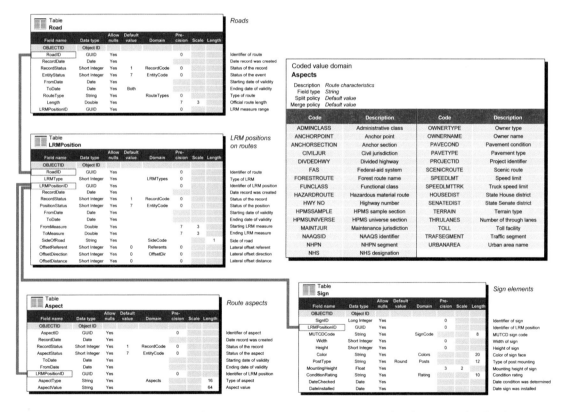

Figure 12.4 **Facilities, aspects, and elements** One change creates a normalized structure for the descriptive characteristics of the transport system's facilities (facility aspects). You can call them events if you want to, but that name has been reserved in the UNETRANS model for classes describing "things that happen." A new Aspects domain includes a subset of the point and linear event types in an earlier example. Other former event types represent facilities and elements of the transport system and will get their own object classes. The Sign table represents a typical element class, while the Road table is a facility class example. To view this figure in detail, go to http://www.esri.com/industries/transport/resources/data_model.html

table, where everything is placed in a one or two tables, aspects and elements are treated as the same thing, a route event. The Aspect table is modeled after the Event table in chapter 7, which follows the legacy design for route events.

What is new here is the introduction of separate tables for facility elements, which will typically need more than one or two descriptive attributes. The distinguishing characteristics of aspects and elements are: (1) an aspect has only one descriptor and cannot be directly observed in the field; and (2) an element will likely need multiple descriptors and has a physical presence that can be observed. Several examples of each are included in the following sections. The Sign table is the first example of how element classes are accommodated in the design.

A suggested domain for aspect types is supplied. Separate classes for point and linear aspects are not required if the LRM position data is stored in its own class. Even when the more traditional integrated design is adopted, with LRMPosition fields being part of the same class as other descriptive data, you could design all classes to accommodate from and to measures, with point aspects and elements having only one field with a useful value.

Mode-specific Inventory Package classes

There are multiple ways to present the many new facility and element classes provided in the new UNETRANS model. No one view can clearly demonstrate all aspects of the Inventory Package data model. The first view, presented here, matches that of the original UNETRANS data model. This version represents inheritance relationships.

The second presentation is by mode of travel. UNETRANS remains multimodal in nature, but since most transport agencies are organized around functional units by mode, it is useful to demonstrate the data model by showing each included mode. The third presentation shows association relationships.

All class diagrams are presented as logical data models. The *Entity* class fields for temporal and editing support are not included. You may add those that are appropriate for your agency. Several classes will appear in multiple modes.

In its logical form, the Inventory Package consists of facilities, their component elements, and their descriptive aspects. All of these classes are derived from ObjectClass, so they do not include geometry.

The distinction between facilities and elements is somewhat arbitrary. A linear facility is a real-world entity that can define an LRM datum; an element cannot. Some facilities provide a way to group elements in order to treat the composite entity as a singularity. For example, a bus terminal (facility) may represent a collection of bus bays (elements), just as a port may be formed from a collection of docks. These could just as easily be complex elements.

Mode-specific Inventory Package classes

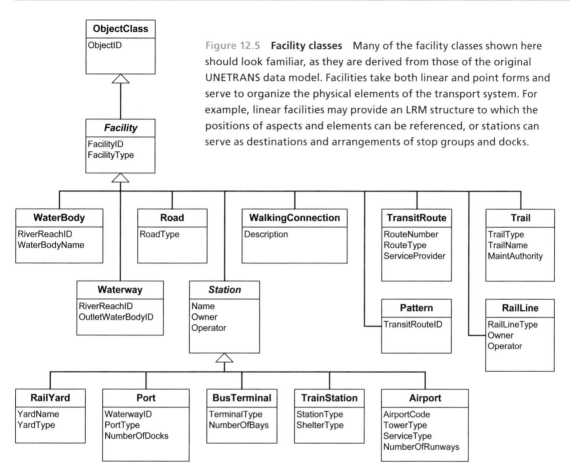

Figure 12.5 **Facility classes** Many of the facility classes shown here should look familiar, as they are derived from those of the original UNETRANS data model. Facilities take both linear and point forms and serve to organize the physical elements of the transport system. For example, linear facilities may provide an LRM structure to which the positions of aspects and elements can be referenced, or stations can serve as destinations and arrangements of stop groups and docks.

Although a facility is a representation of a real-world entity, it need not be a one-to-one relationship. For example, a physical road segment can be part of several routes including a state highway with LRM measures, a local named street with addresses, an HPMS sample section, a National Highway Planning Network route, and a Strategic Highway Network (STRAHNET) military logistics route. A rail line may be owned by one railroad company but could be part of several rail systems through trackage rights. Multiple route systems may include a shared physical entity, each providing its unique identifier, a set of descriptive aspects, and component elements.

A road is a facility that includes such elements as right-of-way, bridges, pavement segments, and traffic signals. A rail line is a facility that can be a collection of such elements as tracks, bridges, signals, and switches. A waterway is a collection of channels, buoys, markers, and locks, and may extend through multiple water bodies. A transit system can be composed of transit routes, each of which consist of a set of patterns defined by an ordered set of transit segments that extend from stop to stop. Modal groupings of facility and element classes are illustrated later.

There will be many more element classes than facility classes. Only a few of the possible element classes are illustrated in figure 12.6. Some, such as SignAssembly and StopGroup, can be composites of other elements. Elements such as RoadProfile and HOVLane can describe some aspect of the roadway over a spatial extent defined by from- and to-measures. They represent a physical part of the roadway but likely do not have their own geometry;

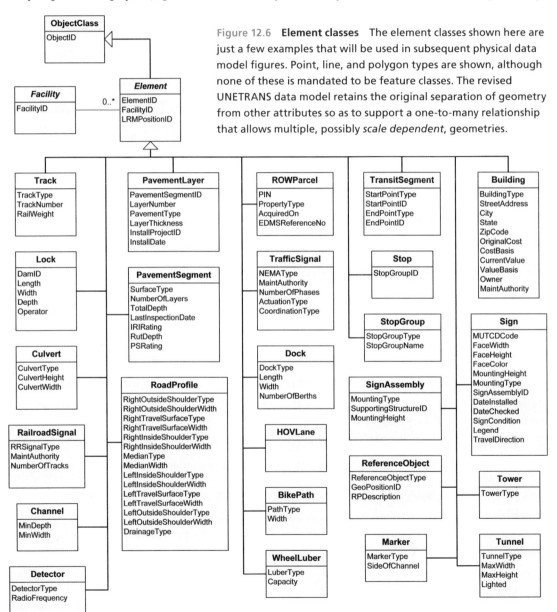

Figure 12.6 **Element classes** The element classes shown here are just a few examples that will be used in subsequent physical data model figures. Point, line, and polygon types are shown, although none of these is mandated to be feature classes. The revised UNETRANS data model retains the original separation of geometry from other attributes so as to support a one-to-many relationship that allows multiple, possibly *scale dependent*, geometries.

they are event tables. In contrast, a ROWParcel can be represented by its own polygon and need not be tied to the facility through an LRM position. A bridge can be either a facility or an element, depending on the needs of your agency, and may be represented by a point, line, or polygon feature.

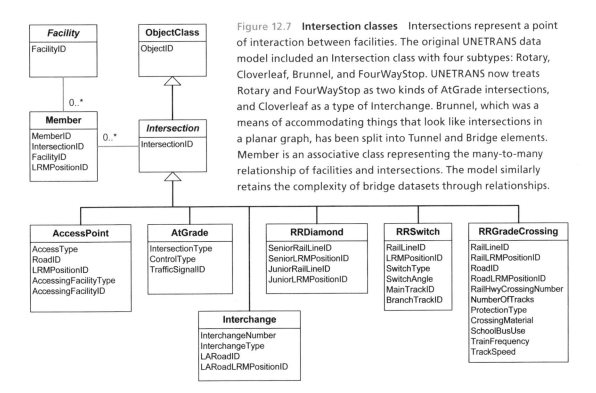

Figure 12.7 **Intersection classes** Intersections represent a point of interaction between facilities. The original UNETRANS data model included an Intersection class with four subtypes: Rotary, Cloverleaf, Brunnel, and FourWayStop. UNETRANS now treats Rotary and FourWayStop as two kinds of AtGrade intersections, and Cloverleaf as a type of Interchange. Brunnel, which was a means of accommodating things that look like intersections in a planar graph, has been split into Tunnel and Bridge elements. Member is an associative class representing the many-to-many relationship of facilities and intersections. The model similarly retains the complexity of bridge datasets through relationships.

Intersections remain part of the Inventory Package but are no longer simple junction feature classes. New types have been added to explicitly support rail travel. Because the model is not based on a geometric network, a brunnel class is not required to represent facility crossings that are not true intersections. The Cloverleaf subtype is now a more generically named Interchange class. The Rotary and FourWayStop subtypes are included in the AtGrade intersection class, which is mode-specific in application. The RRSwitch and RRDiamond classes accommodate at-grade intersections for the rail mode. (A railroad switch is a connection between tracks. A railroad diamond is a crossing of two tracks without the ability to change tracks.) The RRGradeCrossing class represents the intersection of a road and a rail line, and should be graphically represented using a multipoint feature class (one point for each track, all treated as one crossing).

Chapter 12: Improving the UNETRANS data model

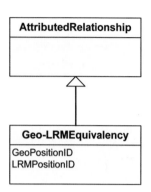

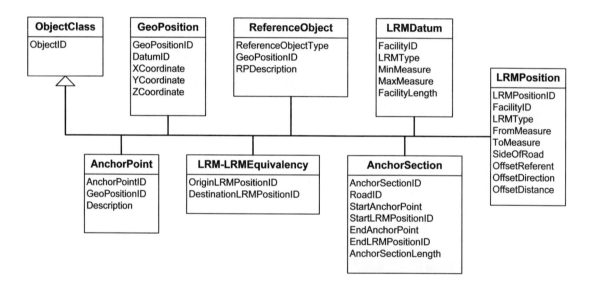

Figure 12.8 **Position classes** There are several new object classes to store position data in the revised UNETRANS logical data model. Both LRM and geographic positions are supported, and each may be defined in multiple datums. LRM positions are tied to the facility that supplies the linear datum information of LRM type, minimum and maximum measure values, and total length. Anchor points and anchor sections, if provided, are accommodated in the model as subdivisions of a road. (Linear datum concepts have not been applied to other facility types.) Note that a single physical facility may be represented in the geodatabase as any number of abstract facilities, each identified by an LRM type. For example, what exists as a city block in one system view could also be represented as a pavement segment, part of a state road, or any other location description approach that involves an LRM. The equivalences between multiple LRMs can be explicitly stored, as can their correlation to geographic coordinates. A point position will have a value only for **FromMeasure** in LRMPosition.

The classes that provide information about LRM and geographic positions and the linear datum are a mix of object and relationship classes. The position of any facility may be stated in terms of geographic coordinates. An LRM datum can be defined by linear facility, and includes the type of LRM, minimum and maximum measures, and total length. Facilities and elements referenced to a linear facility's LRM datum can be located using an LRM position. An anchor section may define the official path and length for any facility with an LRM datum. The official length of an LRM defining facility is the sum of its component anchor section lengths. One or more reference objects may unambiguously define the location of an anchor point.

Modal class groups

The second way to view the example UNETRANS classes is to group them by mode. Groupings for roads, rail lines, and navigable waterways are presented here. The public transit set of classes uses the Network Package and is described in chapter 13 after the discussion of that package. All three modal groupings share a number of classes, such as Aspect, Bridge, and LRMPosition.

Figure 12.9 Road classes The revised UNETRANS data model includes the richest set of example classes for roads. The Facility subclass Road is there, as are the Aspect class for road characteristics, numerous road element classes, and intersection classes. The Road class is the only one that can define an LRM. The LRM datum objects (anchor point and anchor section) are aspects of a road, but may be tied to one or more reference objects that are modeled as road elements.

RoadProfile
RoadProfileID
RoadID
LRMPositionID
RightOutsideShoulderType
RightOutsideShoulderWidth
RightTravelSurfaceType
RightTravelSurfaceWidth
RightInsideShoulderType
RightInsideShoulderWidth
MedianType
MedianWidth
LeftInsideShoulderType
LeftInsideShoulderWidth
LeftTravelSurfaceType
LeftTravelSurfaceWidth
LeftOutsideShoulderType
LeftOutsideShoulderWidth
DrainageType

Bridge
NBINumber
BridgeType
Name
Owner
MaintAuthority
Length
YearBuilt
AcrossFacilityType
AcrossFacilityID
AcrossMaxClearance
AcrossMaxWidth
AcrossMaxWeight
UnderFacilityType
UnderMaxClearance
UnderMaxWidth
LastInspectionDate
NBIRating

Culvert
CulvertID
RoadID
LRMPositionID
CulvertType
CulvertHeight
CulvertWidth

ROWParcel
RoadID
PIN
PropertyType
AcquiredOn
EDMSReferenceNo

Road
RoadID
RoadType

TrafficSignal
NEMAType
MaintAuthority
NumberOfPhases
ActuationType
CoordinationType

ReferenceObject
ReferenceObjectType
GeoPositionID
RPDescription

LRMDatum
FacilityID
LRMType
MinMeasure
MaxMeasure
OfficialLength

AccessPoint
AccessType
RoadID
LRMPositionID
AccessingFacilityType
AccessingFacilityID

Interchange
InterchangeNumber
InterchangeType
LARoadID
LARoadLRMPositionID

Sign
SignID
MUTCDCode
FaceWidth
FaceHeight
FaceColor
MountingHeight
MountingType
SignAssemblyID
DateInstalled
DateChecked
SignCondition
Legend
TravelDirection

Aspect
AspectID
FacilityID
LRMPositionID
AspectType
AspectValue

LRMPosition
LRMPositionID
RouteID
LRMType
FromMeasure
ToMeasure
SideOfRoad
OffsetReferent
OffsetDirection
OffsetDistance

AtGrade
IntersectionType
ControlType
TrafficSignalID

RRGradeCrossing
RailLineID
RailLRMPositionID
RoadID
RoadLRMPositionID
RailHwyCrossingNumber
NumberOfTracks
ProtectionType
CrossingMaterial
SchoolBusUse
TrainFrequency
TrackSpeed

SignAssembly
SignAssemblyID
RoadID
LRMPositionID
GeoPositionID
MountingType
SupportingStructureID
MountingHeight

PavementLayer
PaveSegmentID
LayerNumber
PavementType
LayerThickness
InstallProjectID
InstallDate

PavementSegment
PaveSegmentID
RoadID
LRMPositionID
SurfaceType
NumberOfLayers
TotalDepth
LastInspectionDate
IRIRating
RutDepth
PSRating

LRM-LRMEquivalency
OriginLRMPositionID
DestinationLRMPositionID

Geo-LRMEquivalency
GeoPositionID
LRMPositionID

Building
BuildingType
StreetAddress
City
State
ZipCode
OriginalCost
CostBasis
CurrentValue
ValueBasis
Owner
MaintAuthority

GeoPosition
GeoPositionID
DatumID
XCoordinate
YCoordinate
ZCoordinate

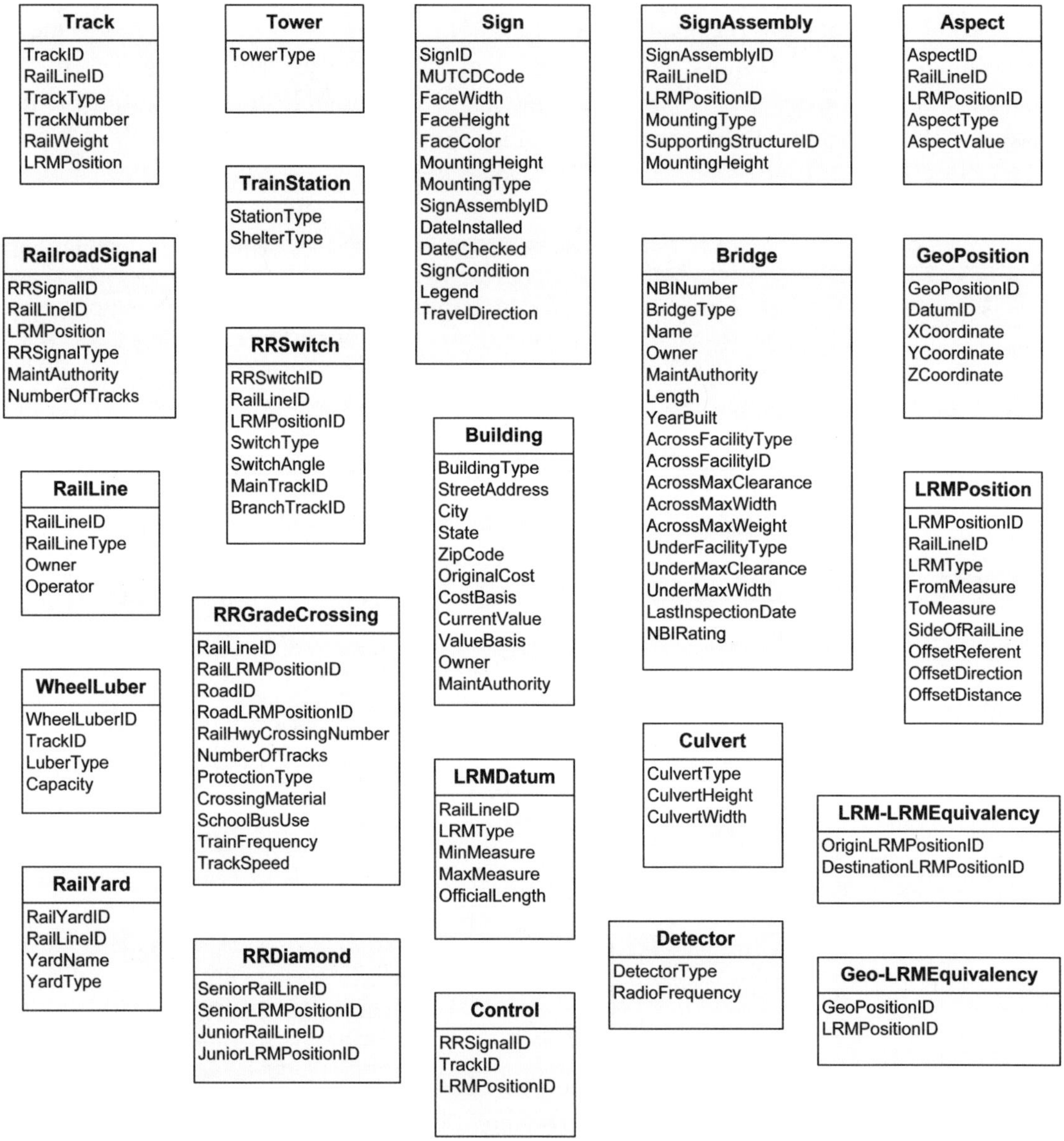

Chapter 12: Improving the UNETRANS data model

Track
- TrackID
- RailLineID
- TrackType
- TrackNumber
- RailWeight
- LRMPosition

Tower
- TowerType

Sign
- SignID
- MUTCDCode
- FaceWidth
- FaceHeight
- FaceColor
- MountingHeight
- MountingType
- SignAssemblyID
- DateInstalled
- DateChecked
- SignCondition
- Legend
- TravelDirection

SignAssembly
- SignAssemblyID
- RailLineID
- LRMPositionID
- MountingType
- SupportingStructureID
- MountingHeight

Aspect
- AspectID
- RailLineID
- LRMPositionID
- AspectType
- AspectValue

TrainStation
- StationType
- ShelterType

RailroadSignal
- RRSignalID
- RailLineID
- LRMPosition
- RRSignalType
- MaintAuthority
- NumberOfTracks

RRSwitch
- RRSwitchID
- RailLineID
- LRMPositionID
- SwitchType
- SwitchAngle
- MainTrackID
- BranchTrackID

Bridge
- NBINumber
- BridgeType
- Name
- Owner
- MaintAuthority
- Length
- YearBuilt
- AcrossFacilityType
- AcrossFacilityID
- AcrossMaxClearance
- AcrossMaxWidth
- AcrossMaxWeight
- UnderFacilityType
- UnderMaxClearance
- UnderMaxWidth
- LastInspectionDate
- NBIRating

GeoPosition
- GeoPositionID
- DatumID
- XCoordinate
- YCoordinate
- ZCoordinate

LRMPosition
- LRMPositionID
- RailLineID
- LRMType
- FromMeasure
- ToMeasure
- SideOfRailLine
- OffsetReferent
- OffsetDirection
- OffsetDistance

RailLine
- RailLineID
- RailLineType
- Owner
- Operator

Building
- BuildingType
- StreetAddress
- City
- State
- ZipCode
- OriginalCost
- CostBasis
- CurrentValue
- ValueBasis
- Owner
- MaintAuthority

RRGradeCrossing
- RailLineID
- RailLRMPositionID
- RoadID
- RoadLRMPositionID
- RailHwyCrossingNumber
- NumberOfTracks
- ProtectionType
- CrossingMaterial
- SchoolBusUse
- TrainFrequency
- TrackSpeed

WheelLuber
- WheelLuberID
- TrackID
- LuberType
- Capacity

LRMDatum
- RailLineID
- LRMType
- MinMeasure
- MaxMeasure
- OfficialLength

Culvert
- CulvertType
- CulvertHeight
- CulvertWidth

LRM-LRMEquivalency
- OriginLRMPositionID
- DestinationLRMPositionID

RailYard
- RailYardID
- RailLineID
- YardName
- YardType

RRDiamond
- SeniorRailLineID
- SeniorLRMPositionID
- JuniorRailLineID
- JuniorLRMPositionID

Detector
- DetectorType
- RadioFrequency

Geo-LRMEquivalency
- GeoPositionID
- LRMPositionID

Control
- RRSignalID
- TrackID
- LRMPositionID

Figure 12.10 **Railroad classes** As with road classes, the revised UNETRANS data model includes a large number of classes for facilities, elements, aspects, and intersections. The two Facility subclasses are RailLine and RailYard, although only RailLine can establish an LRM. (Yards use track numbers/names and other ways to organize their component elements.) Several element classes are shared with the Road mode of travel such as Sign, Bridge, Culvert, and the obvious RRGradeCrossing.

Association relationships

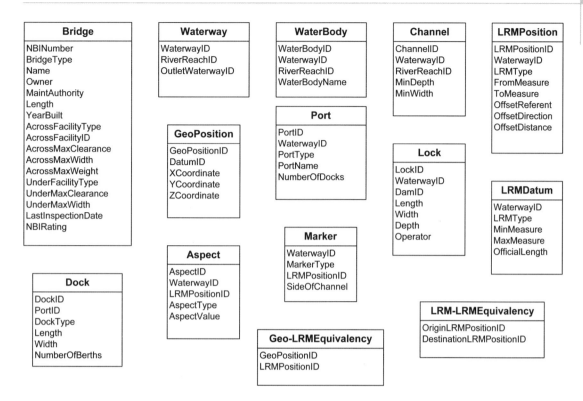

Figure 12.11 **Navigable waterway classes** The original UNETRANS data model included only one class for navigable waterways: WaterEdge. The revised model is a bit more extensive, offering two *Facility* subclasses and several *Element* subclasses unique to the waterway mode of travel. Of course, the ever-present Bridge class is there, too.

Association relationships

The Inventory Package is a logical data model that offers a third way to present the revised UNETRANS structure. This model shows association relationships. Depending on the degree of control you want to exert on the editing process, you may implement some of these relationships using explicit relationship classes. Some of the classes shown here, such as Member and Geo-LRMEquivalency, should be implemented as attribute-relationship classes.

A facility may be formed from one or more elements. This relationship is typically aggregation, not composition; however, it can become a composition relationship for some facilities. For example, you may decide to so construe the relationship of bus stops to the bus station in which they are located. Feature classes may cartographically represent both facilities and elements, but a facility and element class should always be a table. By separating the nonspatial attributes from the representative geometry for a facility or element, you can

Chapter 12: Improving the UNETRANS data model

support multiple cartographic representations and an asynchronous editing process. You will need to create a facility or element identifier in the table class before you create the corresponding feature so that the proper foreign key value may be provided.

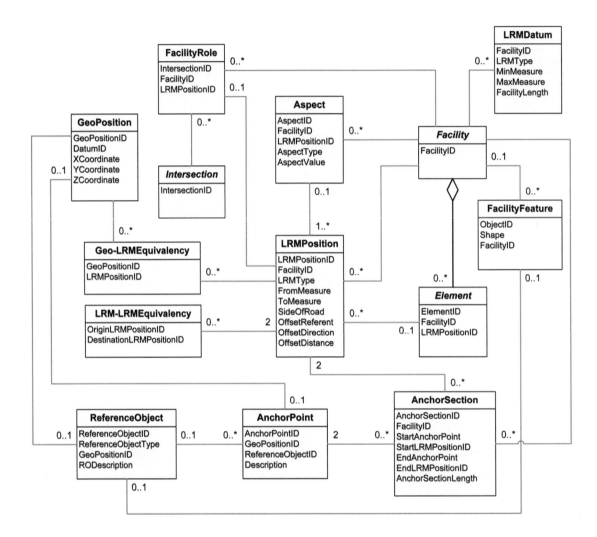

Figure 12.12 **Inventory package** The revised UNETRANS logical data model includes feature and object classes suitable to provide a facility inventory geodatabase. This figure shows the primary object classes, which support facilities, their descriptive aspects and component elements, LRM positions, and geographic coordinates. Facilities, elements, and reference objects may be directly illustrated by feature classes; aspects can be mapped only through dynseg (elements may also be mapped this way). The assumption is that a facility may define an LRM, while aspects and elements may be located according to one or more facility-based LRM positions.

In general, you will use a linear facility to construct an LRM datum for determining and recovering linear positions referenced to the facility origin. The LRMDatum class provides both the LRM metadata and measure domain control fields. Much such data may be applied to a given facility, as identified by LRMType. This means that any query to find the LRM position for a given element, aspect, or subordinate facility will need to specify both the FacilityID and LRMType. Of course, if your agency only has one type of LRM and will never share data with another agency using a different type, then you may choose to eliminate the LRMType field for simplicity.

Facilities, though, need not define an LRM datum. Implementing the optional association relationship between *Facility* and LRMPosition can provide a recursive relationship for the *Facility* class. This would allow a facility of one class to be placed on a facility of another class through an LRM position description, in essence turning the facility into an element. This is a good way to treat bridges as a facility for the structures workgroup and as an element for the highway inventory workgroup.

This general form of the UNETRANS data model does not include a direct relationship between *Facility* or *Element* and GeoPosition. This is because facilities and elements are almost always treated as being linear in their abstract form. As a result, there is no single geographic position that can specify its location on Earth. What is included in the model to provide a geographic location for facilities and elements is a many-to-many relationship between LRMPosition and GeoPosition. The expectation is that the LRM position is the dominant one. This general model does not, however, preclude you from establishing a direct connection between GeoPosition and either or both of *Facility* and *Element*. You may choose to do so when GPS-based surveys are used to compile as-built plans for updating the facility inventory geodatabase, or to store the coordinates of a point element.

Some classes do have a direct relationship with GeoPosition already. These classes are part of the linear datum, which includes the classes of AnchorPoint, AnchorSection, and ReferenceObject. Both AnchorPoint and ReferenceObject can be given geographic coordinate positions. Neither class is assigned an LRM position in the general model, but you may choose to do so for ReferenceObject, which can also have its own feature class representation. The one-to-many relationship for AnchorPoint and LRMPosition is established through GeoPosition-LRMPosition equivalency.

By definition, an anchor point cannot be given a single LRM position, as each intersecting facility will assign its own LRM position description to the location. You can discover anchor point LRM positions by looking at the anchor sections in which they are used as terminal points. Each anchor section is part of one facility. The geographic location of an anchor section can be derived from the geographic positions of its terminal anchor sections.

Chapter 12: Improving the UNETRANS data model

Intersections

One of the more complex and versatile parts of the revised UNETRANS design is its treatment of intersections. In the general model, an associative class (Member) handles the many-to-many relationship that exists between *Facility* and *Intersection*. This allows an intersection to be treated as a singularity with many LRM position descriptions; i.e., at least one for each intersecting facility.

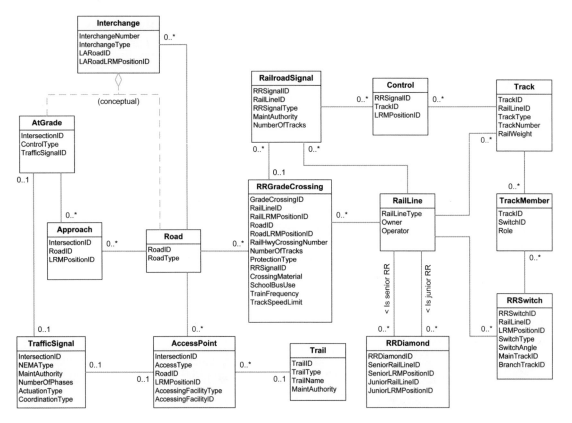

Figure 12.13 **Using intersections** This figure shows how the various kinds of intersection classes relate to facilities and elements. An at-grade intersection must involve at least two roads, although only one road is required for an access point. An interchange could be treated as a collection of at-grade intersections and roads, but we have limited the mandatory relationships to identifying the limited-access road on which the interchange is located. A railroad grade crossing is an intersection between a road and a rail line, while a railroad switch is an intersection between two tracks. Three-way and crossover railroad switches are accommodated.

A more specific data model can be derived from the general form, as shown in figure 12.13, by substituting specific creatable facility and element classes. All creatable *Intersection* classes are present. On the left side is a road system; on the right is a rail system. Connecting the two is the RRGradeCrossing element table. The generic Member class in the previous figure has been instantiated in this more specific data model as RoadMember and TrackMember. These classes could be implemented in a physical geodatabase design as attributed relationship classes or explicit associative tables.

An at-grade intersection is one involving two or more roads. Movement through such an intersection may be controlled by a traffic signal. An interchange may be treated in more complex implementations as an aggregation of at-grade intersections and roads. An access point involves one road and some other facility not included in the dataset, such as a major driveway, or a facility that is not of the road type, such as a recreational trail. Access points may also be controlled by a traffic signal.

A rail line can include one or more tracks. A railroad signal can control train access to a block or indicate the position of a switch. Separate indications may be supplied for each track along a multitrack section of the rail line. A railroad switch and signal is the responsibility of a single railroad, but a diamond crossing normally represents a point where two railroads cross. The junior railroad—the second to arrive and, thus, the one that caused the diamond crossing to be constructed—usually has maintenance responsibility for the facility, while the senior railroad may provide access control to regulate allocation of the right of way to approaching trains.

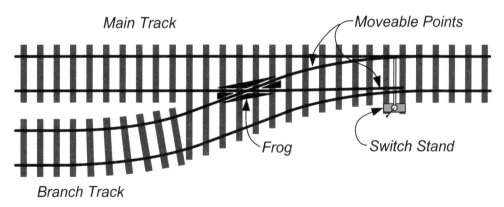

Figure 12.14 **Railroad switch**

A railroad switch allows connections between tracks of a single rail line and is specified in terms of the angle of divergence. Normally, a railroad switch has one input leg and two output legs, but scissor switches and slip switches have multiple input and output legs. Rather than store the relationship of input and output paths in the RRSwitch class, this design uses an associative class (TrackMember) to provide a more flexible structure to support any kind of switch.

A railroad-grade crossing is functionally similar to a railroad diamond, but the intersecting facilities are of different travel modes. The railroad almost always has maintenance responsibility for crossing protection. Both the road and the rail line may supply an LRM position description for the crossing.

Bridges

In some ways, a bridge is like an intersection in that it may be located using multiple position descriptions. As a result, the general model for accommodating bridges will include a BridgeMember class. The Bridge class is one example where it may be appropriate to supply a geographic position, as the National Bridge Inventory requires such information. Any LRM position assigned to the bridge is determined by the perspective of the facility that goes over or under the bridge.

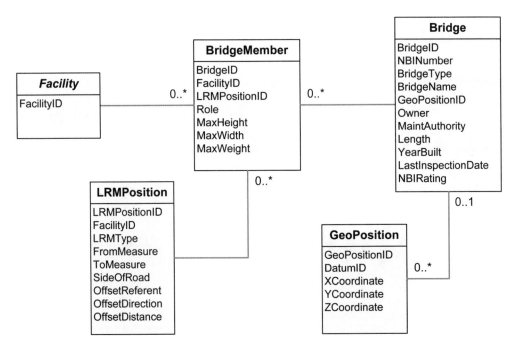

Figure 12.15 **Bridges** The original UNETRANS data model did not do much for the special relationship between bridges and the facilities going across them and passing under them. The revised model allows any number of facilities to cross over or go under a bridge. For example, the model supports situations where a highway and railroad cross a structure, or a road and a waterway go under a bridge. Which facility is doing what and the capacity of the passage for that facility are stored in an associative class, BridgeMember, which also includes a position reference to describe the location of the bridge within the context of each facility's LRM.

Bridges

The model shown in figure 12.15 can accommodate any number of facilities going over or under a bridge. The value you put in the Role field indicates whether the facility goes over or under the bridge. BridgeMember also stores the relevant information about any restriction to travel that may be imposed. Vehicles passing under a bridge may be restricted by height and width clearances. Bridges passing over a bridge may be subject to weight restrictions. In some cases, there may also be width and height clearance issues for travel over a bridge. The design accommodates situations where bridges are stacked on top of each other, as may occur at a complex highway interchange, or arranged side by side, as may occur when facilities are parallel to each other.

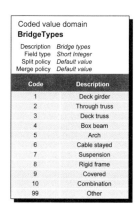

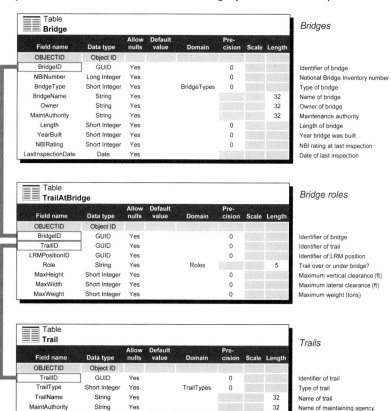

Figure 12.16 **Bridge physical data model—option 1** There are two basic approaches you can use to implement the bridge logical data model. This figure shows the first choice: using an associative table class. It also illustrates the three domain classes used. It is a simple matter to combine the data held in these three tables to produce a trail event table containing information about all the bridges on a trail. The disadvantage of this approach is that you will have only limited control over the referential integrity of the relationship.

Bridges provide a good example for showing how the logical data model might be implemented in a physical geodatabase design. In this first alternative, the BridgeMember class has been instantiated as a TrailAtBridge table to store the relationship between trails and bridges. This alternative allows you to do an on-the-fly join to create a trail event table. You could even support a software switch to modify the symbolization for any included bridge by using the Role field to place the symbol over or under the trail centerline or to select different symbols that provide that appearance.

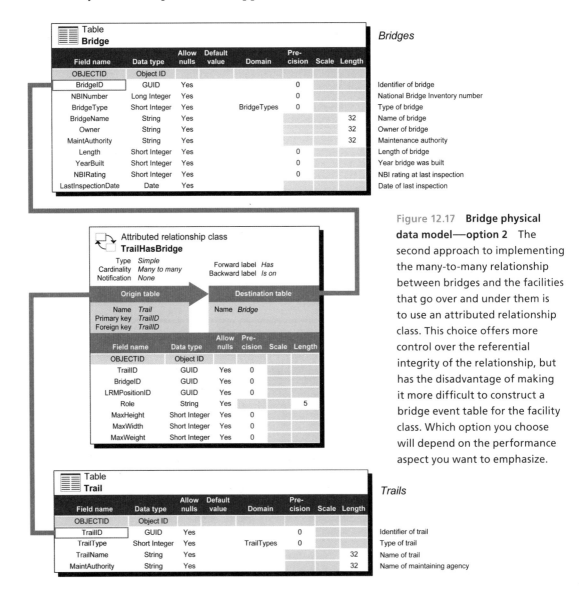

Figure 12.17 **Bridge physical data model—option 2** The second approach to implementing the many-to-many relationship between bridges and the facilities that go over and under them is to use an attributed relationship class. This choice offers more control over the referential integrity of the relationship, but has the disadvantage of making it more difficult to construct a bridge event table for the facility class. Which option you choose will depend on the performance aspect you want to emphasize.

This second alternative implementation of the logical data model uses an attributed relationship class. This more precisely reflects the situation from a class behavior perspective, but it reduces the amount of data-entry control you may be able to exert, such as to supply a Role domain class. It also makes it more difficult to construct a facility event table that includes bridges and the facilities they involve. On the plus side, it allows better control of the cardinality and supports automated messaging between classes.

The latter benefit may be especially useful for movable bridges contained in an operation geodatabase that is used to supply information to staff operating changeable message signs. To support this function, you would add a BridgePosition field to indicate whether the bridge is up or down. You may also want to add a field to indicate when the bridge will next be raised, if it is determined by user request, or to point to a timetable, if it is done on a regular schedule.

Events

An event is something that happens. The Events Package combines the functions of the former Activities and Incidents packages. An event has an LRM position on a facility and can be represented using a feature, either directly or through an event table and dynseg. Common events include work program projects, maintenance work orders, and crashes. All three of these events may result in lane closures, both planned and unplanned.

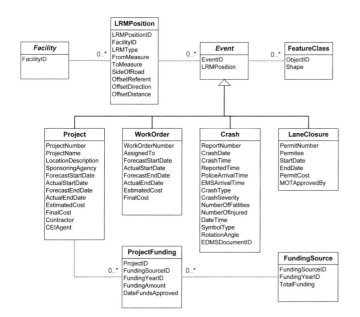

Figure 12.18 **Event classes**
Events are things that happen in UNETRANS, which is a more restrictive definition than the "everything is an event" approach that resulted from NCHRP 20-27(2). Originally called activities and incidents in the original UNETRANS data model, events of all types follow a common design: they happen at a specified location and occur over a period of time. Events are tied to the facilities they impact through the facility identifier in the LRMPosition record, but may be graphically represented by their own feature class. Some samples are shown here to get you started.

Chapter 12: Improving the UNETRANS data model

The revised UNETRANS data model includes an associative class managing the many-to-many relationship between Project and FundingSource, which reflects the fact that many projects receive funds from multiple sources.

Pavement management example

It is useful to illustrate how the Events Package is used in concert with the Inventory Package by proposing a pavement management geodatabase design. This UNETRANS-derived design in figure 12.19 updates the one in chapter 4 based on the short-segment data structure.

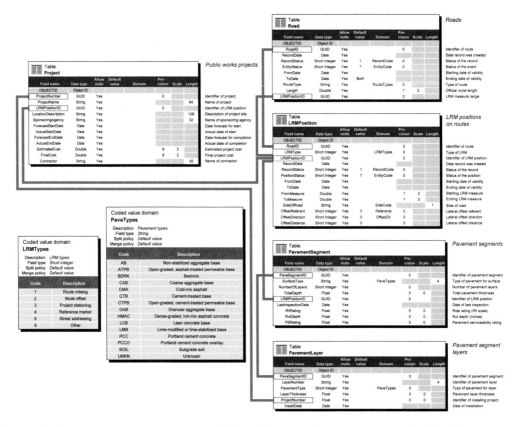

Figure 12.19 **Pavement management system for LRM**   Chapter 4 presented a sample design for a pavement management system where streets are divided into discrete segments. This alternative example is used when routes with LRM measures are employed. Projects (events) and pavement segments (elements) are both tied to the LRM. PavementLayer remains tied to PavementSegment and Project. This example assumes that a project does not span roads. If they do in your agency, then you will want to use an attributed relationship between Project and Road and put the LRMPositionID attribute there. To view this figure in detail, go to http://www.esri.com/industries/transport/resources/data_model.html

Pavement management example

As before, a pavement segment is an aggregation of a set of pavement layers. The difference in this design is the use of LRM positions to define the limits of the pavement segment rather than the extents of member street segments. The many-to-many relationship possible between Project and Road is not shown. It would be structurally identical to that of Project and StreetSegment in the design presented in chapter 4.

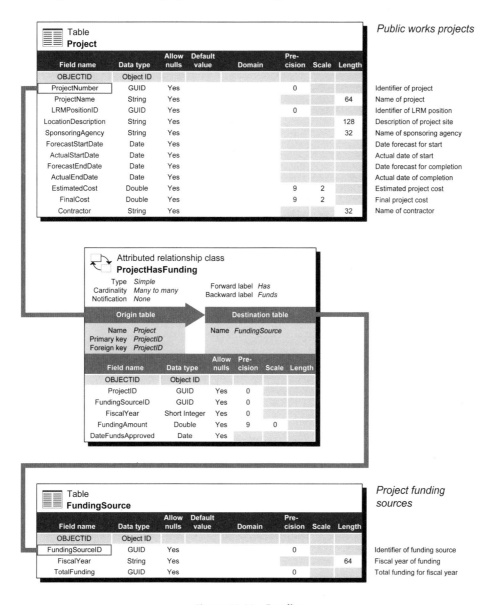

Figure 12.20 **Funding**

Mobile Objects

The revised UNETRANS Mobile Objects Package provides a richer set of classes and relationships, although most were implicit in the original design. A mobile object is not explicitly located on the transport system using an LRM position. Instead, the native form of describing the system user's location is a geographic position. This choice reflects the use of GPS receivers to generate the original location information. An application may convert the geographic position to an LRM position using proximate analysis methods.

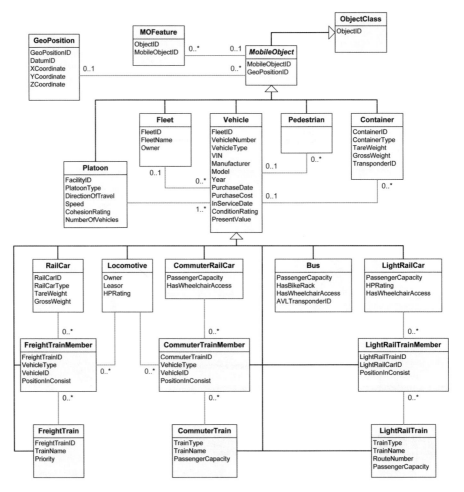

Figure 12.21 **Mobile object classes** Mobile objects are things that use the transport system, such as vehicles, people, and containers. They follow a fairly common structure that provides information on what they are and where they are located. Some also need information about origin and destination. Commercial vehicles can be aggregated to form a fleet, and vehicles of all types can be temporarily part of a platoon.

The Mobile Objects Package is small enough to be able to show all the relationships in one figure. The basic mobile objects are vehicles, pedestrians, and containers. A vehicle may be assigned to a fleet or be part of a platoon, which is a logical grouping of vehicles moving in a coordinated fashion. In some applications, pedestrians may also form a platoon, and an appropriate association relationship will need to be created between Pedestrian and Platoon. Both pedestrians and containers may be conveyed along the transport system by a vehicle.

One new part of the logical data model is explicitly treating how vehicles are arranged to form composite vehicles. Such associations change frequently, such as each time a train is constructed as a set of locomotives and rail cars. The associative Member class also has a use here to manage the various arrangements of basic vehicles into composite forms. The composite vehicle classes are listed along the bottom of figure 12.21. Since light-rail cars are typically self-powered, a light-rail train operated by a transit service is composed simply of one or more light-rail cars. However, freight and passenger trains require separate locomotives.

Using all three packages

A database design for a crash-reporting system offers the opportunity to show how the Inventory, Event, and Mobile Object packages may interact in a single geodatabase. A crash is an event. A crash occurs when a vehicle strikes a fixed object, another vehicle, or a pedestrian. The vehicles and persons involved in a crash are mobile objects. The crash occurs on a facility and may affect some element of the facility, both of which are stored in the inventory database.

A crash may be directly illustrated using a feature class object with a geographic position, which may be determined by a law-enforcement officer using a portable GPS unit. Crashes may also be located on a facility using an LRM position description through post processing. Positions may be established by listing cross streets at an intersection or a distance along a road segment from a known point. Some states have created node numbers to uniquely identify intersections. These numbers may be combined to identify a street segment.

Vehicles and pedestrians involved in a crash typically are assigned a sequence number that ties them to the crash. This sequence number is reflected in the VehicleNumber field of the Vehicle class. It is common practice for Vehicle 1 to be the one that initiated the crash event.

The Pedestrian class shown in the general Mobile Objects model has been generalized in this design to become the Person class. One role of a person is to be a pedestrian, but a person may also be identified as a driver or occupant of a vehicle. The DriverID field in the Vehicle class points to the DriverLicenseNumber field in the Person class. The NumberOfPassengers field in the Vehicle class can be used to ensure that a correct number of Person class rows have been created. Pedestrians are also sequentially numbered, as are drivers, but it is

Chapter 12: Improving the UNETRANS data model

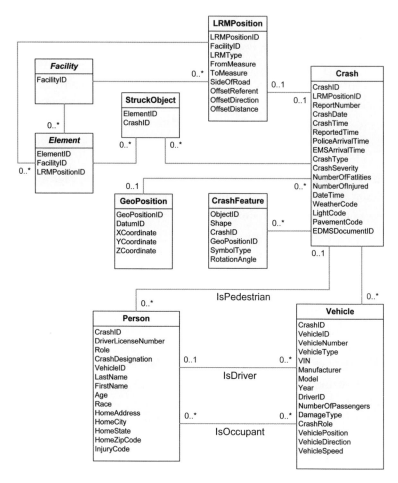

Figure 12.22 **Putting it all together** This example design combines facilities, elements, incidents, and mobile objects in a crash database. A crash can be located using an LRM position, as may be needed by a transport agency, or geographic coordinates, as might be done by an investigating officer using GPS. A crash may involve a vehicle striking fixed object (which could be a road element), a pedestrian, or another vehicle. A person involved in a crash can be given one of three roles.

less common to assign sequence numbers to other vehicle occupants. Accordingly, the VehicleNumber field in the Vehicle class is generalized to CrashDesignation in the Person class, as it may contain something other than a sequence number; e.g., "P1" for Pedestrian 1, "D1" for the driver of Vehicle 1, or "O2-3" for the third listed passenger of Vehicle 2. MobileObjectID in the Person class points to VehicleID in the Vehicle class if the person was a driver or passenger in a vehicle. If not, the value is null.

chapter thirteen

The revised UNETRANS network data model

- Network connectivity
- Network attributes
- Building a network dataset
- Turns
- Alternative methods to accommodate bridges
- Bells and whistles
- The transit data model

Perhaps the biggest change to the revised UNETRANS data model is replacing the original version's geometric network foundation with a transportation-specific network extension. In fact, the source of data for a transport network cannot be a geometric network, which has a different way of managing network elements. The change is the result of ESRI's release of the ArcGIS Network Analyst extension. This chapter provides an overview of how you can use the extension to support pathfinding applications and the inputs such an application will require. It is not a tutorial for Network Analyst extension.

Chapter 13: The revised UNETRANS network data model

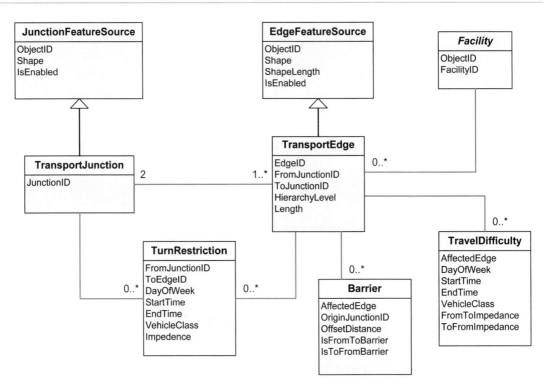

Figure 13.1 **Transport network classes** The revised UNETRANS logical data model continues to support use of a network for pathfinding applications, but it is no longer the underlying data structure and it is not based on a geometric network. The model includes a TurnRestriction class that is used to describe the difficulty of moving within a junction, and a TravelDifficulty class that includes information on the ease with which a specified vehicle class can move along an edge. The relative ability to make a turn or travel down an edge is called impedance or cost. The **IsEnabled** attribute can remove a network element from consideration in pathfinding by setting its value to False. A barrier can block movement on an edge at any point along its length and can apply to only one direction of travel. A **HierarchyLevel** attribute is used to set preferences in pathfinding algorithms.

This logical data model presents a simplified view of the Network Analyst extension in the context of UNETRANS for comparison to the original UNETRANS geometric network design. With the transfer of descriptive characteristics to the Inventory Package, the Network Package's feature classes are substantially simplified. What remains are the basic network components of junctions and edges, with fields that define network connectivity and classes that indicate the relative ease or difficulty of making a turn at a junction or traveling along an edge.

The physical implementation of the logical data model is somewhat different. Network Analyst extension creates the network element classes it needs from a set of three data sources. The only required data source is a feature class that contains route segments, an edge feature source. The process of loading a line feature data source to construct a transport network in the Network Analyst extension will generate network junctions at all line termini and at any point where two lines cross.

If the geodatabase is your data environment, then you may use multiple edge feature sources and also supply junction feature sources and turn feature sources to control connectivity and routing along the network. All sources must reside in a feature dataset. If you are working with shapefiles, then you may supply only one edge source and one turn source. The purpose of feature sources is to supply ArcGIS Network Analyst extension with the input data that it needs to construct and use network elements.

Network connectivity

Edge and junction elements, like those used by a geometric network, are constructed from feature class sources. Constructing the network elements is based on the feature geometry. Specifically, the coincidence of line endpoints and line intersections in edge feature sources determine the basic network. For a line intersection to be perceived as a junction, it must be defined as a vertex in each feature. If this is not the case, then you can create the required vertices using the Integrate tool in ArcToolbox or through applying topology rules. Point features may also be used as an explicit junction feature source to affect the process of creating junction elements.

Connectivity rules may be specified to regulate the overall network-building process. These rules govern the locations where connections may be inferred by the network-building process. Connections may be established at only line endpoints, thereby ignoring all locations where lines cross, or at endpoints and shared vertices on crossing lines. These choices produce networks that function similarly to those constructed by simple and complex edge feature classes in a geometric network.

You may also use elevations to regulate the placement of junctions in the resulting network. As described in chapter 6, elevation fields indicate the relative level of features crossing or meeting at a common location. Features with the same elevation value are judged to connect; those with different elevation values do not connect.

Building a network dataset begins by specifying which feature class will participate in the network. This is done by assigning the classes to connectivity groups. The number of distinct networks equals the number of connectivity groups. Each connectivity group must have at least one edge feature source. Network Analyst extension supports multimodal networks, where different modes of travel may be employed to define a path through the network. Explicit junction feature sources are required to connect edges placed in different connectivity groups. The junction feature source would be in all related connectivity groups.

A typical multimodal network example may involve street and transit segment edge feature sources and a bus stop junction feature source. To ensure that buses traverse only transit segments, the two edge feature sources are placed in different connectivity groups. The bus stop junction feature source would be part of both connectivity groups, thereby serving as a point for travel to move between the two mode-specific networks. Various ways to control connectivity at bridges are presented later in this chapter.

Network attributes

The output of a network-building process is a network dataset. The network attributes can be created for a new dataset by using the New Network Dataset wizard, or added and modified for an existing dataset via the Network Dataset Properties dialog box. (The dialog box can be accessed from the Attributes tab.) The inputs useful to Network Analyst extension in creating and using the network dataset include costs, descriptors, restrictions, and functional hierarchies.

Costs are expressed in terms of time or distance and are the same as impedances or weights, as those terms are used in other network analyses, such as the TAZ example in chapter 5. Pathfinding applications often seek to minimize a cost, such as distance or travel time. Costs are applied in proportion to distance of travel along an edge. If an origin or destination is along an edge, then the cost of traveling along that edge will be based on the percentage used for the trip.

A descriptor applies to an entire network element and cannot be apportioned. Descriptors are typically used to determine costs. For example, a volume-to-capacity ratio can identify the level of congestion and, as a result, travel cost in terms of time to traverse an edge. Vertical and horizontal clearances and maximum safe weight are useful descriptors that can be the input to an evaluator that determines the value of a restriction.

A restriction determines whether travel is possible on a network element. Restrictions may be specific to a given type of vehicle, such as to prevent the passage of large trucks on a residential street, or to eliminate all vehicular travel on a pedestrian mall used as a walking connection to link two bus stops. The value of a restriction is either Restricted or Traversable.

A functional hierarchy indicates the relative desirability for using a particular facility. Network Analyst extension supports three levels: primary, secondary, and local. In general, Network Analyst extension will show a preference to travel on primary facilities. A commonly available functional class aspect in the highway inventory can determine the hierarchy attribute for network elements.

Each type of network attribute has five basic properties: a name, a type of usage, units of measure, a data type, and a default value. The usage type specifies what kind of input the attribute serves (a cost, descriptor, restriction, or hierarchy). Units of measure for a cost may be either distance or time. Other usage types have no units. An input attribute's data type may be Boolean, short integer, long integer, float (single precision), or double precision, except that a restriction is always a Boolean data type, such as on or off, and a cost can never be a Boolean data type. The default value will be applied to an attribute when a new network analysis layer is created. You can set only one cost value for default use. Descriptors cannot have default values because they are to be supplied by the feature source.

Network attributes are typically determined through evaluators that use input parameters supplied by the source class. There are four types of evaluators available: field, field expression, constant, and Visual Basic script. The easiest to construct is the field evaluator, which essentially designates the feature class source field as the network attribute's value.

A field expression will take one or more source fields as an input to determine the network attribute's value. A typical application of this evaluator type is to convert units to a common standard, transform a feature class source field's values to one of the three hierarchy attribute choices, or estimate a cost attribute stated in time by dividing edge length in feet by vehicle speed in miles per hour converted to feet per second using the formula, time = length / (mph x 1.47). You can also generate values for restriction attributes using a field expression. If the expression produces a true result, then the element will be restricted.

A constant evaluator assigns a uniform value to the network attribute for all rows in the feature source. A constant attribute can be a numeric or Boolean data type. You may choose to supply a value of traversable to all restriction attributes until you are satisfied that the network has been properly constructed.

The most complex evaluators use a Visual Basic script, a small computer program, to accept source fields as inputs and to generate the value of a network attribute as the output. For example, if one source you supply is a bridge feature class with a maximum weight field, you can specify that field as a parameter and use a subject vehicle's weight to determine whether the bridge should be considered as a barrier to travel on the connecting edges. You do this by adding a vehicle weight attribute to the network dataset, assigning it a usage of Restriction, and specifying a Boolean data type. If the vehicle is too heavy to be safely supported by the bridge, then the script will produce a result that blocks travel by that vehicle.

To assign a parameter to determine the value of a network attribute, you identify the attribute, name the parameter, identify its data type, and specify a default value. You use the default value when the source input field is null.

Network Analyst extension will itself generate expressions when it discovers certain fields in the feature class source. For example, if you include a Oneway field, Network Analyst extension will assume it has values to indicate the single direction for which travel can occur on the related element and will generate a field expression to determine the value for a restriction attribute. As with geometric networks, the four choices for Oneway are: B (no restriction on direction of travel), which would be a reasonable default value; FT (travel may only occur in the downstream direction); TR (travel may only occur in the upstream direction); and N (no travel permitted).

You can also build custom evaluators that use parameterized attributes to dynamically decide whether to include an edge in a pathfinding solution. This option requires a descriptor and a restriction attribute. The descriptor says what attribute will participate in the analysis, such as, bridge height. The restriction attribute will be the variable input to be compared to the descriptor, such as truck height. If the restriction attribute is outside the limits of the descriptor, such as if the truck height exceeds the clearance under the bridge, then the edge will be restricted, meaning it is eliminated from the solution.

Building a network dataset

This section will present four ways to build a network dataset. Each choice represents a different way to treat overpasses, where crossing facilities do not intersect. The choice you make for your own database design depends primarily on the complexity of the resulting network and the structure of the edge feature class source you can supply. The physical data model you produce by applying the methods and concepts presented in earlier chapters should reflect the selected approach for constructing network elements.

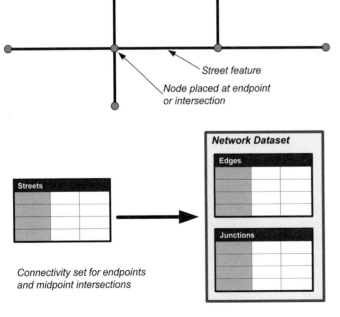

Figure 13.2 **Network dataset built from street feature source—option 1** The simplest way to build a network dataset is to use a feature class as the source and let Network Analyst extension generate the junctions at each endpoint and midpoint intersection. You still need "clean" features, as you would in a geometric network.

Building a network dataset

The easiest way to build a network dataset is to give it an edge feature class source and let Network Analyst extension construct the edge and junction elements using the connectivity setting to place junction elements at line endpoints and shared vertices of overlapping lines. Your input feature class must contain "clean" line work without gaps or overlaps.

Simple feature class: Streets								
Geometry: Polyline								
Contains M values: No								
Contains Z values: No								
Field name	Data type	Allow nulls	Default value	Domain	Precision	Scale	Length	Street features
OBJECTID	Object ID							
Shape	Geometry	Yes						
StreetID	GUID	Yes			0			Identifier of street edge
StreetName	String	Yes					32	Name of street
AltStreetName	String	Yes					32	Alternative name of street
ShieldType	String	Yes		Shields			8	Type of route
ShieldText	String	Yes					10	Route number
RoadClass	Short Integer	Yes		RoadClasses	0			Road classification
Hierarchy	Short Integer	Yes		RoadTypes	0			Street hierarchy level
OneWay	String	Yes		DirectionTypes			2	Indicates allowable direction of travel
TollRoad	Short Integer	Yes		TrueFalse	0			Indicates toll required
Boundary	String	Yes					16	Used to determine boundary crossing
AutoRestriction	Short Integer	Yes		TrueFalse	0			Indicates autos allowed
BusRestriction	Short Integer	Yes		TrueFalse	0			Indicates buses allowed
CarpoolRestriction	Short Integer	Yes		TrueFalse	0			Indicates carpool allowed
SingleAxleRestriction	Short Integer	Yes		TrueFalse	0			Indicates single-axle trucks allowed
DualAxleRestriction	Short Integer	Yes		TrueFalse	0			Indicates multi-axle trucks allowed
SemiTruckRestriction	Short Integer	Yes		TrueFalse	0			Indicates semi-trailers allowed
DeliveryRestriction	Short Integer	Yes		TrueFalse	0			Indicates deliveries allowed
TaxiRestriction	Short Integer	Yes		TrueFalse	0			Indicates taxis allowed
EMSRestriction	Short Integer	Yes		TrueFalse	0			Indicates emergency vehicles allowed
PavedRestriction	Short Integer	Yes		TrueFalse	0			Indicates street is paved
FromElevation	Short Integer	Yes			0			Connectivity level of edge start
ToElevation	Short Integer	Yes			0			Connectivity level of edge end
FromToCost	Double	Yes			0	0		Cost of downstream travel
ToFromCost	Double	Yes			0	0		Cost of upstream travel
CostUnit	Short Integer	Yes		CostUnits	0			Unit of measure for costs
LeftTurnCost	Double	Yes			0	0		Left-turn maneuver cost at end
ThruCost	Double	Yes			0	0		Through cost at end
RightTurnCost	Double	Yes			0	0		Right-turn maneuver cost at end
EdgeLength	Double	Yes			0	0		Length of edge
LengthUnits	Double	Yes		CostUnits	0	0		Unit of measure for length variable
Shape_Length	Double	Yes			0	0		

Figure 13.3 **Street edge class** The primary edge feature source in most transportation networks is for streets. This sample edge feature source contains most of the attributes used by commercial dataset vendors, including information required to control connectivity at overpasses.

This is what such an edge feature class source should look like to provide parameters necessary to generate the full range of network attributes. It is an expanded version of the StreetSegment feature class presented in chapter 4. One difference in the way the features are constructed is there is no need to break features at intersections. While you can certainly use that approach, Network Analyst extension can do the work for you to create edge elements with that structure if there are coincident vertices at all street intersections.

271

Chapter 13: The revised UNETRANS network data model

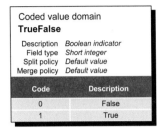
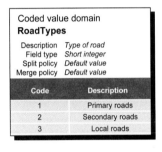
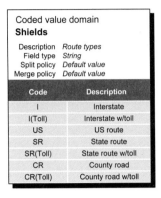
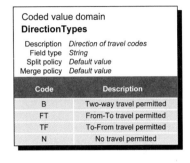
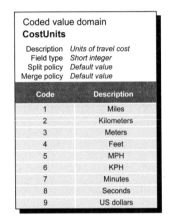

Figure 13.4 **Street edge domain classes** These domain classes are related to the Streets feature class used as the source for an edge feature class in a network dataset. These domains are compatible with those expected by ArcGIS Network Analyst.

Example domain classes will help express the way the Streets feature class source is constructed. Boolean fields are identified by their use of the TrueFalse domain class. Units of measure for travel cost attributes are shown in the CostUnits domain, which includes units for both time and distance. The RoadTypes domain class indicates the functional hierarchy for travel preferences. The DirectionTypes domain implements the travel direction choices recognized by ArcGIS Network Analyst extension. The Shields coded-value domain offers a way to enhance driving directions by providing additional labeling information. (The enhancement of driving directions will be addressed later in this chapter.)

 Although ArcGIS Network Analyst extension infers the junction elements where linear facilities cross, it does so by relying on a common vertex at the location. You will need to add that vertex where you want an intersection to be inferred and a junction added to the network. One way to add the vertices you need is by using the Integrate Tool available in the Arc Toolbox under Data Management > Feature Class.

Turns

The simple network that can be derived from the Streets feature class source presented above includes junctions that allow any direction of travel from a junction, subject to limitations imposed by one-way streets. The four movements possible at a junction of edges are straight, right turn, left turn, and U-turn. A U-turn is also permitted at a terminal junction where no other edges meet, but a U-turn will not be permitted on a one-way edge. All of these types are called two-edge turns, because they connect one edge to another coincident edge at a junction.

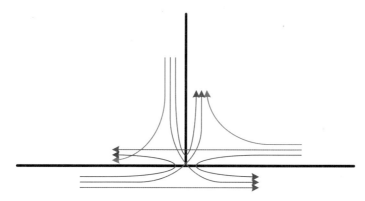

Figure 13.5 **Recognized turns**

All junction features begin by assuming global turn permissions, meaning that all possible ways to move between two edges are permitted. You can, however, affect the cost of global turns by applying a global turn evaluator. A common example is to impose a much higher cost for left-turn and U-turn movements and a slightly higher cost for right turns than for straight movements. Network Analyst extension will be able to determine turn type by looking at the angle described by the movement. Your evaluator script will define the angle range represented by each type of turn. For instance, you may decide that a left turn is any angle between 200 and 300 degrees, a right turn is any angle between 60 and 150 degrees, a U-turn is any angle of 150 to 200 degrees, and all remaining angles are straight movements.

A turn feature class can be defined to restrict certain movements through a junction beyond merely prohibiting wrong-way movements on a one-way street. A turn feature class is also necessary to properly describe any multiedge turns, which involve multiple maneuvers through a complex intersection to complete a turn from an approach edge to a departure edge. A multiedge turn requires movement along interior edges.

Common types of multiedge turns are those that move a vehicle through an interchange along ramps, support movements along divided roads with an edge for each direction of travel, or trace the path of a pedestrian walking between commuter trains in a metro station. All the elements involved in the turn must be on the same elevation and hierarchy level. In figure 13.6, the internal edge must be part of movements through the intersection by the undivided road, a left turn from the divided road onto the undivided road, and a U-turn on the divided road.

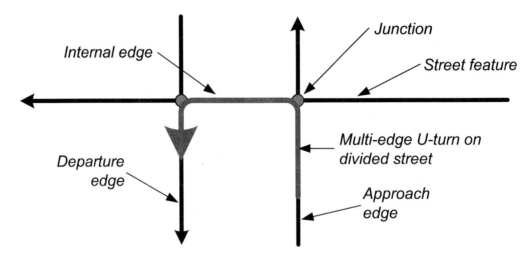

Figure 13.6 **Multiedge turns** If your network involves divided streets and other situations where a person or vehicle must traverse multiple edges to execute a turn, then you will need to define turn feature class entries for these situations. Edges involved in a turn need not be in the same feature class but they must be in the same connectivity group. For example, if walking connections and transit segments are members of the same connectivity group, then the U-turn shown here could start on a transit edge, use a walking connection for the internal edge, and exit on another transit edge, all within a single stop group.

You must properly define turns that differ from the normal expectation of allowing movements in all directions and for all multiedge turns. There are a few rules for defining turns. Any turn except a U-turn must involve at least two different edges. The approach or departure edge of any turn cannot be the interior edge of a multiedge turn. A multiedge turn must be defined as a chain of edges, with each edge connected to only one other edge in a nonbranching sequence. Turns are directional. Each possible movement through an intersection must be explicitly described if any one possible movement is defined. Interior edges can be used by multiple turns defined through a complex intersection. No two turns can have the same approach and departure edges.

Turns are defined in a turn feature class, which is a special version of a plotline feature class that ArcGIS creates. Turn features specify how movements can be made at junctions. A turn feature class has meaning only in a network dataset. Both the shapefile and geodatabase version of the turn feature class are available. The turn feature class is not classified within a connectivity group and cannot include elevation fields. You can, however, use field evaluators in a turn feature class as in other feature class sources. A turn feature must include at least one approach edge and one departure edge, which can be the same for a U-turn. A turn feature class can include up to 18 interior edge fields in a denormalized structure (Edge 1, Edge 2, Edge 3, and so on). The default number of edge fields included when you create a turn feature class is five.

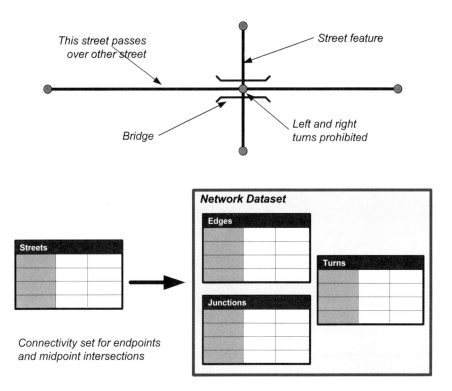

Figure 13.7 **Network dataset built from street feature source—option 2** Letting Network Analyst extension generate the junctions at each endpoint and midpoint intersection results in junctions at bridge overpasses and other places you may not want them. One way to fix this is to add a turn feature class that contains restrictions for movement through junctions. Such restrictions can include prohibiting movements along "turns" that would connect roads on different levels of an overpass, or to preclude movements at legitimate intersections that may be prohibited by regulation.

One way to use a turn feature class is to define permitted movements at a bridge when connectivity is established at line endpoints and shared vertices. You would prohibit right turns, left turns, and U-turns by establishing a turn feature at the junction in the center of the bridge that permitted only straight movements for the approach and departure edges. This option would be useful when the edge feature source does not include elevation data.

Chapter 13: The revised UNETRANS network data model

Turn feature class **Turn**				Geometry Contains M values Contains Z values		*Turn* *No* *No*			*Turn restrictions*
Field name	Data type	Allow nulls	Default value	Domain	Pre-cision	Scale	Length		
OBJECTID	Object ID								
Shape	Geometry	Yes							
TurnID	GUID	Yes			0				Identifier of turn
Edge1End	String	Yes		YesNo	0		1		Turn movement along Edge 1
Edge1FeatureClassID	ClassID	Yes			0				Identifier of Edge 1's feature class
Edge1FeatureID	GUID	Yes			0				Identifier of Edge 1
Edge1Position	Float	Yes			0	0			Position of Edge 1 along feature
Edge2FeatureClassID	ClassID	Yes			0				Identifier of Edge 2's feature class
Edge2FeatureID	GUID	Yes			0				Identifier of Edge 2
Edge2Position	Float	Yes			0	0			Position of Edge 2 along feature
Edge3FeatureClassID	ClassID	Yes			0				Identifier of Edge 3's feature class
Edge3FeatureID	GUID	Yes			0				Identifier of Edge 3
Edge3Position	Float	Yes			0	0			Position of Edge 3 along feature
Edge4FeatureClassID	ClassID	Yes			0				Identifier of Edge 4's feature class
Edge4FeatureID	GUID	Yes			0				Identifier of Edge 4
Edge4Position	Float	Yes			0	0			Position of Edge 4 along feature
Edge5FeatureClassID	ClassID	Yes			0				Identifier of Edge 5's feature class
Edge5FeatureID	GUID	Yes			0				Identifier of Edge 5
Edge5Position	Float	Yes			0	0			Position of Edge 5 along feature
AlternateID1	GUID	Yes			0				StreetID for Edge 1
AlternateID2	GUID	Yes			0				StreetID for Edge 2
AlternateID3	GUID	Yes			0				StreetID for Edge 3
AlternateID4	GUID	Yes			0				StreetID for Edge 4
AlternateID5	GUID	Yes			0				StreetID for Edge 5
CarRestriction	String	Yes		TrueFalse			1		Indicates car movement is restricted
TruckRestriction	String	Yes		TrueFalse			1		Indicates truck movement is restricted
BusRestriction	String	Yes		TrueFalse			1		Indicates bus movement is restricted
WalkRestriction	String	Yes		TrueFalse			1		Indicates walking movement is restricted
AMPeakRestriction	String	Yes		TrueFalse			1		Indicates restriction in AM peak
PMPeakRestriction	String	Yes		TrueFalse			1		Indicates restriction in PM peak
WeekdayRestriction	String	Yes		TrueFalse			1		Indicates restriction on weekdays (M-F)
Shape_Length	Double	Yes			0	0			

Figure 13.8 **Turn class** Network Analyst extension's Turn class employs a denormalized design that provides a set of fields for each edge involved in a turn, up to a predefined maximum number of edges; the default is five. This accommodates one approach edge (Edge 1), up to three internal edges and one departure edge. This example turn class is designed to work with the Streets feature class shown earlier. Edge feature identifiers are foreign keys to **StreetID**. If street edge connectivity occurs only at feature endpoints, then the edge position values should all be set to 0.5.

The fields in figure 13.8 shown with a dark gray background illustrate the default structure for a turn feature class. User-added fields that are also part of the example include pointers to the street identifiers for each edge and various restrictions that may be applicable to the turn feature. The restrictions apply to the entire turn, such that if the restriction applies to any interior edge or junction, then it applies the entire turn movement.

There will be one Turn feature class row for each movement. Thus, a T-intersection that restricts left turns in the afternoon peak will define a maximum of nine movements[1], one of which will carry the PMPeakRestriction field value of True to be used by a restriction evaluator to block the movement at the appropriate time of day.

You may also include fields in a turn feature class that help create costs for various movements through the intersection. For example, you may want to show that a left turn during the afternoon peak will have a higher cost than one made at other times of the day, or that heavy trucks will be delayed more than passenger vehicles. If your dataset includes information about traffic control, then you can use that data to establish the cost for movements controlled by a stop sign differently than those regulated by a traffic signal. A higher cost may be inferred for a traffic signal controlled intersection when right turn on red is prohibited due to pedestrian traffic volumes by time of day or day of week. You may also want to prohibit U-turns at signal-controlled intersections.

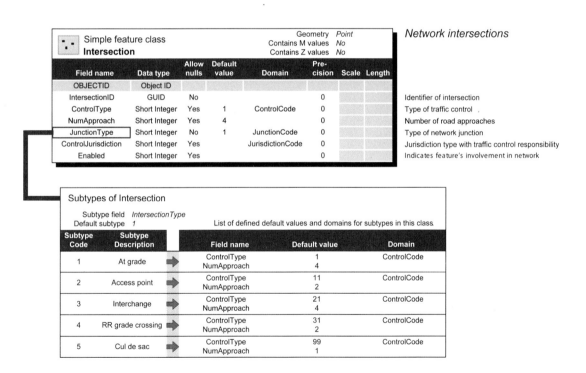

Figure 13.9 **Intersection subtypes** Network Analyst will create a network junction class using the coincidence and/or intersection of the edge feature sources. Having explicit junction class sources, though, provides more flexibility by defining the points where connections can be made in a multimodal network. They can also be used to generate restriction attributes; for example, a stop sign controlled at-grade intersection might generate a different turn class cost than one controlled by a traffic signal.

One good way to provide information about possible turn restrictions and cost factors is to include a multisubtype Intersection feature class. This class was presented in chapter 5 as a strategy to use for geometric networks. It remains a useful approach for transportation networks built with Network Analyst extension.

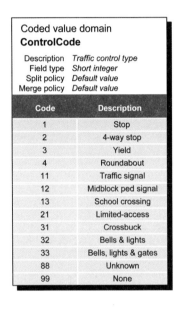

Figure 13.10 **Street intersection domains** Three sample intersection domains are defined here. JunctionCode reflects the value of Intersection class subtypes. ControlCode and JurisdictionCode are repeats from an earlier geometric network example and show the type of traffic control used at an intersection and who is responsible for maintaining that traffic control mechanism.

One change made to the class to improve its usefulness in building a transportation network is to include a cul-de-sac subtype. This option can be used at terminal junctions to infer a cost for a U-turn. A terminus with a large cul-de-sac that allows a vehicle to easily execute a U-turn can produce a lower cost than a dead-end street with no turn-around provision. A field to store the radius of the cul-de-sac could be used by an evaluator to apply different costs by vehicle type, as a small cul-de-sac that works fine for passenger cars may be inadequate for school buses.

Alternative methods to accommodate bridges

There are three common ways to preclude the need for turn features at bridges. The first is to avoid shared vertices at overlapping lines within a bridge structure's extent. This option is rarely used due the frequent need to include vertices that supply a more realistic appearance to the features sources.

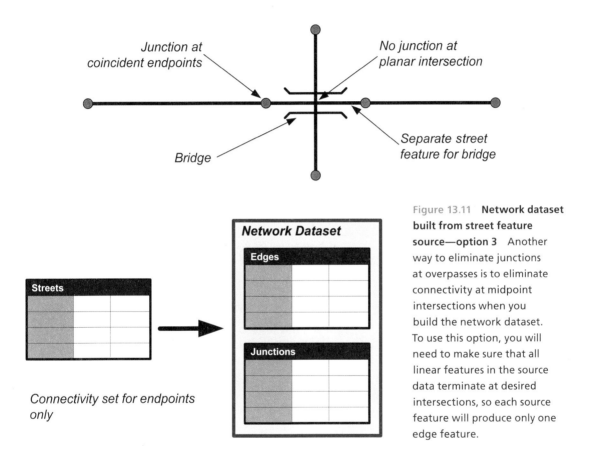

Figure 13.11 **Network dataset built from street feature source—option 3** Another way to eliminate junctions at overpasses is to eliminate connectivity at midpoint intersections when you build the network dataset. To use this option, you will need to make sure that all linear features in the source data terminate at desired intersections, so each source feature will produce only one edge feature.

It is generally preferable to simply turn off shared vertex connectivity and rely only on endpoint connectivity. To facilitate all users getting the same result, regardless of their connectivity setting, you should create a separate bridge segment for the path over the bridge. Endpoint connectivity will ensure that the segment over the bridge is part of the appropriate path and that no junction is established within its spatial extent.

Chapter 13: The revised UNETRANS network data model

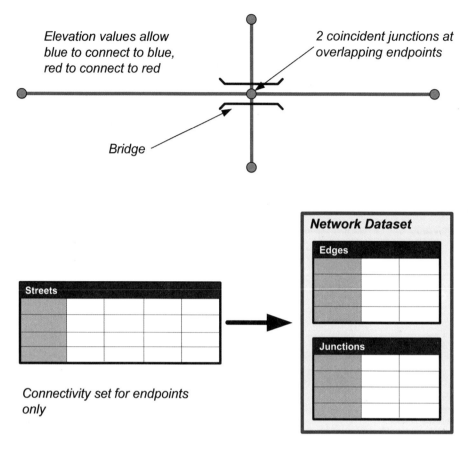

Figure 13.12 **Network dataset built from street feature source—option 4** Yet another way to avoid intersections at overpasses is to provide an elevation attribute for each end of a linear feature **FromElevation** and **ToElevation**. As used here, elevation does not mean height above sea level but height within a vertical structure of overlapping facilities. You could assign, say, an elevation of 0 to the bottom features, an elevation of 1 to the next level, and so on. When the network dataset is constructed, Network Analyst extension will use the endpoint elevations to decide which edges can connect at a given junction, which is defined by coincidence of the endpoints. For this to work properly, you will need to maintain the ratio of one source feature for each resulting edge.

The other way to deal with bridges is to essentially create two overlapping junctions through the use of Elevation field values. Only edges with the same elevation will be involved in turns at such a location.

Bells and whistles

ArcGIS Network Analyst extension includes advanced functionality for generating driving directions. These functions can enhance the directions produced by a pathfinding application so as to provide a more natural and complete report. All such enhancements require you to supply input information.

Supplying a facility name is almost a requirement if driving directions are to be produced by Network Analyst extension. Beyond this, one of the easier enhancements is to supply a segment length attribute that can be used to describe how far someone will travel down an edge before reaching the next decision point. You could also supply a file containing time quantities to determine a travel cost, as many pathfinding applications seek to minimize travel time, distance, or both. As noted above, a road class attribute can be used to set a hierarchy attribute and give a preference to travel on roads supplying more mobility, as most travelers will want. Such an attribute could also be used to identify the most scenic route, if you assign one of the three hierarchical levels to that kind of road.

In addition to a street name, Network Analyst extension can also use route numbers and route type designations. Thus, you can supply a traveler with both a local street name and a numbered route designation when both are available, for example, "Turn left onto Main St. (U.S. 27)." The feature source can contain the shield and route number in one field ("U.S. 27") or two fields ("U.S." and "27").

A signpost point feature class can provide information a motorist would typically see on guide signs at identified points in the transport system. Both shapefile and geodatabase versions of a network dataset can include signpost features. A signpost feature class can accommodate up to 10 branching directions and destinations. Destinations need not be assigned to separate branches. For example, you may want to supply two destinations for a branch that extends a long distance. One destination would be nearby; the other destination would be farther away. The example shown in figure 13.13 supports two branches and four destinations.

A signpost streets table supplies the required connection between signpost features and movement through a turn they support. This table resolves the many-to-many relationship between signposts and the turning maneuvers they may guide. A signpost location can be at any point along a designated edge. A value of 0.0 specifies the start of an edge, while a value 1.0 indicates the end of an edge. Most signposts will be near the start or the end of an edge. A single signpost feature class and signpost streets table combination may support any number of edge feature class sources.

Chapter 13: The revised UNETRANS network data model

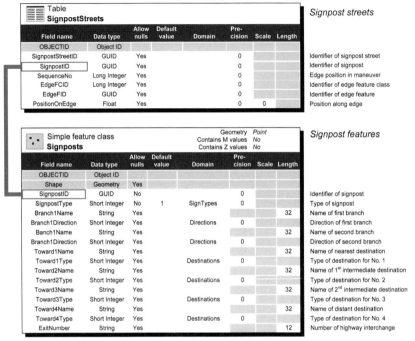

Figure 13.13 **Signposts**

Network Analyst extension also allows a boundary field to indicate when an important boundary is reached in a traversal through the network. If you have, say, a city field, you can designate it as a boundary field so that driving directions will indicate something like, "Entering Anytown" when the path reaches the first segment within the Anytown corporate limits. Network Analyst extension is capable of recognizing changes in a boundary field, such as when the path passes directly from one city into another ("Leaving Anytown. Entering Nextville").

The transit data model

With ArcGIS Network Analyst extension described, now it can be used to present a basic transit database design for the revised UNETRANS data model. A more complete model is presented in chapter 15.

The transit data model

Figure 13.14 **Transit segments** You can construct transit segments as the ordered sequence of transit stops and the paths between them. Transit segments may be constructed as traversals through the street or rail system, but it may be simpler to simply trace the path rather than deal with the overhead of maintaining linear events to describe the portion contributed by each roadway.

A transit segment is a path between two stops, one stop and a time point, or two time points. A transit segment can also be viewed as a sequence of one or more transport facility segments that the transit vehicle travels between time points and/or stops. Normally, a given transit segment will include part of more than one street or railroad track, thus requiring turns to travel the complete distance. In other words, a transit segment usually consists of both edges and junctions in a network dataset.

Another consideration is that most stops and time points are located at midblock locations, which means that a transit segment cannot be formed exclusively of whole street or rail segments that extend from intersection to intersection. Building a transit segment from segments of a street or rail system likely requires some way to subdivide the facility segments at stops and time points.

You must construct transit segments for use in Network Analyst extension before loading the feature class sources. This section will describe end-result specifications for this process and how it can be used in ArcGIS Network Analyst extension.

The logical data model for transit networks includes relatively few classes. A pattern is an ordered set of transit segments. A route is a collection of patterns. (A route class is not shown here because it is part of the Inventory Package.) A transit connection is the beginning or end of each transit segment. A stop group is a collection of transit connections. TransitSegment, StopGroup, and Pattern may be implemented as feature classes. If you do so, you will likely use a polygon for stop groups, so as to define the spatial extent of all included stops, and a polyline for patterns and transit segments.

This model assumes that a transit segment begins and ends at a transit connection, which can be a location where passengers can get on or off a vehicle, or a pseudoconnection, like a transit garage. This model also assumes that passengers may not enter or exit a transit vehicle except at connection points. If you want a model that includes more transit segment termini, it can be developed using the concepts presented here.

Four TransitConnection and five TransitSegment subtypes are included. Subtypes were used to support multimodal movements. Each subtype can be assigned to separate connectivity groups. For example, you could include bus stops in the connectivity groups for bus

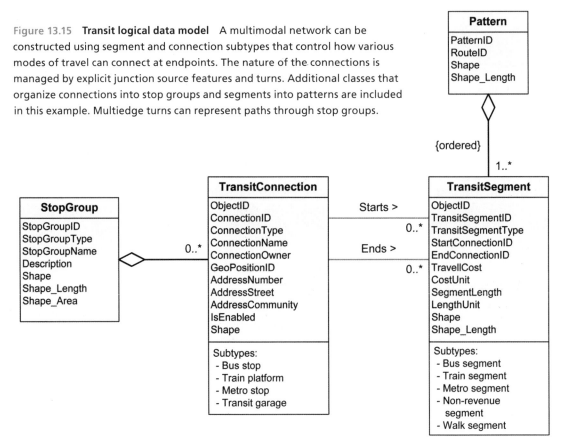

Figure 13.15 **Transit logical data model** A multimodal network can be constructed using segment and connection subtypes that control how various modes of travel can connect at endpoints. The nature of the connections is managed by explicit junction source features and turns. Additional classes that organize connections into stop groups and segments into patterns are included in this example. Multiedge turns can represent paths through stop groups.

segments and walk segments to allow passengers to identify a path to the closest bus stop, or to get from one stop to another to make bus line connections at nearby stops. Train platforms and metro stops, both subtypes of TransitConnection, could be included in other connectivity groups in which walk connection segments could participate to provide directions on how to get from a bus stop to a train platform or a metro stop.

Note that direction of travel is an implicit attribute of TransitSegment. The ordering of transit segments by the pattern of which they are a part will establish a direction of travel for a specific set of origins and destinations, which will also be reflected in the starting and ending connections.

An example transit segment polyline feature class is illustrated in figure 13.16, along with a segment type domain class. Fields are included in the feature class to supply evaluator inputs to processes that determine network attributes. Segment length for transit networks is often stated in time, but if the segment is derived from a street or rail network, that information may have to be inferred from a physical length value. Both cost-attribute inputs are accommodated by this design.

The transit data model

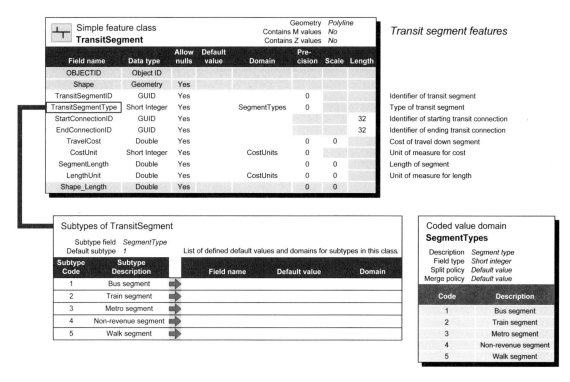

Figure 13.16 **Transit edge feature classes** The example multimodal network dataset includes transit edges derived from transit segments, which begin and end at transit connections. Some jurisdictions use time points to determine the extent of transit segments. The only change required if your jurisdiction follows that practice is to substitute time points for connections and call the terminal fields by names such as **"StartTimePointID"** and **"EndTimePointID."** If your jurisdiction uses both types of segment termini, you will need to add terminal type fields for the starting and ending points.

The generic transit connection point feature class includes geocoding information to put a connection at a physical address, but it does not include any cost-attribute evaluator inputs. This omission is because cost is different on each pattern that includes a connection point. Connection cost is determined by wait time, which is dependent on pattern headways that usually vary by time of day. You should derive a connection cost for each pattern and headway combination that uses a connection.

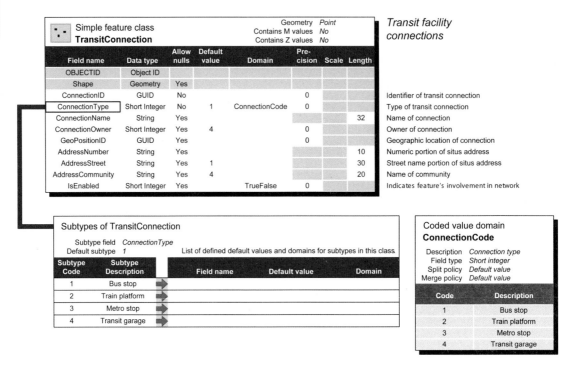

Figure 13.17 **Transit connection subtypes** Transit connections are the equivalent of intersections for the transit network. Connection subtypes can be used to regulate which segments can connect at a junction. For example, a bus segment can only connect to a bus stop or a transit garage. You set the connectivity groups when you define the network.

Notes

[1] Assume the intersection is arranged like the earlier figure illustrating the turns at a T-intersection, with the through portion aligned left to right along the bottom of the figure. The possible turns include: (1) straight left to right; (2) left turn up (the time-restricted movement); (3) straight right to left; (4) right turn up; (5) U-turn from the left; (6) U-turn from the right; (7) right turn down; (8) left turn down; (9) and U-turn from down to up.

chapter fourteen

State DOT highway inventory: Editing

- Refining the conceptual data model
- Route segmentation
- Dealing with route realignment
- Handling divided roads
- Route type issues
- Events for everyone
- Complex elements
- Centerlines
- A special calibration problem
- Migrating to the new structure
- Conclusion

Chapter 14: State DOT highway inventory: Editing

The revised UNETRANS transportation industry data model offers a solid foundation for building a functioning geodatabase to support highway inventory applications at a state DOT. But a foundation is not a complete building. This chapter shows how to apply the principles of the revised UNETRANS data model and other design elements from earlier chapters to construct a complete highway inventory geodatabase suitable for a state DOT using linear referencing to locate points on the highway system. Since all state DOTs already have a highway inventory database, the model must address migrating to the new design.

Chapters 14 and 15 demonstrate how a state DOT with a legacy linear referencing method (LRM) can migrate to the revised UNETRANS data model while preserving the business rules of the agency. Although based on the experiences of actual state agencies, the scenarios in these two chapters exercise some literary license in order to combine the user requirements and business rules of several states.

This chapter presents the editing geodatabase data model with recommendations for how it can be effectively used, showing that one model can be implemented in different ways depending on the agency's business rules. Chapter 15 demonstrates how to publish the geodatabase for analytic applications and the legacy systems a state highway inventory typically supports.

This edit-publish duality, introduced in chapter 6, allows the work units responsible for data maintenance to migrate to an efficient geodatabase structure while preserving the data structures required for legacy applications, especially those maintained outside ArcGIS. The editing application is worthy of its own data structure, one that reduces update workloads. Applying the principles and concepts of prior chapters, this chapter puts it all together as a comprehensive guide for building a state highway inventory using ArcGIS.

Refining the conceptual data model

Chapter 12 offered a general conceptual data model for a state DOT facility inventory. That model is centered on linear referencing positions as a means of connecting facilities to the elements from which they are physically composed, their aspects, measurement datums, and geometric representations. The general model can be used to build any type of linear facility data structure. This chapter refines the general model to create a highway inventory geodatabase, as shown in figure 14.1.

Refining the conceptual data model

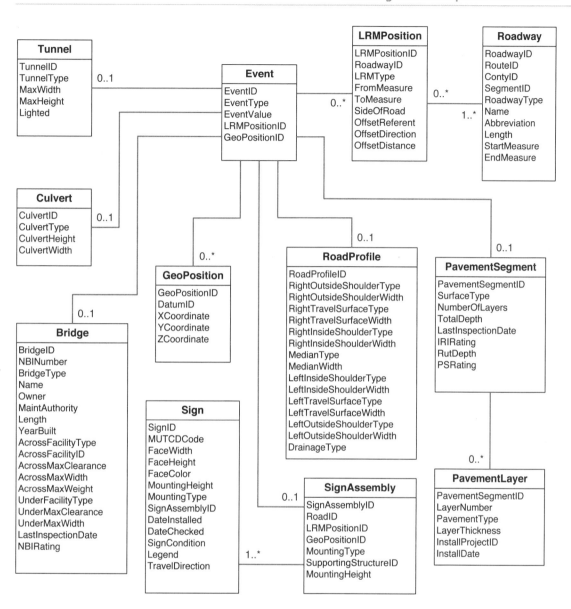

Figure 14.1 **Highway inventory conceptual data model** The conceptual data model for a highway inventory geodatabase is based on the revised UNETRANS model presented in chapter 12. The central class is Event, which represents every aspect of a highway, whether it is an aspect, element, or occurrence. There are really only four classes in the event model: Roadway, Event, LRMPosition, and GeoPosition. All the other classes represent elements, which are located on a roadway using Event as an associative entity to connect the elements to their various geographic and LRM position descriptions. There is one Event record for each aspect, element, or occurrence. Each event can have any number of geographic and LRM position descriptions. Geographic positions are defined by datum. LRM positions are defined by LRM type.

The highway inventory conceptual data model adopts the "everything is an event" view described in chapter 8 and commonly implemented in state highway agencies. Elements, attributes, and things that happen are all viewed as events occurring on the linear facility at a point (point event) or along a segment of the facility (linear event). The example elements included in figure 14.1 are Tunnel, Culvert, Bridge, Sign Assembly, and Pavement Segment. Sign and Pavement Layer are element components. The nonelement events are all embodied in the Event class except for Road Profile, which is a multiaspect table describing a cross-section of the roadway at a given point.

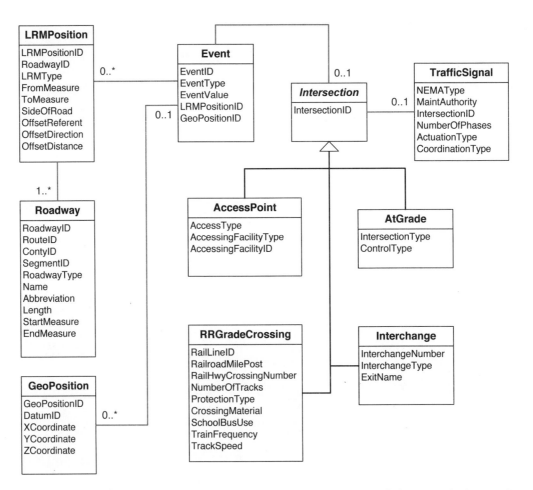

Figure 14.2 **Conceptual model extension for intersections** A special type of element is the intersection. Chapter 12 presented an overview of this model in a more general context. The abstract *Intersection* class has four subtypes. An intersection is related to one event, but is expected to have many position descriptions, at least one for each intersecting roadway. Some intersections will be related to a traffic signal instance.

Figure 14.2 supplements the general model of figure 14.1 by showing the additional classes and relationships required to support intersections. This part of the inventory data model is taken from chapter 12. It shows that an event may be an intersection, of which there are four subtypes described by instantiable classes. At this point in the modeling process, it does not matter whether *Intersection* is implemented as a feature class with geometry or just a table. The most important design aspect represented by the *Intersection* entity may be the fact that it exists at all. Many state highway agencies only implicitly include intersections in their inventory databases, like those used by highway safety. An explicit *Intersection* entity gives you the foreign key you need to tie crashes, traffic counts, and other supplemental information to the highway inventory. Figure 14.2 includes a Traffic Signal entity to accommodate inventory data to manage these devices.

Route segmentation

A linear system of highways has no boundaries. Some way of creating discrete entities—highway segments—and establishing a linear referencing system is required. Our hypothetical state DOT has traditionally used a route-based LRM with a resolution of 0.001 mile. Routes extend from state line to state line and are designated using a four-character name: three characters (right justified with leading zeroes) for the route number and one character for a route suffix, such as 075A for SR 75 Alternate. If the route suffix is not required, then an extra leading zero is inserted. Segments are defined for the extent of each route within a county. In addition to the route name, you need a field to store a route abbreviation for labeling, a route type field to distinguish between the various kinds of highways in the geodatabase (e.g., state highway, county road, city street, etc.), and a route length stated in miles.

An RCLink field is typically included to provide a public key for connecting roadway characteristic events to the routes on which they related. RCLink is a character string consisting of the four characters of the route name, a three-digit FIPS code for the county, and a three-digit sequence number for the segment within that county. The three components are separated by a hyphen. A typical RCLink value would be "0027-046-003."

As noted in chapter 3, an intelligent key is great for human use but is not a good candidate for a primary key, which must be unique and not subject to the potential for human error in entry. This means you cannot reliably use RCLink as a unique identifier. You will instead use a combination of RouteID, CountyID, and SegmentID to define each roadway segment, which you will combine into an RCLink public key field. Since you cannot use a complex key in ArcGIS, RoadwayID is a simpler computer-generated, one-part key.

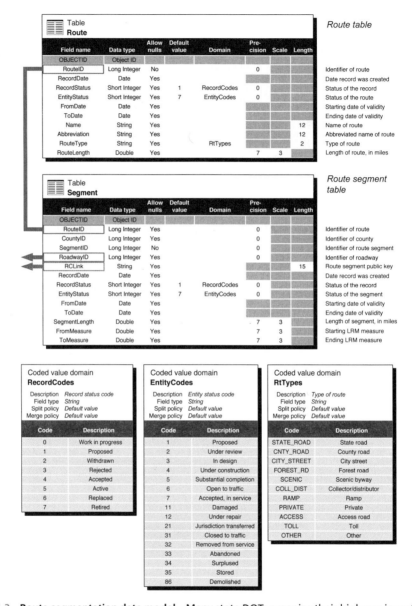

Figure 14.3 **Route segmentation data model** Many state DOTs organize their highway inventory around the concept of named routes, which are composed of multiple segments. Each segment may initially represent the extent of a highway within a given county or DOT district, but realignments and other changes result in the proliferation of segments. The included Segment class has a complex candidate primary key (**RouteID, CountyID,** and **SegmentID**), a computer-generated simple candidate primary key (**RoadwayID**), and an **RCLink** public key that is constructed from the complex primary key as a string field. Of these, **RoadwayID** is used as a foreign key for relating the Segment class to classes storing events and geometric representations. Both tables are shown with the fields that support record-level metadata. Useful field domains are also included.

Figure 14.3 shows the two tables and three domains required to implement the conceptual design using the standard class templates presented in chapter 6 for edit management and history recovery. The implicit association cardinality between Route and Segment is one to many; which means zero, one, or more segments may describe one route. The zero multiplicity could reflect either a proposed roadway or one for which segments have not yet been defined. Most initial geodatabases will have a one-to-many cardinality, with each route being composed of at least one segment for each portion of the route within a county. Some short routes with an extent contained within a single county will likely only need a single segment.

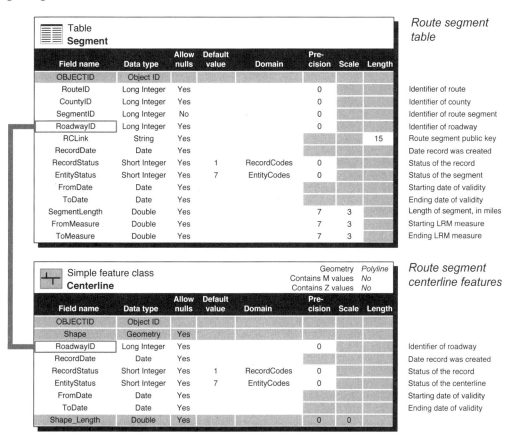

Route segment table

Route segment centerline features

Figure 14.4 **Segment centerlines** **RoadwayID** is used as a foreign key to link each segment to its geometric representation as a centerline feature. This implementation of the general conceptual model includes explicit from- and to-measure values for a single LRM type rather than an **LRMPositionID** field. Embedding measure values in the Segment class does not preclude the use of multiple LRM types. You simply need to include **LRMPositionID** in the Event class. The Centerline feature class has no measure data fields nor is it M-aware. That information is not supplied until the geodatabase is published. (See chapter 15.)

Of course, one of the primary advantages of ArcGIS is its ability to graphically depict the roadway system. The design of this part of the geodatabase is shown in figure 14.4. A centerline feature should be created for each roadway segment, which means the cardinality between Centerline and Segment is one to one. However, both centerlines and segments can evolve independently, so there can be many versions of each within the geodatabase. The one-to-one cardinality just reflects that, at any point in time, there is only one centerline feature and one segment table record with an active status for a given piece of roadway. For instance, you might need to change the official length of the segment, but not need to do anything to the geometry in Centerline. Or, you might need to update the geometry with new aerial photography and not need to make any updates to the segment record. RoadwayID is the foreign key link between the two classes.

Look closely in the class header and you will see that Centerline feature class has no M values, indicating no measures. This is because you want to be able to change them to support multiple LRM types without having to edit multiple copies of the Centerline feature class. We will revisit this topic when we address the addition of measure values to the centerline features. Chapter 8 covered the basics of this process.

Dealing with route realignment

Regardless of how many segments are required to represent the extent of a route, the LRM is established separately for each segment. In the initial state, each route segment's measures begin at 0.000 and end at the measured length of the segment. Realignments and other structural changes to the highway system will alter that situation over time. Figure 14.5, an example of such a sequence of events, illustrates three points in time for Route 27 in County 56.

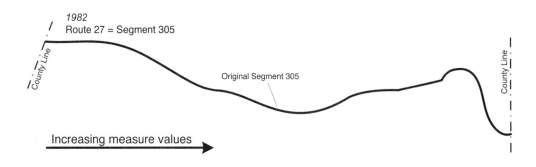

Figure 14.5 **Evolution of a route (continued on next page)**

Dealing with route realignment

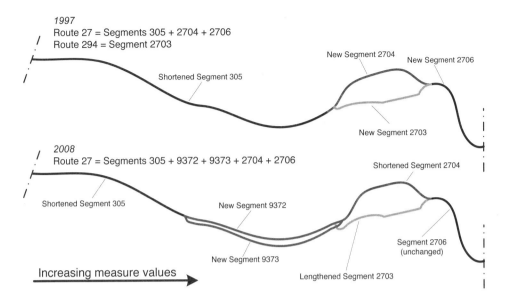

1997
Route 27 = Segments 305 + 2704 + 2706
Route 294 = Segment 2703

2008
Route 27 = Segments 305 + 9372 + 9373 + 2704 + 2706

Increasing measure values →

Figure 14.5 **Evolution of a route (continued)** This figure illustrates the typical sequence of segmentation changes over the life of a route within the highway inventory. Three points in time are shown. In 1982, there is only one route segment. In 1997, a bypass constructed around a town creates a new segment for Route 27 while part of the original segment for this route is reassigned to a new Route 294. Route 27 now consists of three segments and the total length is increased. Some years later, a divided portion of Route 27 and a reconstructed Route 27/292 changes segment lengths. Two new segments are created for the divided-highway portion of Route 27 because they are significantly different in length and do not follow parallel paths. Route 27 now consists of five segments.

In the top view, representing 1982, Segment 305 defines SR 27 in County 56. The measure values start at 0.000 and end at 17.580. Figure 14.6 shows what the records for this portion of the inventory geodatabase would look like. (The record-level metadata fields are omitted for clarity.)

Figure 14.6 **Route and Segment tables—1982** A representative set of records is provided for the initial states of the Route and Segment tables. One Route record is related to one Segment record. Route 27 is 17.580 miles long.

Route (1982)

RouteID	Name	Abbreviation	RouteType
25	SR 25	SR25	State Road
27	SR 27	SR27	State Road
29	US 29 Connector	US29Conn	US Route
37	SR 37	SR37	State Road
38	I 38	I-38	Interstate

Segment (1982)

RouteID	CountyID	SegmentID	RoadwayID	RCLink	FromMeasure	ToMeasure
25	56	297	90	0025-56-001	0.000	24.108
27	56	305	42	0027-56-001	0.000	17.580
29	56	109	274	029C-56-001	0.000	2.093
37	56	921	107	0037-56-001	0.000	42.901
38	57	836	99	0038-57-001	0.000	23.164

A new bypass is constructed around a town in 1997. The DOT's business rules say that such a realignment will be handled by creating new roadway segments. SR 27 in County 56 now consists of three segments: the western part of Segment 305, a new Segment 2704 defining the path of the bypass, and an eastern portion of Segment 305 now called Segment 2706. (The agency does not allow multipart segments.) The middle portion of the original SR 27 is now called SR 294. With the route realignment, the portion of former SR 27 (Segment 305) that now defines SR 294 must get a new segment identifier.

Route (1997)

RouteID	Name	Abbreviation	RouteType
25	SR 25	SR25	State Road
27	SR 27	SR27	State Road
29	US 29 Connector	US29Conn	US Route
37	SR 37	SR37	State Road
38	I 38	I-38	Interstate
...	...	...	...
294	SR 294	SR294	State Road

Segment (1997)

RouteID	CountyID	SegmentID	RoadwayID	RCLink	FromMeasure	ToMeasure
...	...	...	...	...	...	...
27	56	305	42	0027-56-005	0.000	12.019
27	56	2704	9744	0027-56-008	0.000	4.302
27	56	2706	9745	0027-56-009	13.605	17.580
...	...	...	...	...	...	...
294	56	2703	9746	0294-56-010	12.019	13.605

Figure 14.7 **Route and Segment tables—1997** Realignment in 1997 results in three additional Segment records required to fully describe Route 27. A portion of the original Segment 305 has been redefined as Segment 2703, which represents the path of Route 294. The original measures for Segment 305 have been retained for the remaining portion of Segment 305 and the portion that has been used to create Segment 2706. Retaining the original measures precludes reassigning measure values to events on these segments. Route 27 is now 20.296 miles long.

Figure 14.7 shows the new state of the geodatabase. The original Route table record for SR 27 is unchanged, but there are now three segment records. There also is a new Route table record for SR 294. The initial portion of Segment 305 is retained so the LRM measures still start at zero, as do the measures for the new Segment 2704, the bypass. The old SR 27 measures are "reused" on Segments 2703 (SR 294) and 2706 (SR 27). Doing so avoids the work

Dealing with route realignment

of revising all the measures downstream of the point where SR 27 starts the realignment around the bypass. Since each segment's LRM is completely detached from that of other segments, even for the same route, this labor-saving business rule really has no drawbacks, except that you have to remember that the length of SR 27 in County 56 is not equal to the highest measure value. It is now the sum of Segments 305, 2704, and 2706, or:

(12.019 - 0.000) + (4.302 - 0.000) + (17.580 - 13.605) = 20.296 miles.

Part of Segment 305 is converted to a four-lane divided highway in 2008 and the intersection of SR 27 and SR 294 is reconstructed. Since the divided portion of SR 27 has dissimilar lengths on each carriageway, the agency decided to treat them as separate segments so field measures can be determined and applied for each direction of travel. The western end of Segment 2704 is included in the divided highway, which results in a shortening of that segment, as Segments 9372 and 9373 subsume a portion of it. Segment 2703 is lengthened by the project, as the intersection of SR 27 and SR 294 is moved to the west to take advantage of the divided highway approaches. The resulting record changes are shown in figure 14.8.

Route (2008)

RouteID	Name	Abbreviation	RouteType
25	SR 25	SR25	State Road
27	SR 27	SR27	State Road
29	US 29 Connector	US29Conn	US Route
37	SR 37	SR37	State Road
38	I 38	I-38	Interstate
...	...	...	...
294	SR 294	SR294	State Road

Segment (2008)

RouteID	CountyID	SegmentID	RoadwayID	RCLink	FromMeasure	ToMeasure
...	...	...	...	...	...	...
27	56	305	42	0027-56-005	0.000	7.109
27	56	2704	9744	0027-56-008	0.196	4.302
27	56	2706	9745	0027-56-009	13.605	17.580
27	56	9372	10534	0027-56-010	0.000	5.163
27	56	9373	10535	0027-56-011	0.000	5.204
...	...	...	...	...	...	...
294	56	2703	9746	0294-56-010	11.857	13.605

Figure 14.8 **Route and Segment tables—2008** In this snapshot of the database, the number of active-status Segment records required for Route 27 has grown to five. Upgrading a portion of the route to divided-highway status added two Segment records. Segment 2704 was shortened; it no longer begins at 0.000. Segment 2703 was lengthened, as indicated by a lower **FromMeasure** value than in figure 14.7. Route 27 is now either 20.353 or 20.394 miles long, depending on how you traverse the divided-highway portion.

Figure 14.8 shows only the records as they would exist after the changes in 2008 are applied; however, if you have chosen to implement the status fields and continuous versioning, then all the older records are also included in the tables. This means you will be able to recover the state of the system for any point in time since 1982. There are still some outstanding issues to settle, though. We address them in the next section.

Handling divided roads

One of the problems with data modeling is that some very different business rules can produce the same data structure. For example, the direction in which measure values numerically increase on the two divided highway segments is not indicated. Measures typically increase in a single direction, with phenomena recorded by side of road for divided highways where parallel, closely spaced carriageways exist. One of the reasons for defining separate directional centerlines is to be able to properly record and recover the location of elements and characteristics applicable to each direction of travel when it is difficult or impossible to see both sides, or when lengths are not equal. Just looking at the Segment table will not tell you that Segments 9372 and 9373 are directional centerlines; direction of travel is really an event.

Alternate 1

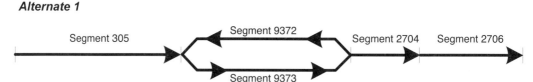

Segment (2008)

RouteID	CountyID	SegmentID	RoadwayID	RCLink	FromMeasure	ToMeasure	MLDirection
...	...	...	...	...	...	...	...
27	56	9372	10534	0027-56-010	0.000	5.163	HiLo
27	56	9373	10535	0027-56-011	0.000	5.204	LoHi
...	...	...	...	...	...	...	...

Figure 14.9 **Segment direction and ordering** This state DOT's general business rule is that the direction of increasing measures is from west to east (left to right in this drawing) or south to north. The divided highway portion of Route 27 presents a potential problem in that the direction of increasing measure values must be assigned. If both are defined as left to right, then measure values will increase in a direction opposite that in which field crews will compile data; i.e., with traffic flowing right to left. Alternate 1 shows the records of the Segment table from figure 14.8 with the measure values increasing in the direction of traffic flow.
An **MLDirection** field has been included to indicate that the direction of increasing measure values for the westbound segment (9372) is in a reverse direction (High to Low) compared to general business rule (Low to High).
(Figure 14.9 continued on next page.)

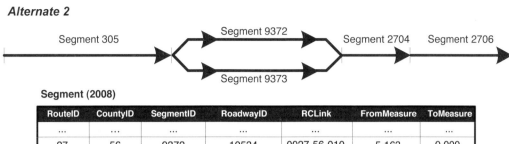

Figure 14.9 (continued) **Segment direction and ordering** Alternate 2 shows how to solve the problem by ordering the from- and to-measures. Under this alternative, field crews would continue to compile data in the order of increasing measures, but dynseg would use decreasing measure values.

Figure 14.9 shows the two possible database design alternatives to indicate something special is going on with Segments 9372 and 9737. In Alternate 1, the Segment table gets an extra field, MLDirection, which indicates whether the direction of data compilation is from low measure to high measure (with the measure direction), or from high measure to low measure (against the measure direction). Alternate 2 does the same thing by reversing the order of FromMeasure and ToMeasure. ArcGIS and the geodatabase are indifferent as to whether measures increase or decrease along a line segment. Field crews, though, are typically constrained by the directional flow of traffic.

Direction of travel is of potential interest for other applications besides inventory. Neither alternate solution to the problem of indicating the direction of data compilation provides a means of linking the various segments to provide a continuous path. As it stands, the Segment and Route tables tell us which segments form which routes, but not their geographic sequence. A simple field addition will not solve the problem. A different approach is required.

That approach is an explicit table that stores the relationship of routes and segments. Our new SegmentSequence table includes a PositionOnRoute field that allows us to state the ordering of segments across a route. PositionOnRoute is a string field so we can insert values without renumbering downstream segment positions. For example, with an alphabetic sort, you could put Segment 1g between Segments 1 and 2 and still have the opportunity to place more segments on either side of the new segment. In order to solve the problem of what to do with divided roads where separate directional centerlines are provided, you would append "L" or "R" to indicate the roughly parallel left and right carriageways. (Direction for left and right is determined by increasing measure direction.) Or you could use a Direction field and list segment sequence separately for each direction of travel, although this would double the number of records required.

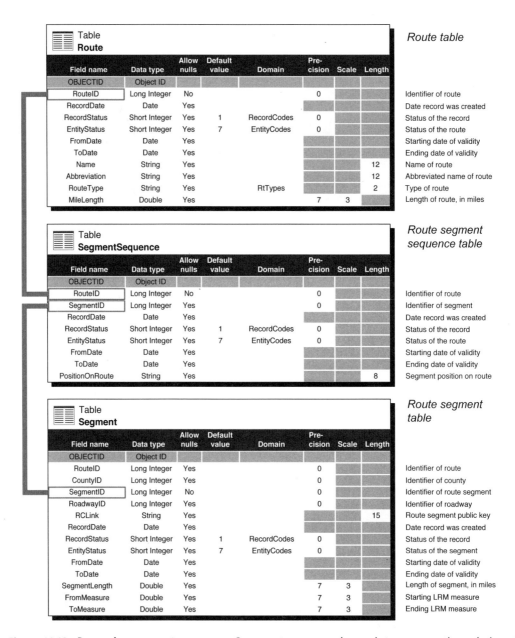

Figure 14.10 **Preserving segment sequence** One way to manage change is to preserve the ordering of segments within a route. Besides helping to identify those portions of the inventory geodatabase that may require edits after realignments and other changes at the segment level, the structure shown here would facilitate a route-level LRM separate from the one used at the segment level. (An alternative solution involving traversals is offered later in this chapter.) The **PositionOnRoute** field in the SegmentSequence table indicates the relative position of each segment along the route.

Route type issues

One common editing task we have not addressed is a change in route type. It is relatively uncommon for route type to change between state and local jurisdiction. The workload for changes between city and county jurisdiction, though, is likely to be much more common and more complicated. Given the current data structure, you would have to create a new set of relationships, if not entirely new route and segment records.

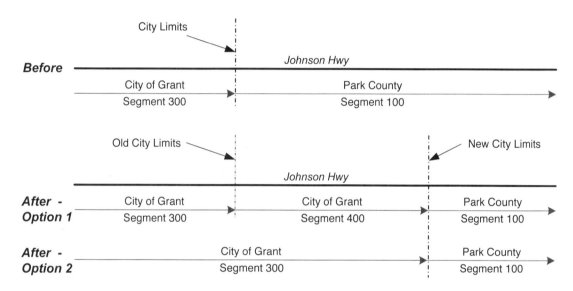

Figure 14.11 **Annexation impacts on route segmentation** The City of Grant has annexed a portion of unincorporated Park County. Since jurisdiction is being changed for a portion of Johnson Highway, it is necessary to make revisions at the segment and route levels. You can either create a new Segment 400 for the annexed portion to preserve the action in the segmentation schema, with the appropriate shortening of Park County Segment 100, or you can simply change the lengths of the existing segments. Either way, you must reassign and recalculate measures for all included events. A better alternative is to omit jurisdiction and route type from the segmentation rules and treat them as event types.

In the example of figure 14.11, the City of Grant has annexed a portion of Johnson Highway previously in unincorporated Park County. The route type for this portion of the highway must be changed from county road to city street. To accomplish this task, a new Segment 400 must be derived from the western end of Park County Segment 100. The result will be a new Segment table record for the annexed portion and a shortened Segment 100. The alternative would be to make Segment 300 longer by the same amount that Segment 100 was shortened.

The bad result of both actions is the need to reassign events previously tied to County Segment 100 to City Segment 300 or 400 because the RoadwayID and RCLink values have changed along with the route type. This is a lot of extra work.

So, how do you avoid all this work? Don't do it! Treat RouteType as an event type, not a way to define routes. Building an attribute into the way you segment road systems is a bad idea. You can use route numbers and street names to help segment the system, but do not include that segmentation rule in the business process that maintains the segmentation schema.

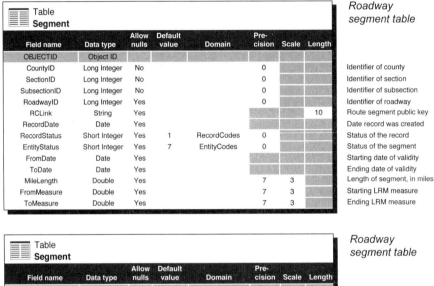

Figure 14.12 **Alternative Segment tables** There are several ways to eliminate routes as an organizing element at the edit level of the geodatabase. (Routes can still be constructed through route name linear events.) The top example takes us part of the way by restricting changes to a single county, with subsections being components of road sections to accommodate realignments. The bottom example offers a complete separation of the segmentation schema from the routes that compose the highway system. In this version, which offers the most efficient Segment table design, a segment is just an arbitrarily defined piece of the highway system. The bottom alternative also changes from the one-LRM type structure of the top example to accommodate multiple LRM types. It is perfectly fine to use numbered routes as the original method of defining segments in a highway inventory, but once that segmentation is completed, treat route number as an event type, not a segmentation rule.

The two alternative Segment table examples in figure 14.12 show a couple of ways some state DOTs are avoiding route-based difficulties. The top example implements a county-based segmentation schema with sections and subsections defining roadways and their measures. RoadwayID remains as an internal foreign key and RCLink remains a useful public key. This state has only one LRM based on miles, so it records measure ranges for each segment directly.

The bottom example in figure 14.12 completely ignores the segmentation process and says simply that its identifier and terminal LRM position statements define a roadway. This example allows multiple LRM types by employing the LRMPosition approach initially discussed in chapter 8. Only the RoadwayID field is included as a foreign key. If you can use this design, do so. Simple is better. It also helps avoid any confusion regarding the assignment of county identifiers for roads that form part of a county boundary or include short sections of road in an adjacent county.

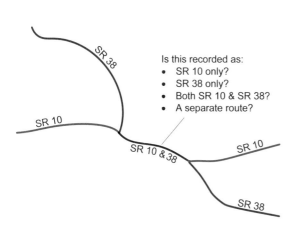

Figure 14.13 **Dealing with route overlaps** Almost every state will face the situation of multiple route numbers assigned to all or part of a given segment. In this example, SR 10 and SR 38 both occupy part of the roadway system. A common way to treat such a situation in the inventory is to include the segment's events only on to the lowest numbered route; i.e., SR 10. Some states choose to ignore the distance of the overlap for the other state roads, resulting in two positions on the segment having the same measure value for the higher numbered route(s). Other states also provide a gap in all events but increment the measure values for the overlapping route by an amount equal to the distance of the gap in that route. A few states have chosen to create an exception event for the overlapping routes that points to the route number where event data is stored.

There is another reason to avoid the route-based segmentation schemas, at least to the point that you retain the route basis as the system evolves. So far, we have talked about routes as being singular and separate and have looked only at route evolution as a motivator for business rules. Consider, though, the example shown in figure 14.13, where two routes share a single segment. This is a common occurrence.

Many states have adopted a general rule that route overlaps are assigned to the lowest numbered route. In this example, where SR 10 and SR 38 share a segment, that common segment would be part of SR 10 within the inventory geodatabase. In some states, the gap in the SR 38 record would be completely ignored, with the ending measure of the first SR 38 segment being exactly the same as the beginning measure of the last segment. Other states add the intervening SR 10 segment distance to the measures, such that the ending measure of the first segment is substantially different from the beginning measure of the last segment on SR 38.

Yet another approach is to create an exception event for SR 38 that points to the equivalent segment of SR 10 where the descriptive data can be found. That exception event is the only one recorded for SR 38 over the distance of the overlap; however, it will present a potential problem of double-counting. You will not be able to simply sum the length of all roadways to find the length of the highway system. You will need to also delete the length of all exception events.

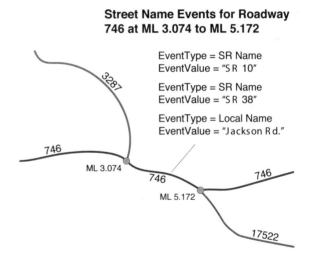

Figure 14.14 **Treating route overlaps with events** A better approach is to treat route number as a linear event. If you chose this option, then you will only have segments and use something like the bottom Segment table design from figure 14.12. You can record any number of SR Name events for the route-overlap portion of Segment 706. Organizing segments by route is denormalization and will both increase the workload of editing and create difficult situations for users as LRM positions change but the locations they describe do not.

The best approach is shown in figure 14.14, which is to ignore the route designation when creating segments and to look instead at the underlying structure. Route type, route name, and all other descriptive aspects of the roadway are treated as events. Query the geodatabase to find all events where EventType = "SR 10" and EventValue = "SR 38", and ArcGIS will highlight all the records meeting that criteria. It does not really matter that the software highlighted 10 or 15 line segments, or that they may be tied to many different RoadwayID values. Such queries can be perpetuated when you publish the geodatabase by constructing traversals for each state road. This process is described in chapter 15.

Events for everyone

Our fictional state DOT has decided to treat everything as an event, which is a fairly typical approach for legacy systems. Figure 14.15 shows what this part of the geodatabase design looks like. It includes four tables and a number of supporting domains. Events are defined by type, with location on a roadway specified by a related LRMPosition table record. A geographic position may also be provided. This design differs from the normal legacy database structure in that the LRM data is moved to the LRMPosition table rather than treated as an attribute in the Event table. This change is intended to support multiple LRM types. In this case, the other LRM type is street addressing, which the agency believes will facilitate data exchange with local governments.

The overall structure for this part of the geodatabase is essentially the one discussed in chapter 8, with the history-recovery template of chapter 6 used as a guide. Nevertheless, a little review may be useful.

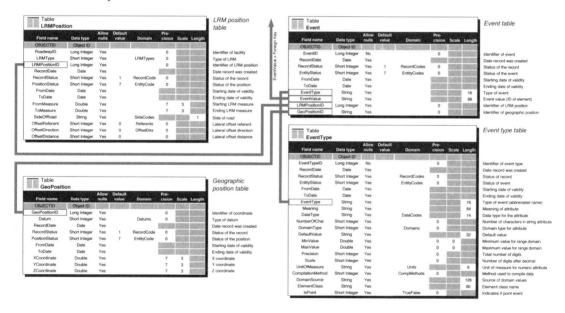

Figure 14.15 **The event logical data model** These four tables form the core of the highway inventory data model. Three of the tables store relationships. Event includes the equivalency relationship between GeoPosition and LRMPosition. LRMPosition stores the relationship between events and the roadway segments on which they exist, and between the measure and a linear datum. GeoPosition includes the relationship between a position and a geographic datum. When the event type represents an element of the roadway, **EventValue** stores the identifier of that element, which serves to construct a relationship between the element and its various LRM and geographic position descriptions. In addition to documenting the event records, EventType also provides the higher-level metadata suggested in chapter 6. To view this figure in detail, go to http://www.esri.com/industries/transport/resources/data_model.html (Figure 14.15 continued on next page.)

Chapter 14: State DOT highway inventory: Editing

Coded value domain
LRMTypes

Description *LRM types*
Field type *Short integer*
Split policy *Default value*
Merge policy *Default value*

Code	Description
1	Route milelog
2	Node offset
3	Project stationing
4	Reference marker
5	Street addressing
9	Other

Coded value domain
OffsetDirs

Description *Lateral offset direction*
Field type *Short Integer*
Split policy *Default value*
Merge policy *Default value*

Code	Description
0	None
1	In
2	Out
3	North
4	East
5	South
6	West
7	Toward
8	Away from

Coded value domain
Referents

Description *Lateral offset referent*
Field type *Single Integer*
Split policy *Default value*
Merge policy *Default value*

Code	Description
0	None
1	Edge of pavement
2	Face of curb
3	Back of curb
4	Back of sidewalk
5	Edge of right of way
6	Centerline of road
7	Centerline of ditch

Coded value domain
SideCodes

Description *Side of road*
Field type *String*
Split policy *Default value*
Merge policy *Default value*

Code	Description
L	Left side only
R	Right side only
B	Both sides & street
S	Street only
N	Both sides & not street

Coded value domain
TrueFalse

Description *Boolean logic value*
Field type *Short integer*
Split policy *Default value*
Merge policy *Default value*

Code	Description
0	False
1	True

Coded value domain
Datums

Description *Geographic datums*
Field type *Short integer*
Split policy *Default value*
Merge policy *Default value*

Code	Description
1	State Plane
2	UTM Zone 16
3	WGS 84
4	NAD 83
5	NVAD 88

Coded value domain
CompMethods

Description *Method of data collection*
Field type *Short integer*
Split policy *Default value*
Merge policy *Default value*

Code	Description
1	Field measurement
2	Office review
3	Consultant
4	Publication
5	Aerial photography
6	Remote sensing
7	SCADA
8	Run report
88	Other method
99	Unknown

Coded value domain
Units

Description *Unit of measure codes*
Field type *String*
Split policy *Default value*
Merge policy *Default value*

Code	Description
IN	Inch
FT	Foot (unspecified type)
US FT	US survey foot
INT FT	International foot
MI	Statute mile
KTS	Nautical mile
CM	Centimeter
M	Meter
KM	Kilometer
LAT	Latitude
LONG	Longitude
LB	Pound
TONS	US ton (2,000 lbs.)
KIP	18,000 lbs.
EACH	Each (count)

Coded value domain
DataCodes

Description *Data type code*
Field type *String*
Split policy *Default value*
Merge policy *Default value*

Code	Description
Short Integer	Short integer (2 bytes)
Long Integer	Long integer (4 bytes)
Single	Single-precision floating point
Double	Double-precision floating point
String	String of characters
Date	Date & Time
BLOB	Binary large object
GUID	Globally unique identifier
Raster	Raster image
Geometry	Vector geometry

Figure 14.15 **(continued)** The event logical data model

First of all, you will see no distinction in table design between point and linear events, nor between point and linear LRM positions. In practice, a point event will have only a FromMeasure value in LRMPosition. The cardinality between Event and LRMPosition is one to many, in that one event may relate to any number of LRM position descriptions, all with the same LRMPositionID value. This is a bigger statement than it may first appear, as illustrated in figure 14.16.

Figure 14.16 shows a few rows of the LRMPosition table. The four highlighted rows all have the same LRMPositionID value, 3295. At least one Event table record would contain this LRMPositionID value. That event would be an intersection of two state roads. Both state roads are also included in the geodatabase as locally named streets with addresses. The result is two LRM position descriptions for the state road milelog measures and two for the local street address measures. You should have expected this.

LRMPositionID	LRMType	RoadwayID	FromMeasure	ToMeasure
4723	StateMilelog	67	23.905	24.108
3295	StateMilelog	42	17.58	
3295	StateMilelog	96	2.093	
3295	LocalAddress	42	1400	
3295	LocalAddress	96	300	
6429	StateMilelog	64	7.664	9.633
2084	StateMilelog	65	3.902	

Figure 14.16 **LRMPosition class** The key to supporting multiple LRM types is the ability to put multiple records in the LRMPosition table for a single **LRMPositionID** value. In this example set of rows, a single event—an at-grade intersection—stores an **LRMPositionID** value of 3295. The LRMPosition table includes four rows with that value, two for the state milelog LRM type and two for street addressing, which is a second LRM type. Because it is an intersection, each of the two involved roadways defines a linear datum for stating the event's position. Thus, **LRMPositionID**, **LRMType**, and **RoadwayID** form the candidate primary key for the LRMPosition table. Sharing the **LRMPositionID** value means that all the rows are equivalent position descriptions for the same location.

What you may not have realized is that you only need one event record for this intersection. Almost every state DOT legacy system would have two event records for this one intersection—one for each involved roadway—because they embed the LRM position data in the event table. But two event records with two LRM position descriptions makes it look like there are two intersections when there is only one. Plus, that design would mean you will have to edit two event records if anything changes at the intersection. More work!

If you recall, chapter 8 showed a geodatabase design with an LRM-to-LRM equivalency table. Such a table is required if each LRMPositionID is unique; i.e., only one row with each value. The alternative business rule applied here, where there can be multiple rows with the same LRMPositionID value, but different RoadwayID or LRMType values, eliminates that need. Now, you just find all the rows with the same LRMPositionID value; they are equivalent in that they are position descriptions for the same location using different LRM types.

Look a little closer at figure 14.15 and you might notice it lacks the attributed relationship class in the chapter 8 design for tying LRM positions to geographic positions. That relationship now is part of the Event table, which lists both GeoPositionID and LRMPositionID among its attributes. This means you can go into the Event table with a GeoPositionID value and discover any equivalent LRMPositionID values through a simple SQL SELECT statement. (More on this process in the next chapter.)

The LRMPosition table does more than support the use of multiple LRM types. It also allows you to travel down a roadway and discover events. As the next chapter on publishing the inventory geodatabase demonstrates, a query into the geodatabase as to what may exist at a point along a roadway begins by finding the related LRMPosition table records and then going to the Event table to find those records that store the selected LRMPositionID value(s).

One business rule that could be adopted is to restrict the proliferation of LRMPosition table records that point to the same physical location. There is no intrinsic geodatabase requirement that says each linear measure can be contained in only one LRMPosition table record. If, say, speed limit, street name, and jurisdiction all change at a common location, such as a city limit, all those Event table records could point to the same LRMPosition table record, or they could point to separate, coincident LRMPosition table records that describe the same location. It would be easier to discover that all these events occur at the same location if they shared a single LRMPositionID value, but having multiple values will not alter the final result of dynamic segmentation (dynseg) operations and other analytical processes that are based on the measure value.

Figure 14.17 shows how events can relate to roadway elements, those physical parts of the roadway that are likely to have their own descriptive tables. An event that covers an aspect of the roadway, such as speed limit and number of lanes, can be adequately described with a single value, but elements need more information. Elements include bridges, culverts, pavement segments, and, as shown here, tunnels. When an external table provides the descriptive data for an element, EventValue stores the foreign key for that table, which is identified by ElementClass in the EventType table.

Notice that the Tunnel feature class does not have any absolute position description, only a map location that is part of its polygon geometry. The Event table supplies position descriptions for Tunnel through the Event table's foreign key references to the LRMPosition and GeoPosition tables. An acceptable alternative would be to put a geographic position description in the Tunnel feature class, either by adding GeoPositionID as an attribute or including explicit geographic coordinate fields. However, by limiting the position information to the

Events for everyone

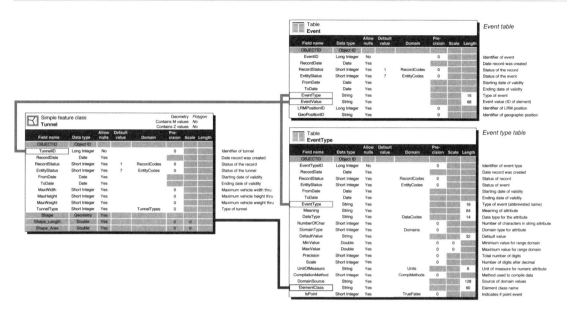

Figure 14.17 An example element event data model Events provide location references for physical components of the transport system. Unlike an aspect, which describes one characteristic of a portion of the highway system, an element can be described by many attributes. The event type of "Tunnel" tells you that the information stored in **EventValue** is a foreign key pointer to the tunnel at this location. The **ElementClass** field in the EventType table points to the class or table that contains that matching foreign key value. This structure supports automation of a process linking elements to events. As the manager of relationships, the Event table connects the element class to its various position descriptions. Conversely, an application that scanned a segment for possible restrictions to travel could follow the event-element connection to extract vehicle weight, height, and width restrictions for pathfinding. To view this figure in detail, go to http://www.esri.com/industries/transport/resources/data_model.html

Event table, you avoid having duplicative and potentially conflicting data in the editing geodatabase. You could publish the event location information as part of an expanded Tunnel feature class by finding the correct value of TunnelID in the Event table and then following the LRMPositionID trail to the appropriate LRMPosition table record. This process is described more fully in chapter 15.

There probably are some nonelement events at your agency that also need external tables, such as the traffic example shown in figure 14.18. These tables are taken from chapter 10. EventValue stores the foreign key of TrafficSectionID when EventType is "Traffic Section" (a.k.a., traffic break). Multiple traffic characteristics are stored in the AnnualEstimate table. The source of the traffic data may be traced to the count site where the data was compiled and, subsequently, to the actual counts. However, you could also treat count sites as an event type and use CountSiteID in EventValue to record the locations where traffic counts are conducted.

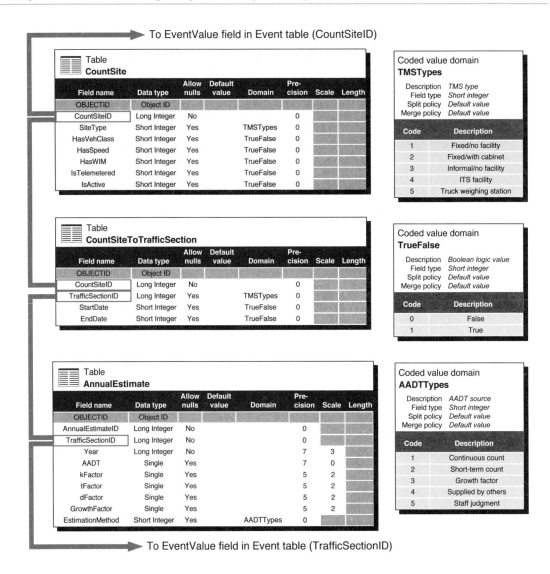

Figure 14.18 **Traffic sections look like elements** Some complex events look like elements. In this example, **EventValue** stores **TrafficSectionID**, which points to the AnnualEstimate table. This table includes one row for each traffic section and year combination. Descriptive characteristics typically provided for each traffic section include annual average daily traffic (AADT), peak hour factor (k), percent of trucks (t), and directional bias during the peak hour (d). The estimation method says how the descriptive statistics were derived, by applying a growth factor, direct observation, or some other method. The relationship of traffic sections to the counts used to generate the descriptive statistics is stored in the CountSiteToTrafficSection table. The relationship needs an attributed relationship table because a single traffic section could get data from multiple count sites, and one count site might be used for multiple traffic sections. Count sites may themselves be stored as point events. This could also be done with an attributed relationship class.

Events for everyone

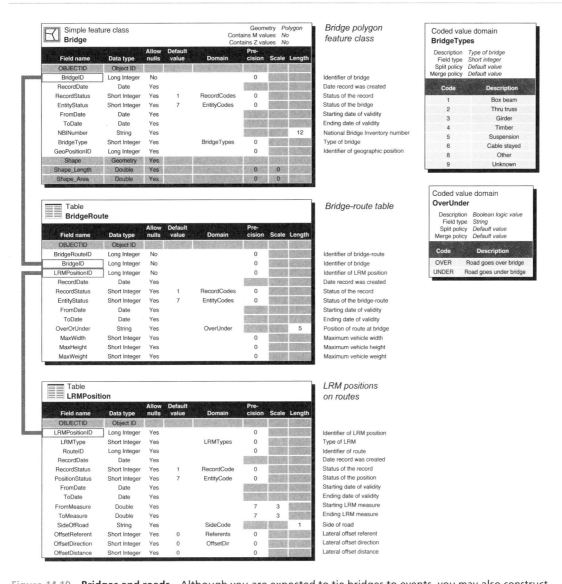

Figure 14.19 **Bridges and roads** Although you are expected to tie bridges to events, you may also construct a separate bridge-route geodatabase that forms a special standalone highway inventory for structures. In this example, the Bridge feature class contains polygons that describe the outline of each structure from an overhead view. These polygons can be overlaid on a roadway map layer to provide the proper appearance of one road passing over another. The BridgeRoute event table stores a direct connection from the Bridge feature class to the involved roadways by including an **LRMPositionID** field. The LRMPosition table includes a **RouteID** field that defines the LRM datum and connects the bridge to each route. Unlike the Tunnel feature class, which represents a restriction only for vehicles going through it, a bridge imposes different restrictions on traffic going over or under it. Thus, each roadway faces its own set of passage restrictions, all of which are stored in the associative BridgeRoute table.

Complex elements

Some elements typically found in state DOT geodatabases, such as bridges and intersections, are more difficult to accommodate. Conceptual models and design approaches were discussed generally in chapter 12. This chapter will cover implementations more suitable for a state DOT highway inventory.

The bridge element has a potential relationship with many roadways. Usually, only one roadway goes over a bridge but many could pass beneath it. You can even have bridges over bridges, as in complex interchanges. Figure 14.19 shows how a state highway inventory could be constructed to handle these various relationships using an associative table, BridgeRoute.

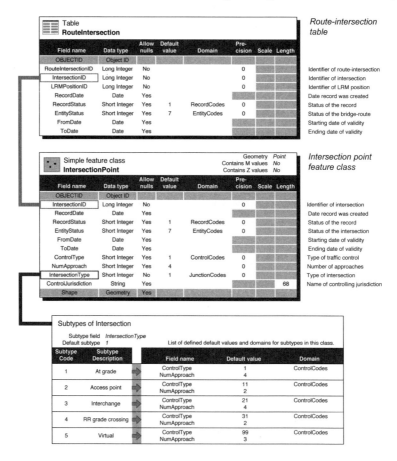

Figure 14.20 **Intersections and roads** As with bridges, intersections usually include more than one roadway. This figure repeats the Intersection feature class and its subtypes initially introduced in chapter 12 and adds a RouteIntersection table to store the many-to-many relationship between intersections and roadways. **LRMPositionID** in the RouteIntersection table points to the related roadways in the LRMPosition table.

In this example, Bridge is a feature class, but the same structure would work with a Bridge table. Part of the relationship's attributes is a field, OverOrUnder, that indicates whether the route involved in this relationship goes over or under the bridge.

The associative table that stores the relationship of bridges to the roadways that go over or under them also includes the weight and dimensional limits that would be needed to route commercial loads through the road system. EventValue in the Event table stores Bridge ID when EventType = "Bridge," which means it also allows a direct connection to the BridgeRoute table. Alternatively, you can start in the Bridge table, find the BridgeID value, go into the BridgeRoute table with that value to find all associated LRMPositionID values, and then use those values to locate all the included routes in the LRMPosition table. More guidance on this type of data analysis process will be provided in chapter 15.

Intersections form another group of complex elements. Like bridges, intersections typically involve multiple roadways. Figure 14.20 shows that a geodatabase design similar to that used to manage the relationships of bridges and roadways is suitable for intersections. The basic concepts come from chapter 3, which introduced subtypes; chapter 4, which described specific intersection subtypes; and chapter 12, which provided a conceptual model for intersection data. The design shown here for the RouteIntersection table is simpler than that suggested by the Approach entity in the conceptual models of chapter 12. All that would be required

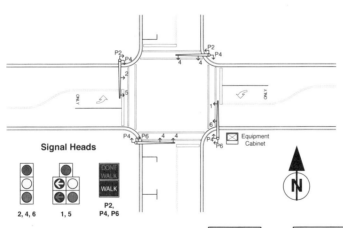

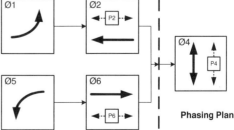

Figure 14.21 **Traffic signals** A traffic signal installation is typically described by a plan, a description of signal heads, and a phasing or timing plan. Agencies that include traffic signals in their highway inventories normally track expensive electronic components, such as controller units, conflict monitors, and vehicle detectors, most of which are located in an equipment cabinet.

to implement the approach concept, where you had a separate record for each direction of travel on a roadway at the intersection, is to add a DirectionOfTravel field. However, the reason to implement the approach concept is to store additional information, such as turning movement counts and signal timing, so more fields would be necessary.

There is, though, a simple way to add traffic signal information typically found in highway inventories, and that is to treat the traffic signal as a component of an intersection. A common four-legged intersection is shown in figure 14.21. It includes mast arm-mounted traffic signals, pedestrian signals, signal phasing, and an equipment cabinet that contains a signal controller, vehicle detectors, a conflict monitor, and a system coordinator. Figure 14.22 shows the conceptual data model for an extension to the highway inventory to retain this information.

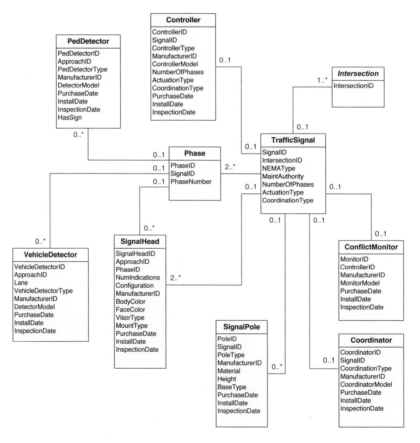

Figure 14.22 **Traffic signal conceptual data model** The intersection data model shown in figure 14.2 included a single TrafficSignal class. That entity has been expanded here to show the components of the traffic signal installation. Some components are tied directly to the TrafficSignal class; others are organized through the Phase class because they are specific to the movement that phase controls. All would be illustrated on a map using the intersection point feature.

Some agencies have chosen to create a feature class for every part of a traffic signal they need to manage, partially due to the structural preferences of their inventory application. However, that solution imposes a huge editing workload on staff by making the abstraction too close to the real world, resulting in a requirement for very-large-scale mapping in order to distinguish among all the features. This simple intersection alone would require 16 signal head point features, four mast arm symbols (point or line), and a controller cabinet point or polygon symbol (at this scale either could be used). The really expensive components, the electronic units, are not visible. You may be able to place all those point features on top of the cabinet, but that is certainly not an elegant solution.

A motor turning a cam with electrical contacts to supply power to each signal indication operating electromechanical controllers is an increasingly rare traffic control system due to high maintenance costs and limited functionality. A 24-hour timer may select from among as many as three timing plans, each with its own camshaft. For example, the time provided to green indications on some approaches can be different in the morning and evening peaks. The National Electrical Manufacturers Association (NEMA) adopted the TS-1 standard for controller operation and connection to the wiring cabinet in the mid-1970s. In the early 1980s, New York and California adopted the 170 standard for controllers that were really small computers. The NEMA standard was updated to version TS-2 in 1992. This standard was then expanded in 1998 to form the National Transportation Communications for ITS Protocol (NTCIP), which allows the interoperability and interchangeability of traffic control and other intelligent transportation system (ITS) equipment. The most recent standard, for the advanced traffic controller (ATC), was also promulgated under the NTCIP.

Semiactuated controllers remain in main street green until a vehicle or pedestrian is detected, at which time they cycle to the minor movements to provide green indications to side streets and protected turn movements. An actuated signal is one that responds to detectors for all main street and side street movements. In an actuated signal, the cycle is controlled by the detection of vehicles and pedestrians. Vehicle detectors are installed in the pavement (inductance loops) or over the intersection (video cameras) and connected via a cable to an electronic unit in the cabinet that tells the controller when a vehicle is present. Pedestrian detectors, typically pole-mounted push buttons, are used to indicate the presence of a pedestrian seeking a WALK signal to cross one of the intersection approaches. The controller and any detectors that may be provided determine the capabilities of the traffic signal installation. A fully actuated controller can be used to run a semiactuated installation.

A conflict monitor observes the operation of the controller to ensure that a green indication is never supplied to conflicting movements through equipment failure or that all red bulbs are inoperative on a given approach. The occurrence of one or both conditions will place the signal into flashing operation. A coordinator unit may also be present in the cabinet to synchronize the local signal with a larger system of interconnected controllers and, possibly, to a central office.

A lot of electronic equipment is being added to traffic signal installations and at other points along the roadway. For example, those traffic cameras shown on the morning news program are part of the ITS infrastructure being installed in metropolitan areas around the country. So, too, are changeable message signs placed on limited-access highways, emergency call boxes, and other devices that seek to supply more information to traffic control staff and motorists.

The alternative presented here suggests a different direction: tie the traffic signal installation to the intersection(s) it serves and create tables, not feature classes, to store all the components. The intersection point feature can illustrate the location of a traffic signal using a control type field that includes "Traffic Signal" as one of its values. The vehicular traffic signal heads and detectors can be organized by phase. A phase is an atomic unit of the signal timing plan and usually represents a green indication followed by a yellow and all-red clearance interval, possibly with a concurrent pedestrian WALK indication followed by a flashing DON'T WALK indication. Pedestrian signals and detectors should be tied to the related signal timing phase.

The included entities and attributes are representative of what you may have in a traffic signal inventory. Each agency may care about different things. Price is a typical determinant of the things to track in a highway inventory. For example, some states may only track LED indicators (expensive) in their inventory and ignore light bulbs (cheap). In all cases, the design allows for the location of spare parts at non-intersection locations. This portion of the design is reflected in the multiplicity of 0..1 at the TrafficSignal end of the relationships shown for Controller, Coordinator, and ConflictMonitor. This equipment may be in a warehouse, on a truck, or off for repairs. It is likely that a unit removed from one location will be reinstalled at another, so the design treats each movable piece of equipment as a separate entity rather than attributes of TrafficSignal.

The TrafficSignal entity is related to the *Intersection* entity through a one-to-many relationship, reflecting the possibility that one traffic signal installation might control multiple intersections but each intersection has only one controlling traffic signal installation. A single piece of equipment for a traffic signal, such as a controller and conflict monitor, is tied to the TrafficSignal entity. So, too, are mast arms and other signal poles that cannot be assigned to a particular traffic movement. Everything else is related to a signal phase, which is an organizing entity within the timing plan and the conceptual data model.

A traffic signal may consist of two or more phases (the typical upper limit is eight phases). Each phase controls the signal display for a given set of approaches, such as the main street through movement or the side street left turn. Some traffic signals operate on a fixed-time basis. Others are demand actuated and depend on inputs from vehicle and pedestrian detectors to "change the signal" so that a green aspect is shown to a different approach or traffic movement.

In a geodatabase implementation of this logical model, you would create only tables for the entire traffic signal inventory. Their geometric representation would be through an intersection point feature class. If you provide a traffic control type field in the *Intersection* class, then you can use its value to select the appropriate point symbol to illustrate the intersection feature.

Centerlines

Earlier in this chapter we noted that there is a one-to-one relationship between Segment and Centerline, with one centerline feature representing the extent and shape of a roadway segment, although both may independently evolve over time. While this rule is technically true, you may want to consider maintaining multiple centerline features for each segment. Figure 14.23 shows what we mean.

You may want to create scale-specific versions of the Centerline feature class. Doing so can greatly improve performance by reducing the number of vertices on a centerline through the generalization process when working in smaller scales. It also can be part of a general scale-specific cartography plan for applying the new drawing capabilities of ArcGIS. While you could technically put all the various centerline versions into one class and then extract

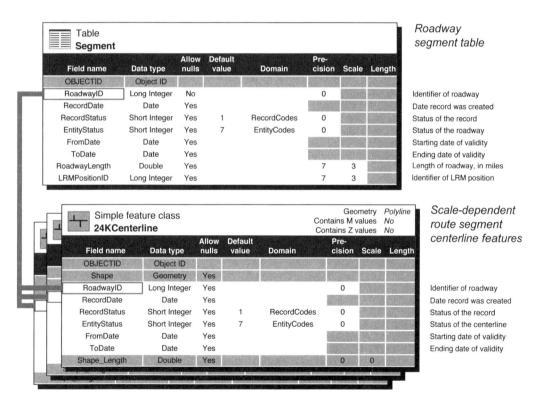

Figure 14.23 **Multiple Centerline feature classes** Many state DOTs produce maps at multiple scales due to the large geographic extent of a state. Accordingly, the inventory geodatabase must support multiple, scale-specific Centerline feature classes. A given roadway may be represented by a detailed centerline at large scales and a more generalized centerline—or none at all—at smaller scales. One Segment table record would relate to all the various Centerline feature classes in which that segment was present.

Figure 14.24 **Interchange representations change with scale** An interchange consists of a cluster of small roadways serving as on- and off-ramps. At larger scales, you will probably want to include all these ramps, but at smaller scales it is common to replace the ramps with a point symbol. Most interchanges are symmetrical, making it easy to decide where to place the point symbols. Other interchanges with complex, asymmetrical ramp arrangements are more problematic.

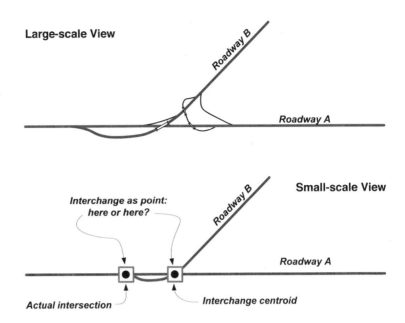

the ones you desire by using a scale attribute, it is probably better to keep them in separate feature classes since edits must be separately applied for each scale. You may even want to omit some roads or change cartographic representations at smaller scales.

Figure 14.24 shows one such example. At large scales, you will likely want to include all the ramps, connectors, service roads, and other components of interchange geometry, but at smaller scales, point symbols may suffice. In this example, you will need to carefully consider the location of the point symbol. You should place the point symbol where it matches the Roadway A measure for the interchange event, perhaps by using dynseg.

If you only have one LRM type and work at a fairly small scale, you could just store terminal measure values as attributes of the Centerline feature class. However, you need more specificity than that for any sort of larger scale mapping to avoid getting things put in the wrong place during dynseg. It is also likely that you will need to support multiple LRM types. The best way to meet these requirements is not to include measures in the Centerline features you edit.

It was noted earlier that the Centerline feature class did not include measures (the Contains M values parameter is set to No). But measures must be provided for dynseg operations to occur. A polyline can contain only one set of measure values, so it is not possible to store measures for multiple LRM types in a single feature class. What the Centerline feature class relies on is a CalibrationPoint feature class, shown in figure 14.25, that allows you to select the measures associated with the LRM type of interest, and then to apply that set of measure

Centerlines

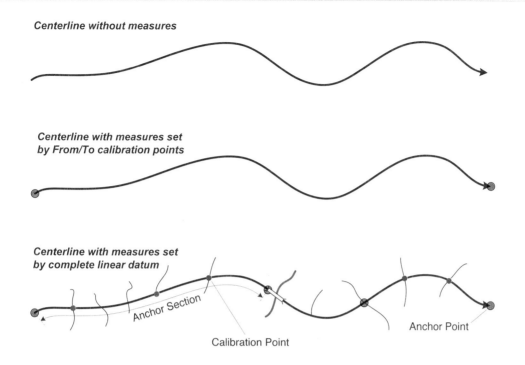

Figure 14.25 **Using calibration points to assign measures** The Centerline feature class provided in the highway inventory geodatabase design does not include measures in the Edit Geodatabase. Measures are added when the geodatabase is published, with one set of centerlines being generated for each LRM type. The measures are derived from a CalibrationPoint feature class, which stores a measure value for each LRM type. At a minimum, calibration points must be established for the start and end of a centerline feature. Other calibration points may be placed along a centerline to provide additional measure calibration, such as at intersections, bridges, and other discernable fixed points. Such calibration points can define a formal linear datum consisting of anchor points and anchor sections.

values to the Centerline feature class. Using the Calibrate Route Feature task in ArcMap, you can add measure value controls to each centerline feature by placing a calibration point at the start, end, and any number of intermediate vertices along the line. Each calibration point must coincide with a centerline vertex. Since the calibration process affects only one feature at a time, calibration points snap to the selected centerline feature. The calibration process will uniformly use the option to interpolate between points. A simple script can automate the entire process, one segment/centerline combination at a time.

Dynseg works by interpolating positions between vertices. When you load measures into a polyline shape, ArcGIS calculates the measure value for all the vertices that exist between the ones for which you supply explicit values. With measures recorded to the nearest 0.001 mile, your field measure resolution is about 2.5 feet. You must have very good geometry to be off by less than that over the length of a 20- or 30-mile roadway segment. It doesn't take

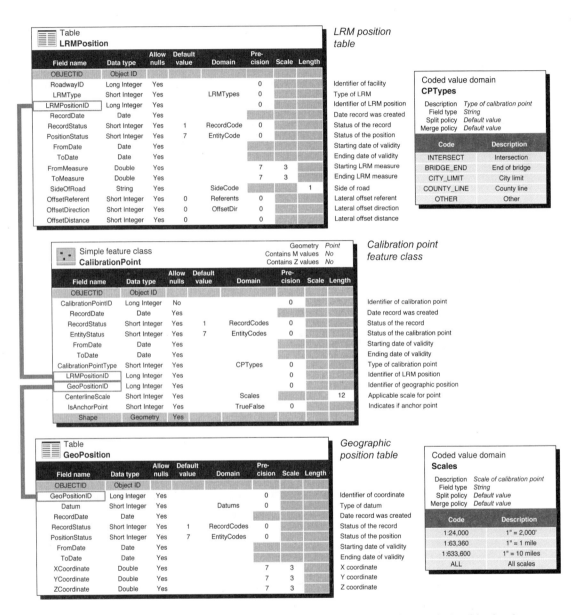

Figure 14.26 **Calibration point classes** The Event table serves as an equivalency relationship class between LRMPosition and GeoPosition in the core event data model. The CalibrationPoint feature class serves a similar function for the geometry portion of the highway inventory geodatabase. Calibration points may be separately defined for each scale Centerline feature class included, as generalization of centerlines may move a roadway so that it is too far from the calibration point at the snapping tolerance used.

much for events near an intersection or a bridge to be placed on the wrong side of a feature that has its own geometry. You do not have to define an explicit measure value for every vertex, and not every centerline feature will require the same density of calibration points. You just need to provide enough intermediate calibration points to provide a linear datum for the geometry, the same way you would use anchor points and anchor sections as a linear datum for fieldwork. In fact, as shown in figures 14.25 and 14.26, calibration points can serve to represent actual anchor points.

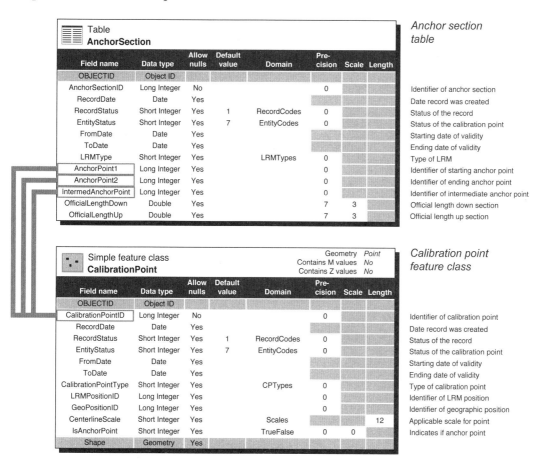

Anchor section table

Calibration point feature class

Figure 14.27 **Anchor section class** The CalibrationPoint feature class is adequate to accommodate anchor points, which are a type of calibration point. However, a separate AnchorSection table is required for each anchor section defining the official length for part or all of a given centerline. Three foreign key fields in the AnchorSection table all point to the same **CalibrationPointID** field. Anchor points 1 and 2 define the start and end of each anchor section. Calibration point ordering indicates the direction of increasing measure values. A third anchor point field is available to provide an optional intermediate calibration point reference when there are multiple paths between the listed terminal anchor points.

Just as the Event table serves as a type of associative entity between the LRMPosition and GeoPosition tables, the CalibrationPoint feature class can serve the same purpose with the geometry portion of the geodatabase. As indicated in the data model of figure 14.26, the CalibrationPoint feature class has two other useful attributes. The first of these is **CenterlineScale**, which indicates the scale-specific centerline feature class for which the calibration point should be used. You could have scale-specific calibration point feature classes, but putting them together allows the class to serve as a single point of reconciliation between the LRMPosition and GeoPosition tables.

The other useful attribute, **IsAnchorPoint**, is a Boolean flag that indicates whether the location is also an anchor point for fieldwork. Such a calibration point must correspond to a physically identifiable, field recoverable element, such as an intersection or the end of a bridge. If you really wanted to provide the full linear datum, as shown in the new UNETRANS Inventory package of chapter 12, you will need to also create a reference point feature class and an associative table to tie reference points to anchor points.

The AnchorSection table included in figure 14.27 uses three association relationships with the CalibrationPoint feature class. All three relationships terminate at **CalibrationPointID**. The use of anchor points at the start and end of anchor sections is obvious. Less obvious, perhaps, is the use of an intermediate anchor point, which allows you to distinguish between two paths connecting the same terminal anchor points. Such a situation could exist with one-way streets, divided highways, and looping streets. Divided highways are also the reason behind two official length fields, one for going down the anchor section (direction

Calibration point feature class for milelog measures

Simple feature class
MilelogCalibrationPoint

Geometry: Point
Contains M values: No
Contains Z values: No

Field name	Data type	Allow nulls	Default value	Domain	Precision	Scale	Length	
OBJECTID	Object ID							
CalibrationPointID	Long Integer	No			0			Identifier of calibration point
RecordDate	Date	Yes						Date record was created
RecordStatus	Short Integer	Yes	1	RecordCodes	0			Status of the record
EntityStatus	Short Integer	Yes	7	EntityCodes	0			Status of the calibration point
FromDate	Date	Yes						Starting date of validity
ToDate	Date	Yes						Ending date of validity
CalibrationPointType	Short Integer	Yes		CPTypes	0			Type of calibration point
CenterlineScale	Short Integer	Yes		Scales			12	Applicable scale for point
IsAnchorPoint	Short Integer	Yes		TrueFalse	0			Indicates if anchor point
RoadwayID	Long Integer	Yes			0			Identifier of facility
AtMeasure	Double	Yes			7	3		LRM measure for point
Shape	Geometry	Yes						

Figure 14.28 **Simplified calibration point feature class** You can use the MilelogCalibrationPoint feature class where there is only one LRM type based on milelog measures and geographic coordinates are not needed. **RoadwayID** and **AtMeasure** have replaced the **LRMPositionID** field in the multi-LRM type version.

of decreasing measure values) and up the anchor section (direction of increasing measure values). In order to support multiple LRM types, you must specify the type to which this anchor section applies.

The alert reader will have noticed one glaring problem with the design for CalibratePoint: it does not include any measures. Instead, it stores an LRMPositionID value. You must go to the LRMPosition table with that value as a foreign key to find the actual measure values you need for a given LRM type. You need to join the LRMPosition table to the CalibrationPoint feature class using the LRMPositionID field. This is really part of the process of publishing the centerline features for analytical uses, like dynseg operations. The next chapter talks more about relational joins and publishing the centerline features. Figure 14.28 gives you a sneak preview of what the published CalibrationPoint feature class would look like for milelog measures.

The join process takes a few fields from the LRMPosition table and adds them to the original CalibratePoint feature class. The output of that process is the MilelogCalibrationPoint feature class shown in figure 14.28. The RoadwayID field is brought over unchanged. Since a calibration point has only a single milelog measure, the FromMeasure field's values move from the LRMPosition table to an AtMeasure field in the MilelogCalibrationPoint feature class. Although the desired LRMPosition table rows are selected using LRMType = "State Milelog", that field is not needed in the output class. Instead, it is in the feature class's name. Now you can use the MilelogCalibrationPoint feature class and the Calibrate Route Feature task in ArcMap to publish a Centerline feature class that has measures.

A special calibration problem

One state DOT has an LRM measure problem that calibration points can solve. This state uses a route-based segmentation schema, with routes traversing the entire state and no breaks at county lines. These routes are subdivided into inventory segments of zero length to one mile, but measure values are continuous across all segments along the route. Although variable-length segments are the fundamental inventory unit, the reference points that define the limits of inventory segments are really the key to the legacy database design and the business rules it implements.

The central inventory table, shown in figure 14.29, consists of roadway segments specified by beginning and ending referencing points along each numbered route, plus an official length. Every reference point must be used as the start of a segment, which means the last record for a given route has the same starting and ending reference point value and a 0.000 length. The ReferencePoint value of the last record is also the value of EndReferencePoint of the next-to-last record. Segments that include a point event also have zero length. Thus,

Chapter 14: State DOT highway inventory: Editing

while presented as a set of linear events, it is possible to view the table as a set of point events representing the locations where some attribute changes, a point event exists, or an element begins or ends. The Description field explains what the reference point is. This is why the last segment has a zero length: so the end of the route can be the beginning of a segment (i.e., the value in ReferencePoint) and thus have a Description field entry.

Table InventorySegment								
Field name	Data type	Allow nulls	Default value	Domain	Precision	Scale	Length	
OBJECTID	Object ID							
RouteName	String	No					4	Identifier of route
ReferencePoint	Double	No			7	3		Beginning segment measure
EndReferencePoint	Double	No			7	3		Ending segment measure
Length	Double	No			7	3		Field-derived segment length
UpdateYear	Short Integer	Yes			0	0		Year of last update
Description	String	Yes					30	Description of segment

Legacy state route inventory segments

Figure 14.29 Denormalized inventory segment class One state DOT uses a fully denormalized database model that creates a segment whenever one of the included attributes changes. Where an attribute change occurs is called a reference point. Each inventory segment must begin and end at a reference point. In applying this rule, the last inventory segment always has the same beginning and ending reference point, resulting in a zero length. This rule is necessary because the **Description** field applies only to the beginning reference point. The state wants to move away from this denormalized design for data editing and adopt the normalized structure described in this chapter. The denormalized form would be generated when the geodatabase is published.

Since inventory segments are created wherever a linear attribute changes or a point event, such as an intersection or boundary, exists, the highway inventory database is denormalized, a term introduced in chapter 3. This means that data redundancy is present, as all the attribute values except the one that changes at a reference point is duplicated in at least one more record. A change in a single event value typically results in edits to many inventory segments.

Another source of inventory maintenance is a route realignment, which causes changes in downstream measures from the point of realignment. The new route designation includes a suffix letter that indicates the number of times it has been realigned, such as "A" for the first time, "B" for the second time. Otherwise, measures are never changed—even if discovered to be incorrect—to preserve the historical locations of features upon which other units of the agency rely. Their databases use route number and measure as a form of location identifier. For example, instead of an intersection identifier, the state safety staff uses a combination of route number and measure to identify an intersection for recording crashes. At the intersection of two state routes, the lower numbered route is used to record crashes. Changing a

A special calibration problem

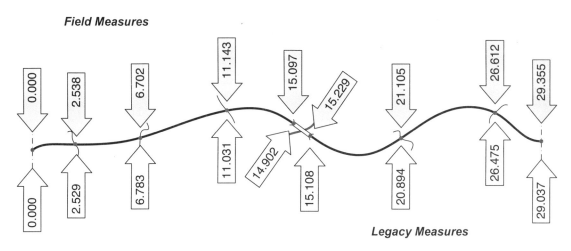

Figure 14.30 **A legacy of error** One of the reasons the state DOT wants to get away from using reference points and inventory segments is because reference points are not field recoverable. The original route measure locations assigned to the reference points were determined by an inaccurate method. The distance measuring instruments and procedures now employed cannot duplicate the legacy measure values. Since the locations defined by the reference points cannot actually be found in the field with the required precision, updates to the highway inventory cannot occur reliably.

measure would result in future crash reports not being found at the same location. So, too, would a change in route numbers. History is preserved in most cases, but there is no ability to improve the spatial accuracy of the database. Thus, reference points are not really field-recoverable measures along the route.

Although reference points cannot be edited because they are used as element identifiers, segment length can be altered. Since the measures associated with reference points cannot be changed, corrections properly reflecting the extent of the segment can only be expressed in the official length attribute. As a result, a segment's official length is not the mathematical difference between the beginning and ending reference points but the actual field-derived length. The length of a route is the sum of all route segment lengths, not the difference between the highest and lowest reference point measure values. There is no point along a route where the field length and the official length are reconciled and the difference reduced to zero, so the measure difference grows in a non-linear manner along the facility's complete length. To prevent errors from getting too large, segments are limited to one mile.

The agency preserves history by copying the database at the end of each calendar year, with the historical copies receiving a calendar year suffix. Although data maintenance occurs throughout the agency, the central office staff in the capital city does most of it, using data collected by contractors and staff in the field. Distance measuring instruments (DMIs) locate positions in the field using the legacy measure values. Field-measured distances are tied to the database position with an equivalent reference point. However, differences between the legacy values and the field measures observed by a DMI are poorly understood and rarely documented. Thus, data collected at a field measure equivalent to the legacy measure is often compiled in the wrong place, and data compiled in the field can be applied to the wrong inventory segment.

To assure compiling data where intended, field staff is directed to reset the DMI to zero at the start of each inventory segment. Unfortunately, some segments are very short, and field staff may not reset the DMI at every closely spaced segment break, particularly in urban areas where the sudden stop required in the middle of every intersection could be dangerous and disrupt traffic. There is also the problem of long rural segments that would run for miles except for the fact that the DOT subdivided them at implied mileposts. The agency even went so far as to install a numbered sign at each milepost location, although the signs are not always accurately placed.

Data is compiled in positive (primary, or low to high) and negative (secondary, or high to low) measure directions on divided roadways based on measures along a logical centerline. In mountainous regions of the state, it is difficult to see from one side of the road to another

Simple feature class **ReferencePoint**					Geometry	Point		
					Contains M values	No		
					Contains Z values	No		
Field name	Data type	Allow nulls	Default value	Domain	Precision	Scale	Length	
OBJECTID	Object ID							
ReferencePointID	Long Integer	No			0			Identifier of calibration point
RecordDate	Date	Yes						Date record was created
RecordStatus	Short Integer	Yes	1	RecordCodes	0			Status of the record
EntityStatus	Short Integer	Yes	7	EntityCodes	0			Status of the calibration point
FromDate	Date	Yes						Starting date of validity
ToDate	Date	Yes						Ending date of validity
CalibrationPointType	Short Integer	Yes		CPTypes	0			Type of calibration point
RoadwayID	Long Integer	Yes			0			Identifier of roadway
AtMeasure	Double	Yes			7	3		LRM measure for point
GeoPositionID	Long Integer	Yes			0			Identifier of geographic position
CenterlineScale	Short Integer	Yes		Scales			12	Applicable scale for point
IsAnchorPoint	Short Integer	Yes		TrueFalse	0	0		Indicates if anchor point
LegacyRouteID	String	Yes					1	Identifier of legacy route
LegacyMeasure	Double	Yes			7	3		Legacy LRM measure
Shape	Geometry	Yes						

Reference point feature class

FIGURE 14.31 **A multilingual reference point** It turns out that many workgroups had adopted reference points as a proxy for a unique identifier. Thus, the continued use of the legacy—and wrong—measure values is a business necessity, at least until all the workgroup databases and the applications they support are modified to use unique identifiers. Since that day may never come, an efficient way is needed to provide the "real" measure values for data management and the legacy measures for other uses. Enter the ReferencePoint feature class. This derivative of the CalibrationPoint feature class stores the actual LRM position (**RoadwayID** and **AtMeasure**) and the legacy LRM position. Serving as a sort of Rosetta Stone, the ReferencePoint feature class can generate centerline features using either set of measures. The former reference point identifier is preserved as the candidate primary key for the class.

due to intervening topography and land cover. It is difficult to tell exactly where you are along the imaginary centerline, particularly when both paths around the median obstacle are different lengths. There is also the problem of great changes in elevation across the state along many routes. The result is that the recorded length of many routes is substantially different from the length implied by geometry stored in the Centerline feature class.

The problem presented is how to resolve the inherent differences between the legacy measures stored as reference points and the field measures. From the perspective of the database design project, the easiest choice is to discard the legacy measures and accept the field measures. The agency rejected this choice because to undergo a wholesale change in measures would move almost every event in the inventory, thus requiring the same maintenance process in all other facility and project management databases. It is easier, from an organizational perspective, to limit the effect of LRM translation to the highway inventory maintenance and field staffs.

What the state DOT decided to do was to maintain both LRMs while migrating to the field measures over several years. This choice requires the agency to maintain two LRMs for the same routes and to construct a crosswalk between them so that field staff can find the actual location inferred by a position stored in the database. As shown in figure 14.31, a ReferencePoint feature class supplies that crosswalk. ReferencePoint is really a two-LRM version of our earlier CalibrationPoint feature class. This decision is possible through the construction and editing of centerline features that do not include measures and the use of calibration point features that represent reference points. Once the flexible-measure centerline reference point features are constructed, geometry updates will be required only when errors are discovered, or a road is created, realigned, or extended.

The highway inventory workgroup also provided the new element identifiers that allow other workgroups to migrate over time to an explicit intersection and bridge entities, thereby removing their dependence on nonchanging location references as a proxy for such identifiers. As this migration occurs, the agency can stop preserving the erroneous measures of the legacy LRM and eventually have only the one that is field recoverable.

Legacy users who seek to do dynseg to display their data will need route polylines with measures corresponding to those of legacy reference points. Field data collection crews and other users will need centerline features containing the measure values derived from field observation. These two views of the highway system inventory are made possible through the separation of the editing and publishing environments, and the use of calibration point features as a universal translator.

Migrating to the new structure

As this chapter illustrates, migration to a new geodatabase cannot happen overnight. Indeed, the basic premise of this entire chapter has been the need to support legacy data structures and business rules, allowing each to evolve at its own pace. Where appropriate, we have suggested changes to those business rules that will impact only the data editors while preserving the data views of other workgroups. Clearly, editing data in its denormalized form imposes extra work on the data editors and increases the risk of errors through data duplication and a lack of connectivity across the database. Separating the editing geodatabase from the databases required by other applications offers the only logical solution, one that confines change to the editing workgroup.

What we are talking about is how to manage change for that workgroup. Change is always difficult for a large organization. Change presents risks to the organization and the individuals who impose and implement the new methods and tools. Thus, an effective migration strategy manages change and mitigates risk. People in the organization know how to do their jobs and know where they stand within the group. New methods mean that people will initially have to do new things. They will worry about not performing the new tasks as well as the old tasks, thus altering their status in the organization.

The only effective solution is to reduce the risk of failure for the individuals involved, which will lower the risk to the organization. You can do this through incremental migration by supporting the old and new ways of editing for different parts of the highway system and combining them when you publish the statewide geodatabase. You should have already done a proof-of-concept pilot project to test the geodatabase. It is now time for full-scale deployment. So, how do you get your data into the new structure?

Most likely, you hired a consultant to generate the new geodatabase structure through the data modeling exercise in chapter 2. If your experience is like that of most agencies, the data modeling process was an extra duty, something you had to make time for while doing your "real" work. The first test comes when you realize the consultant's geodatabase must become your geodatabase. It is better to realize this early in the project, as noted in our discussion of agile system development techniques.

States and provinces consist of multiple counties, parishes, or other political units. State DOTs are typically subdivided into districts or regions. You can use one of these structures to split off data maintenance for the system within a group of these political units for trial deployment of the new normalized editing environment. Consider it a full-scale pilot project or a small-scale deployment, but do not allow parallel use of the old and new editing processes for the same data. Ask for one or more volunteers to make the move. Size the pilot in proportion to the number of volunteers you get. If you do not get any volunteers, you need to evaluate the entire project. You missed something along the way if everyone is afraid of the end result. Now is the time to fix it. If you have been employing the agile development process, then you have been getting frequent deliverables of functioning software and making small changes as you went, using feedback from the pilot deployment users. That feedback means ownership is already transferring from the consultants to the users.

Once the pilot deployment is under way, let the early adopters who volunteered to do the initial migration solve their own problems with the consulting team or whoever is doing the geodatabase and edit application development. People will talk about their experiences. If the expected benefits are occurring, everyone will find out. (If not, everyone will know that, too.) Give others the opportunity to migrate to the new system as they feel comfortable, but not everyone will. Ultimately, you will need to set a date certain for full migration once the unit is generally assured that the pilot has been successful, meaning the big problems have been solved, not avoided. This full-deployment decision point need not wait until the system is completely delivered. Indeed, it is better if the entire workgroup is involved in designing and implementing the final functional tweaks to the system. This is the point where the workgroup owns the new editing process and the tools and geodatabase upon which it relies.

As for the physical process of migrating data, ArcGIS offers tools to assist. The new geodatabase includes the same data as the legacy database, just in a slightly different structure, so there is a place for everything you want to retain. If you have a denormalized roadway event data structure, then you will need to normalize the data. We showed in chapter 8 how ArcGIS dynseg functions can take a denormalized event table, extract the individual events, and place them in a normalized event table. Use these methods to create a set of one-event tables that can be fed into the single all-events table used by the new geodatabase, one event type at a time. Extract the measures embedded in your legacy event tables to create LRMPosition table records.

If you now have state-length routes and want to use shorter segments as a strategy for deployment at a county or district level, then you will need to subdivide those routes at county lines. It should be a fairly simple matter to devise a script that automates this process by intersecting county polygons and roadway centerlines and generating all the required identifiers. Assuming the legacy database includes county as a linear event or county line as a point event, you should be able to readily extract the from- and to-measures required to create segment termini measures for calibration points. You can do the same thing with any bridges or intersections that are already in the database as events because they should also already have geometric representations that can generate calibration points for populating the new centerline features with measures. At the very least, you should be able to create intersection points using centerline crossings.

The key to a successful data migration is to incorporate the existing business rules into the new geodatabase design to the extent possible or desirable (remember, this is the time to make changes). As we have seen in this chapter, one data model can implement a variety of business rules. You can keep using many of the old rules and presenting the old data structure to users while utilizing the new geodatabase, as we will demonstrate in the next chapter. Of course, you cannot get all the benefits of the new technology with the old work processes. After all, many of the design requirements inherent in our inventory data model are intended to help you work smarter, not harder. You have to eventually use the new tools in the new ways to get the maximum benefit, but you do not have to jump in with both feet all at once. The geodatabase design offered in this chapter is intended for incremental implementation. Once you have moved the main body of inventory data—the events—over to the new structure, other changes, like the additional support for bridges and intersections, can be brought online at a later date. You can even keep the long routes and move to shorter segments later; you're just changing rows, not columns. Give the workgroup time to get used to the new technology and you should be able to migrate the business rules without having to change the geodatabase design. You will just be extending it to provide additional functionality.

Conclusion

This chapter has shown you how to create a normalized geodatabase for editing a highway inventory at a state DOT. Such geodatabases typically coexist with legacy systems from which they may originally or periodically receive data loads. This means that some of the structure of the legacy system will likely need to be included in the geodatabase design or, at least, be translatable from one to the other, possibly in both directions. Fortunately, many state DOTs have transaction-oriented legacy inventory systems that are compatible with a normalized geodatabase. Even when such is not the case, as with our reference point agency, significant benefits can be gained by using the concepts presented in this chapter.

There is, however, one drawback of a normalized editing geodatabase: it presents some additional burden for data analyses. Spatial analyses—the bread and butter of GIS applications—typically prefer denormalized geodatabases. As shown in chapter 6, a "publish" geodatabase for spatial applications can be derived from an edit-oriented geodatabase. Chapter 15 will show you how to denormalize and publish the data contained in a normalized highway inventory geodatabase.

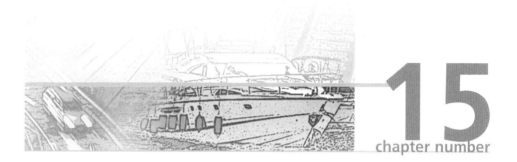

chapter number 15

State DOT highway inventory: Publishing

- Certifying the Edit Geodatabase
- Publishing centerline features for dynseg
- Creating event tables
- Publishing traversals
- Getting started with SQL
- Using relational joins
- Proceed with caution
- Conclusion

This book's general principle of geodatabase design is to separate the editing geodatabase used by data maintenance staff from the published version used by everyone else. All the preceding chapters dealt with data editing and geodatabase design strategies for optimizing that process. They showed you how to prepare your editing geodatabase for publishing the data so that others may use it as input to various spatial applications. This chapter presents the publishing process, where you will take the normalized editing geodatabase and publish it as a set of denormalized tables and feature classes. It also completes the two-chapter focus on constructing and using a state highway inventory geodatabase.

The editing geodatabase relies on a high degree of normalization to do its job. The publishing geodatabase relies on a high degree of denormalization to meet its functional requirements. In between is the transformation process presented in this chapter. That process requires you to go into a big stack of data, such as contained in the Event table, and extract one or two event types for further analysis. Another part requires you to combine data from multiple tables and feature classes into a single table or feature class. The publishing transformation process involves basic tasks that form a sequence of record selections and table joins.

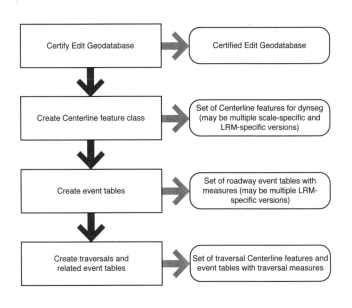

Figure 15.1 **Publication process flowchart** Publishing the highway inventory geodatabase has four basic tasks. First, certify the Edit Geodatabase, which is an input to all subsequent processing. Second, generate the Centerline features with measures. One Centerline feature class is published for each unique combination of scale and LRM type. Third, generate the event tables. Initially, you will extract each event type and join it with LRM position records to supply the required event tables for dynseg operations. Some of these event types will also be combined to form various denormalized multievent tables. The optional last step is to create traversals equivalent to the path of named routes and then to translate events to a traversal-specific LRM.

Four basic tasks are required to publish the highway inventory. Figure 15.1 presents a high-level view of the process. The process steps are listed on the left; the result of each process is shown on the right. Collectively, the four products shown in figure 15.1 comprise the Published Geodatabase.

The first task is to certify the Edit Geodatabase. The certified geodatabase is the input for the remainder of the process and is itself a published product. The certification process consists of completing the editing process, removing remnants of that process, and performing quality-control checks.

The second task is to prepare the Centerline feature class for dynamic segmentation (dynseg) operations. This means adding the measure values. You may have multiple Centerline feature classes, such as one for each commonly used mapping scale. For now, you cannot store multiple measures for a single vertex, so you will need one set of Centerline features for each LRM type that your agency uses.

The third task is to generate the event tables. You will have one set of event tables for each relevant LRM type. Not all events will necessarily be published in each LRM type, as they may be specific to a particular field of application and its unique LRM type. For example, crashes may be published using an intersection node reference, while pavement sections are published using a pavement management LRM. Published event tables will include those imported from nongeodatabase sources that describe elements to which you will need to add LRM position references.

The final task, which is optional, is to generate static traversals that represent the path of each state road. Each traversal will establish a new LRM. You will also likely need to transform select event types to use traversal measures. Not every agency will want to create traversals or transform events to their route-specific LRMs, although doing so may make it is easier to provide assistance to the public by tying events to a route milepost signing system.

The following sections provide guidance on how to accomplish these four tasks. There is also a presentation of SQL, the language with which you can communicate with a relational database. SQL may be useful in accomplishing some tasks involving external datasets, such as preparing them for event table publication.

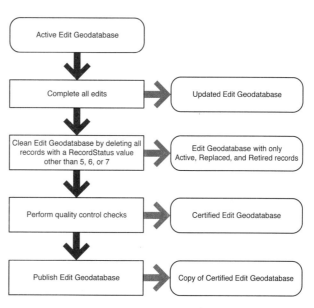

Figure 15.2 **Certifying the Edit Geodatabase** Certifying the Edit Geodatabase starts with completing all edits. This task includes applying all accepted edits (**RecordStatus** = 4). The second task is to remove all the edit records left over from the editing process, the rows with a **RecordStatus** value less than 5. After you strip the geodatabase of unnecessary records, you conduct quality-control checks for internal inconsistencies. You then publish the scrubbed geodatabase by applying its updates to the Publish Geodatabase using the process described in chapter 6. This process involves replacing existing records when the new version has a **RecordStatus** value of 6 or 7. You then add all new records with a **RecordStatus** value of 5.

Certifying the Edit Geodatabase

Since the Edit Geodatabase is the input for all other publication steps, it must be certified first. The certification process is intended to periodically create a complete, internally consistent version of the geodatabase used by the editing workgroup. This version is the first dataset placed in the Publish Geodatabase. It represents a complete view of the highway inventory at that point in time; however, it also contains all the edits committed since the Edit Geodatabase was created so you can recover the state of the inventory from any historic timeframe.

The first task is to complete all edits for this maintenance cycle. Next, you need to remove all the edits that never made it into the production dataset, which are all records with a RecordStatus value other than 5 (Active), 6 (Retired), or 7 (Replaced). You might think you would need to retain only the active records, as they show the current state of the system plus the retired records that describe how the system looked at a previous time. But you also need the replaced records that tell other users which records to delete from their copies of the geodatabase. By retaining the replaced records in the published version, you are allowing other users to extract these records from the certified copy, find them in their own geodatabases, and remove them in a batch process. Users can select records with an active status when they want to use the latest view of the system. Or you could publish a subset of the Edit Geodatabase that contains only these records. Remember, too, that a search on RecordDate will not find any of the replaced records as they have null values for this field. Only Active and Retired records have validity date entries (Active records have only a start date).

Once you have removed all the "sausage-making" records from the geodatabase, it is time to apply your quality-control checks. There should not be any null values in foreign key fields. Look for gaps or overlaps in linear events that should apply to every part of the highway system (speed limit, route type, functional class, etc.), measure values that are outside the range allowed for the underlying roadway segment, and internal inconsistencies (such as roadway segment assigned to both city and county jurisdiction). You should pay particular attention to divided roadway sections to ensure that all the appropriate event types are recorded with left- and right-side values and that a median type event exists.

The end result of the quality-control process is a Certified Edit Geodatabase. This version should be adopted as the new default geodatabase for subsequent editing. A copy of the certified geodatabase should be the first dataset placed in the Publish Geodatabase so that other users can construct data products that are not included in the published highway inventory and perform any internal updates within their own data resources. You may want to formally notify users that the Publish Geodatabase has been updated with a new dataset. All subsequent publication steps should occur in the Publish Geodatabase so editing can continue without interacting with the publication process.

Publishing centerline features for dynseg

Figure 15.3 illustrates the task required to publish a set of Centerline feature classes for dynseg and other mapping applications. The sequence of steps must be completed separately for each supported LRM type.

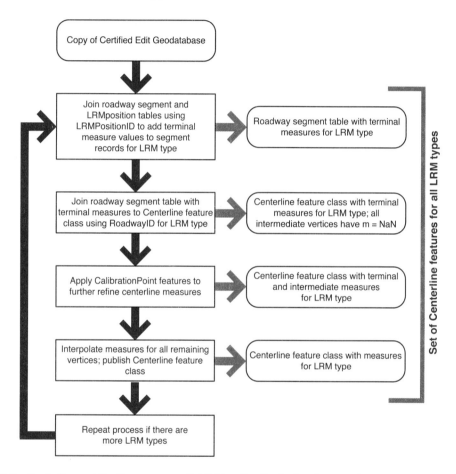

Figure 15.3 **Centerline publication process** Publishing the Centerline feature class represents a cycle of four tasks to complete for each version. First join the Segment and LRMPosition tables to generate a Segment table with measures appropriate to the LRM type. (Skip this step if you already have a Segment table with linear measures.) Next, join the Segment table with measures and the Centerline feature class, which adds the route identifier and terminal LRM measures to centerline features. Use the Create Route function in ArcGIS to make the Centerline feature class M-aware and to transfer the beginning and ending measures to each feature. All the intermediate vertices would have an m-coordinate of NaN. Finally, apply the calibration points to add more vertex measures along the feature and have ArcMap interpolate the remaining vertices to complete the process of providing an m-coordinate value for all vertices.

As you will recall from chapter 14, the native Centerline feature class maintained in the Edit Geodatabase has no measures. The tasks required to publish a set of Centerline feature classes for the user community begins with creating a roadway segment table suitable for adding start and end measures for each LRM type to the Centerline feature class. If you are, as recommended, using a roadway segment table devoid of any segmentation rules and supporting multiple LRM types, such as shown in figure 15.4, then you will first add the LRM data from the LRMPosition table to the Segment table. This table join will have to happen separately for each LRM type. You should name each version of the Segment table to reflect the LRM type it provides. If you only support one LRM type and directly store terminal measures in the Segment table, then you can skip this task.

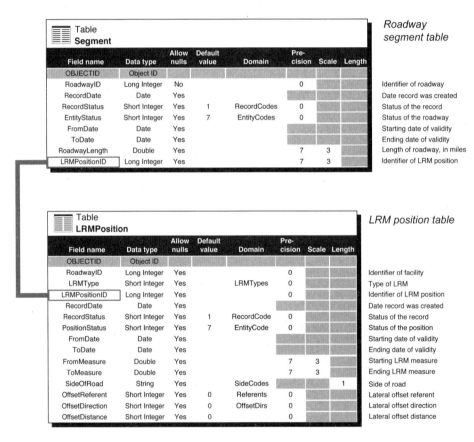

Figure 15.4 **Joining the Segment and LRMPosition tables** If you have used the most flexible form of the Segment table, as shown here, then you must convert the **LRMPositionID** field to an actual measure range before joining the Segment table and the Centerline feature class. The output of this process is a SegmentLRM table. If you have more than one LRM type, then you must do this for each type and name the output table appropriately.

Publishing centerline features for dynseg

The next task is to apply the terminal measures stored in the SegmentLRM table to the related members of the Centerline feature class using RoadwayID as the foreign key. This process is illustrated in figure 15.5 through a foreign key relationship using RoadwayID. The Create Routes dialog in ArcMap or a similar process allow you to change the Contains M values parameter setting to Yes, and set all the intermediate vertices to NaN (not a number). The book, *Linear Referencing in ArcGIS* from ESRI Press explains how to perform this process manually. You are encouraged to develop a script that automates this process although the workload of applying measures to Centerline features after the first publication process may not be great. You really only need to reestablish or add Centerline measures for updated segments. If you select Segment and Centerline records with a RecordDate value reflecting changes since the last publication cycle, then you will have much less work to do.

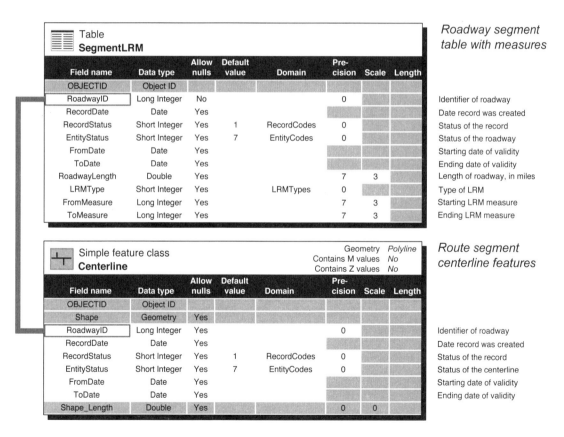

Figure 15.5 **Joining the SegmentLRM table and the Centerline feature class** Joining the SegmentLRM table and the Centerline feature class uses **RoadwayID** as a foreign key to add terminal measure values to centerline features. Each LRM type with terminal measures has one output CenterlineLRM feature class.

In fact, updating the Publish Geodatabase does not necessarily involve a massive replacement of everything that has previously been published. You could structure it like the Edit Geodatabase, where only updates are applied instead of doing a global drop and replace. The Edit Geodatabase tells you (and all other users) what has changed since the prior publication date. You could simply select those records with a RecordDate value after the last publication cycle and perform the publication steps only with those records. You would then apply those records to the published Centerline feature class and event tables. Records with a status of Active would be added to the published feature classes and tables. Records with a status of Retired or Replaced would replace the same records already in the published dataset. Procedurally, you would read through all the records in the updated dataset, delete all the matching records in the Publish Geodatabase, and add the update dataset to the Publish Geodatabase. Such a process can be readily automated with a script.

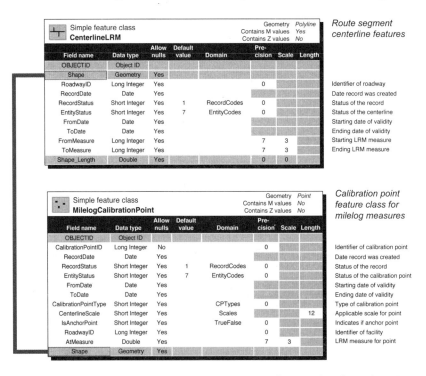

Figure 15.6 **Applying calibration points** With the CenterlineLRM feature class from the prior step, apply the appropriate calibration points for the LRM type. The CalibrationPoint feature class includes **LRMPositionID**, not actual measures, so you will need to produce an expanded calibration point feature class for this task by joining CalibrationPoint and LRMPosition using the desired LRM type. This figure repeats the MilelogCalibrationPoint feature class from chapter 14 as the template for the output of that process. Note that the relationship between these two feature classes is spatial due to the nature of the process. You could generate a list of calibration points for each centerline feature using **RoadwayID** as a foreign key.

Once you have applied the terminal measure values, which make the Centerline feature class M-aware, the next task is to apply the intermediate linear datum measures supplied by the CalibrationPoint feature class for the appropriate LRM type. This process is illustrated in figure 15.6, which demonstrates that the process relies on a spatial relationship rather than an explicit foreign key association, processing one Centerline feature at a time. However, you can use RoadwayID as a selection mechanism to provide the correct calibration points for a given roadway segment. Doing so will allow you to select the appropriate LRM reference for calibration points that correspond to intersections, bridges, and other features that may be located along multiple roadway segments. Incidentally, you can calibrate just a portion of a single route, which supports the type of incremental update process described earlier.

The calibration point is applied to each related roadway segment in measure order geographically along the Centerline feature. You can set a tolerance for how close the calibration points must be to a centerline feature for them to be applied. Once all the calibration points have been applied, you will use ArcGIS to complete the process by interpolating measure values for all remaining vertices. ArcGIS can interpolate between calibration points or extrapolate before or after a calibration point. Again, it is a good idea to create a script to automate this process.

If you have decided to use an inventory data model based on segmentation rules, such as route type, then you will use a process more like that shown in figure 15.7. You first join the Route and Segment tables using RouteID as the foreign key. The result is the top table shown in figure 15.7. Since the segmentation schema includes LRM type, meaning that no route has more than one LRM type, you can join the Route-Segment table directly to the Centerline feature class. You then apply the calibration points separately for state roads and local streets because they define different LRM types. Besides having the one Centerline feature class with all route types, you could create two Centerline feature classes, one for state roads and one for local streets. You can also publish subsets of the M-aware Centerline feature class by selecting on the county portion of RCLink to provide a geographic subset. You can publish all these same feature classes if you did not adopt the route-segment structure by applying the appropriate event tables later in the publishing process.

The process goes through certification, just as with the Edit Geodatabase, before you publish the final result. ArcMap functions can aid the process. For example, you can use the Layer Properties dialog in the Linear Referencing toolbox to show all the measure anomalies. Such anomalies include areas where there are no measures (NaN values remain), where measures do not increase in the digitized direction, where measure values repeat, and other conditions that you specify.

Chapter 15: State DOT highway inventory: Publishing

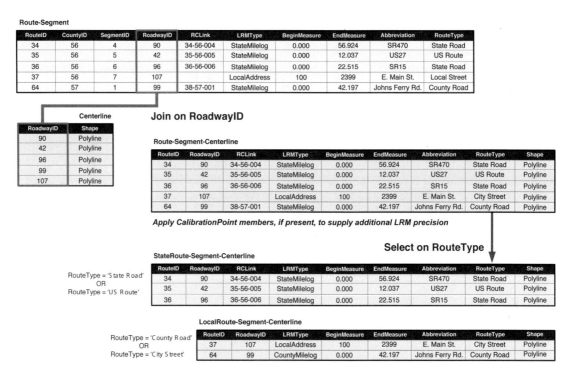

Figure 15.7 **Centerline publication at the record level** This figure shows what happens when you publish centerline features for state roads separately from those depicting local roads from the viewpoint of the involved tables and feature classes. This chapter later describes how to construct SQL statements that do this work.

Creating event tables

What you have at the end of the process of publishing the normalized Centerline feature class is one or more denormalized Centerline feature classes with measure values ready for dynseg. Now you need some event tables, which are the core of the published geodatabase. An event table suitable for dynseg requires three things: a roadway identifier that matches the values in the Centerline feature class, measure values that correspond to the same LRM type and range of measures included on the Centerline feature with that roadway identifier, and at least one event column.

Figure 15.8 shows the sequence of tasks required to produce event tables. As with the roadway segment centerline publication process, the first task is to join the LRMPosition and Event tables using LRMPositionID as the foreign key. This will add measure values to records in the Event table. In creating the published event tables, you will rely upon the foreign key relationships established in the Edit Geodatabase, which are shown in figure 15.9. The top of the figure would be different if you adopt the "no-rules" version of the inventory design in which there is only a Segment table. Otherwise, the process is identical. RouteID is the foreign key to link the Route and Segment tables. Combining these two tables through a join denormalizes the Route table, as it would generate multiple Route-Segment table records with the same Route table field values.

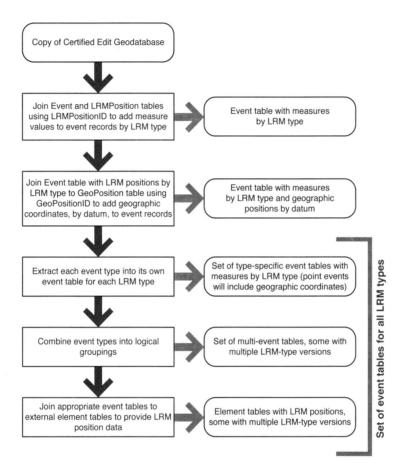

Figure 15.8 **Event table publication process flowchart** You must complete five tasks to publish event data in a form suitable for dynseg and other analytical operations. First, join the Event and LRMPosition tables, which results in an output row for each combination of event type and LRM type. Next, add geographic position data with a second table join, producing one record for each geographic datum. The output of these two tasks will have one row for each event, LRM type, and datum combination, so the next task is to extract events by each desired unique combination. The result will be a set of event tables for each distinct event type and duplicated for each LRM type and datum. You will then likely need to combine event types for specific applications, such as to join the tables containing speed limit, number of lanes, and traffic section. You will also want to publish element tables with position references by joining the appropriate event table to the element table.

Chapter 15: State DOT highway inventory: Publishing

Route

RouteID	Name	Abbreviation	RouteType
34	SR 470	SR470	State Road
35	SR 27	SR27	State Road
36	SR 15	SR15	State Road
37	East Main Street	E. Main St.	Local Street
38	Johns Ferry Road	Johns Ferry Rd.	County Road

Segment

RouteID	CountyID	SegmentID	RoadwayID	RCLink
34	56	4	90	34-56-004
35	56	5	42	35-56-005
35	56	109	274	35-56-006
37	56	7	107	
64	57	1	99	38-57-001

Event

EventID	EventType	EventValue	LRMPositionID	GeoPositionID
4590	Sign	39021	3109	9318
4591	SpeedLimit	35	265	
4592	AtGradeInt	2763	3295	2106
4593	ParkRestriction	Loading Zone	2874	
4594	Bridge	410028	5731	

LRMPosition

RoadwayID	LRMType	LRMPositionID	FromMeasure	ToMeasure
67	StateMilelog	4723	23.905	24.108
42	StateMilelog	3295	17.58	
96	StateMilelog	3295	2.093	
107	LocalAddress	3295	1400	
63	LocalAddress	3295	300	
64	StateMilelog	6429	7.664	9.633
65	StateMilelog	2084	3.902	

Figure 15.9 **Foreign keys supporting event table publication** The event table publication process relies on foreign keys that exist in the Edit Geodatabase. Although included in this typical structure, you will not need to join the Route and Segment tables if your LRM is based only on segment measures.

An association exists between the Segment and LRMPosition tables using RoadwayID as a foreign key but you do not really need to join these tables now. The dynseg process will use this foreign key without any other action at this step in the publishing process since LRMPosition also includes RoadwayID. Note also that the process does not need to be accomplished for each LRM type independently. The presence of multiple rows for each value of LRMPositionID, should you include more than one LRM type, will be reflected in the output Event table by multiple rows for each event, as in figure 15.10, which uses an at-grade intersection event to illustrate the process.

Creating event tables

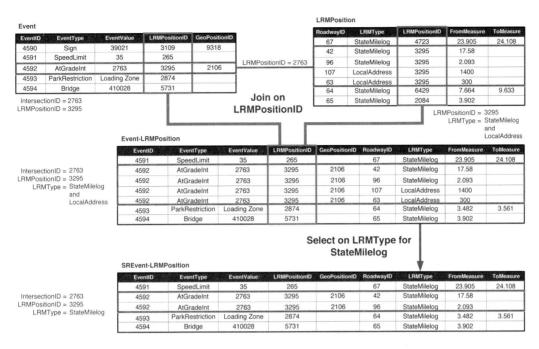

Figure 15.10 **The event table publication process at the record level** The Event and LRMPosition tables are joined on **LRMPositionID**. The highlighted event shows an at-grade intersection occurring at LRM position 3295. Given two LRM types and two intersecting roadways, the output is a group of four rows in the output table. You can then separate the rows by route system to generate a table containing only state road events. GeoPosition table records can be incorporated for point events.

At the top of the figure, there is one Event table record for this intersection. To the right, you can follow the LRMPositionID value to find four LRMPosition table records with this value, two for each intersecting state road and two for the intersecting named city streets. Physically, there are only two roads intersecting at this location, but the presence of two LRM types doubles the number of rows required to describe the location from the perspective of each route. (You should have already created two Centerline feature classes, one with milelog measures and another with street addresses.)

Joining the tables through the LRMPositionID foreign key will generate four event records describing the intersection. You can then select by **LRMType** value to produce separate Event tables for state roads and local streets. Both subsets will include two rows for the intersection event, one for each involved roadway. This structure allows the intersection event to be "discovered" when traveling along either facility.

Of course, you may apply additional selection criteria, such as eliminating all Event table records with a RecordStatus value other than "Active." You will also need to decide what RecordDate and FromDate values to retain in the result set. You should include the date of

Chapter 15: State DOT highway inventory: Publishing

publication as the result set's RecordDate and retain the input Event table record's FromDate and ToDate values in the result set. You can facilitate queries that travel along a segment by building an index on FromMeasure. You can also select records by RoadwayID and order by FromMeasure to print a geographically ordered list of events along a segment. (More information is provided on how to do this in the SQL section later in this chapter.)

Table: RoadwayEvent — *Roadway event table*

Field name	Data type	Allow nulls	Default value	Domain	Precision	Scale	Length	
OBJECTID	Object ID							
RoadwayID	Long Integer	No			0			Identifier of roadway
EventID	Long Integer	No			0			Identifier of event
RecordDate	Date	Yes						Date record was created
RecordStatus	Short Integer	Yes	1	RecordCodes	0			Status of the record
EntityStatus	Short Integer	Yes	7	EntityCodes	0			Status of the event
FromDate	Date	Yes						Starting date of validity
ToDate	Date	Yes						Ending date of validity
EventType	String	Yes					16	Type of event
EventValue	String	Yes					68	Event value (ID of element)
LRMPositionID	Long Integer	Yes			0			Identifier of LRM position
FromMeasure	Double	Yes			7	3		Starting LRM measure
ToMeasure	Double	Yes			7	3		Ending LRM measure
SideOfRoad	String	Yes		SideCodes			1	Side of road
OffsetReferent	Short Integer	Yes	0	Referents	0			Lateral offset referent
OffsetDirection	Short Integer	Yes	0	OffsetDirs	0			Lateral offset direction
OffsetDistance	Short Integer	Yes	0		0			Lateral offset distance
GeoPositionID	Long Integer	Yes			0			Identifier of coordinate
Datum	Short Integer	Yes		Datums	0			Type of datum
XCoordinate	Double	Yes			7	3		X coordinate
YCoordinate	Double	Yes			7	3		Y coordinate
ZCoordinate	Double	Yes			7	3		Z coordinate

Figure 15.11 **The event table prototype** The RoadwayEvent table shown here is a general prototype for all event tables. It includes all the fields contained in the Event, LRMPosition, and GeoPosition tables. No actual event table would include all these fields. For example, only point events would contain the geographic position fields, and only linear event types will find the **ToMeasure** field useful. Point event types only need the **FromMeasure** field, possibly renaming it as the **AtMeasure** field.

Figure 15.11 presents the data dictionary view of the data generated by the process of figure 15.10 persisted as a RoadwayEvent table. The table has additionally joined on GeoPositionID, so you will see the geographic coordinates of the event in addition to the LRM position. You now have all the events combined with their LRM and geographic positions in one table. You may prefer, though, to add geographic positions to event data later in the process. Geographic positions are, by their nature, point position statements and are generally limited to point events. The typical highway inventory contains relatively few point events, so you may choose to add geographic coordinate fields to the event tables after you have extracted the appropriate event types.

There is one more step to get a published event table in the form most users will need. To use your massive Event table, users will need to select specific records based on event and LRM types. Rather than require all your users to perform this function, you should separate the Event table's records by event type. Some users will want single-event tables. Other users will want multiple-event tables. Figure 15.12 is an example of a typical single-event table you will want to publish.

Table: NumberOfLanes

Field name	Data type	Allow nulls	Default value	Domain	Precision	Scale	Length	
OBJECTID	Object ID							
RoadwayID	Long Integer	No			0			Identifier of roadway
EventID	Long Integer	No			0			Identifier of event
RecordDate	Date	Yes						Date record was created
RecordStatus	Short Integer	Yes	1	RecordCodes	0			Status of the record
EntityStatus	Short Integer	Yes	7	EntityCodes	0			Status of the event
FromDate	Date	Yes						Starting date of validity
ToDate	Date	Yes						Ending date of validity
NumberLanes	Short Integer	Yes			0			Number of lanes
FromMeasure	Double	Yes			7	3		Starting LRM measure
ToMeasure	Double	Yes			7	3		Ending LRM measure
SideOfRoad	String	Yes		SideCodes			1	Side of road

Roadway number of lanes event table

Figure 15.12 **An example linear event table** This sample linear event table containing events depicting the number of lanes for each portion of a highway illustrates how few fields may actually be required. The **EventValue** field has been renamed as **NumberLanes**. **SideOfRoad** has been retained to serve divided roadways; other offset fields have been eliminated.

You can readily see that, in comparison to the table shown in figure 15.11, there are fewer fields in this table. Only those fields that apply to this event type, Number of Lanes, are selected. The event type has been converted into the name of the table. EventValue is renamed NumberLanes, the geographic position information (which does not apply to linear events) is eliminated, and all the LRM position attributes are dropped except RoadwayID, the measure values, and SideOfRoad. Apply the NumberOfLanes table to the published Centerline feature class with measures in the dynseg process and you will generate a map with roadway centerlines segmented by number of lanes. Through a series of selections on EventType, you would be able to publish all the events as a set of single-event tables. Such a process can be readily automated. As with Centerline feature updates, you only have to process the event records created since the last Publish Geodatabase update.

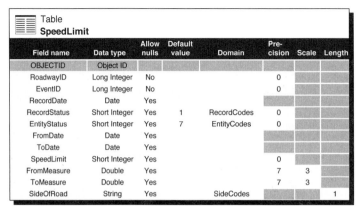

Figure 15.13 **Building a multievent table** Each of the top two tables has one event type. They are joined to produce the two-event type table at the bottom.

Number of lanes and all the other published event tables you have produced to this point in the process contain a single characteristic. Chapter 8 showed that you can handle up to three characteristics when mapping events, one each for the three presentation controls available. For a line feature, they are color, width, and line style. For a point feature, they are color, size, and point symbol. You can combine the single-event tables to produce a multievent table using a single SQL statement that performs an inner join, such as the several examples presented later in this chapter, or through the line-on-line overlay function available in ArcMap. If you use the line-on-line overlay function, the union option will perform an inner join to produce an output that has all the rows and columns in both input tables, even if one of the tables has no data for part of the spatial extent of the other. If you instead use the intersect overlay option, then you will get event data only where both input tables provide information. ArcGIS also includes a line-on-point overlay process that can add the linear event properties to the point event table's members.

Figure 15.13 offers a sample multievent table with traffic volume (AADT) and speed limit. You would likely want to publish a number of such multievent tables for a variety of uses. Indeed, this is how you will generate the denormalized tables needed by various inventory-based applications throughout the agency and beyond, such as the Highway Performance Monitoring System (HPMS). But single-event tables are likely to be the most useful to ArcGIS users, who can create event-specific map layers through the dynseg process, which allows them to construct any of the many possible combinations including many you might not anticipate.

On the other hand, it does not really matter if you publish some event types only as part of a multievent table. The ArcGIS concatenate and dissolve functions inside ArcMap's Linear Referencing Toolbox can select a single event type from a multitype table. You can use other ArcGIS functions to publish a wide variety of data products for others to use. You can even use some methods to generate event tables where none exist, such as by overlaying polygons representing city boundaries to generate linear events that show which portions of a roadway are within cities.

Chapters 14 and 15 show you how to prepare tabular data and feature classes for successful dynseg operations. To find out more about the dynseg process itself, see chapter 8 of this book, *Linear Referencing in ArcGIS* from ESRI, and ArcGIS Desktop Help at the ESRI Support Center (http://support.esri.com). These references also provide guidance on the measure calibration process and various analytical methods available in the Linear Referencing Toolbox.

Chapter 15: State DOT highway inventory: Publishing

Publishing traversals

A traversal is a path through the highway system. A static traversal is a path that will have repeated uses and is determined by some characteristic of the component roadway, like route name or functional class. A dynamic traversal is a one-time path usually determined by the requirements of a trip between an origin and a destination. Network Analyst extension for ArcGIS creates dynamic traversals based on the origin and destination, vehicle characteristics, and other parameters of the trip. This section deals with static traversals.

This book recommends that you not include route type as part of the highway system segmentation process, mostly to avoid major edits when route type changes. This section demonstrates how to create traversal centerlines that correspond to the path of named routes and to transform events to a new LRM so they can be defined relative to a traversal's origin.

Table SRName — *State road name event table*

Field name	Data type	Allow nulls	Default value	Domain	Precision	Scale	Length	
OBJECTID	Object ID							
RoadwayID	Long Integer	No			0			Identifier of roadway
EventID	Long Integer	No			0			Identifier of event
RecordDate	Date	Yes						Date record was created
RecordStatus	Short Integer	Yes	1	RecordCodes	0			Status of the record
EntityStatus	Short Integer	Yes	7	EntityCodes	0			Status of the event
FromDate	Date	Yes						Starting date of validity
ToDate	Date	Yes						Ending date of validity
SRName	Short Integer	Yes			0			Number of state road
FromMeasure	Double	Yes			7	3		Starting LRM measure
ToMeasure	Double	Yes			7	3		Ending LRM measure

Figure 15.14 **Route event table** If you have adopted the most simple and flexible of designs for the Segment table, then you would treat route name as a linear event. This sample table represents a linear event storing state roads by route name. It could be used through the dynseg process to generate a map with state roads depicted as groups of unconnected line segments, or it could serve as the start of a traversal-building process.

The simplest way to create a static traversal for route name is to create a RouteName event table. Figure 15.14 shows one that includes all the state road segments and their names. You can select any subset of rows by SRName value, apply it to the Centerline feature class through dynseg, and get a map showing the path of that one route. The map would look the same whether there are 10 event segments or 1,000.

But what if you wanted more than a map of that route? What if you wanted to create a new, route-specific LRM with its own measures? Figure 15.15 shows the basic sequence of tasks required to create an SRTraversal feature class and to transform useful events to use the new traversal measures that you generate from the segments that form each traversal feature. It starts with the state road name event class to generate a set of centerline features that correspond to the extent of each named road through the dynseg process. Note that some portions of the highway system will generate multiple state road segments. This occurs when more than one state road uses that section of roadway.

You next would sum the length of all the component sections to get the total length of the traversal. This number is the ending measure value for the new, traversal-centric LRM. The starting measure for the traversal is zero. These two values are applied to a whole-traversal centerline feature you will create by merging all the sections into a single feature. You could calibrate the traversal LRM by intersecting each section with the traversal centerline and using simple math to calculate the equivalent measure in the traversal's LRM. You could even permanently store traversal measures as one of the LRM types in the CalibrationPoint feature class.

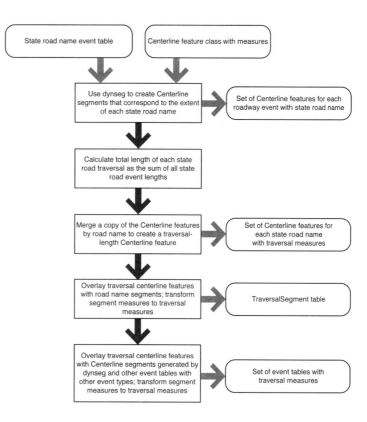

Figure 15.15 **Flowchart of the traversal publication process** The traversal publication process depends on spatial functions available in ArcGIS. First, use the SRName event table from figure 15.14 and the appropriate published Centerline feature class in the dynseg process to generate a group of line segments for each state road name. Then, sum the length of each segment by road name to derive the terminal measure for the traversal. (Assume a starting measure of zero.) Next, use ArcGIS to merge the segments by road name to create a new centerline feature depicting the full extent of each route. Some traversal centerlines will overlap where there is more than one road name assigned to a given section of highway. If you overlay the calibrated segment centerlines and the traversal centerlines, you can transfer the calibration results to the traversal LRM. You can then overlay line segments generated by individual event tables through the dynseg process and the traversal centerlines to use ArcGIS to transform roadway events to traversal events. This overlay process translates original event values to the measures that match the traversal LRM.

Another way to generate a new traversal feature derived from all the route name segments is with the Make Route command in ArcMap. This process starts with the Select by Attributes option in ArcMap to pick all the route segments that have the correct EventType and EventValue entries. The Make Route command will then merge the segments into a single route and transform the segment measures into route measures. This approach retains the original centerline feature calibration.

However you decide to create the new traversal routes, you would overlay linear and point event features generated via dynseg to translate their roadway segment measures to the traversal LRM measures using the Transform Route Measure capability of ArcMap. This process, like all others, can be automated through a script.

A nongeometry method of creating and preserving traversals is shown in figure 15.16. Each row in the SRName table represents a segment of the total traversal. A new RouteTraversal table stores the information about each traversal and a SegmentSequence table stores the relationship of SRName event records to the traversal of which they are a part. An associative table is required to store the location of the event-derived segment along the traversal, which is given in the PositionOnRoute field, along with the from- and to-measure values (using the traversal's LRM).

This structure is almost exactly the same as that in figure 14.10 in the preceding chapter for ordering Segment records along a Route with one difference: SegmentSequence additionally stores the traversal measures for each component event. The reason you need to explicitly store these LRM values is the difference in measure datums. In the Route-to-Segment version, segments have the same measure values as the route they form. In the Traversal-to-Event version shown here, the event-derived segments have a different LRM datum—the one supplied by the roadway segment—while the traversal has its own LRM datum.

You might wonder why go to all the trouble to create this data structure in the geodatabase when you can reconstruct traversals using geometry operators. The reason is that those geometry-driven methods can require a lot of processing time. Using the explicit ordering mechanism allows you to examine the component events to see if they have changed prior to publishing the next edition of the highway inventory database. You will have to process only those traversals where the component segments have changed.

Publishing traversals

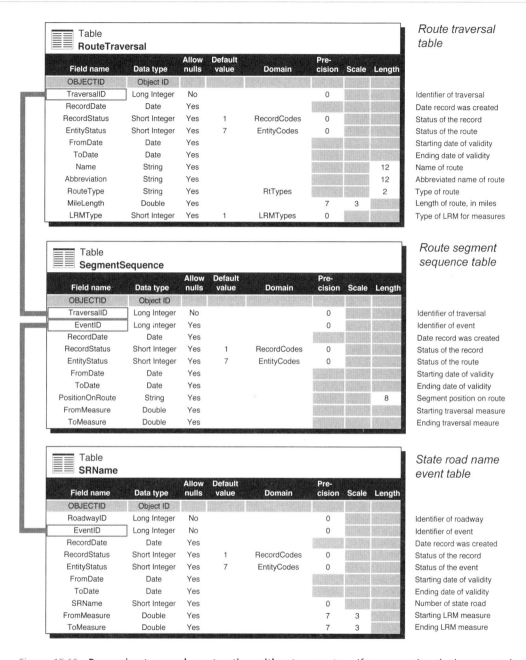

Figure 15.16 **Preserving traversal construction without geometry** If you are using the incremental publication process that relies on the **RecordStatus** and **RecordDate** fields to insert updates rather than doing a drop-and-replace process, it may be useful for you to store the relationship between state road name traversals and the highway sections from which they are formed. This would let you quickly identify those traversals that need to be reconstructed so you can avoid rebuilding all the others.

Getting started with SQL

This chapter has mentioned things like selecting rows and joining tables as ways to get useful information out of our normalized geodatabase or to publish the data it contains in a denormalized form. Chapters 2 and 3 provided a general foundation on the structure and operation of relational database management systems (RDBMS) and the geodatabase at a conceptual level. This section will show how to apply those concepts.

In a way, you can view the components of the Edit Geodatabase as stacks of bricks and bags of mortar, and the Publish Geodatabase as various structures constructed from those building materials. You can construct almost anything from the normalized bricks and mortar. A denormalized table is like a brick wall—quite useful when part of a building but hard to reuse somewhere else.

SQL is the language of relational databases and ArcGIS. SQL used to be an abbreviation for Structured Query Language, but now it is just SQL, which some people pronounce like the word "sequel." SQL is the way you turn bricks and mortar into brick walls. One useful aspect of SQL is that after you build something, you still have all the original bricks and mortar so you can build something else. The results of an SQL query do not alter the source of the data unless you tell it to. Another neat thing about SQL is that you do not have to figure out how to do the work; you just tell SQL what you want. The RDBMS will determine how to do the work.

While there are many words in the vocabulary of SQL, you only need to know a few to create and use the Publish Geodatabase for most applications. An SQL query, or statement, is fairly simple. The basic form of an SQL statement tells the RDBMS what fields and rows of data you want, where they come from, and how to present the data. Here is the basic form of an SQL statement:

```
SELECT    fieldname1, [fieldname2], ..., [fieldnamen]
FROM      tablename1, [tablename2], ..., [tablenamen]
```

The verb SELECT tells the RDBMS the fields or columns you want included in the result set of rows, while the FROM part tells it the tables where those fields are located. For example, if you wanted to see the contents of the Route table, you would say:

```
SELECT    RouteID, Name, Abbreviation, RouteType,
RouteDesignator
FROM      Route
```

What the RDBMS would return as an answer would be something like this:

Getting started with SQL

RouteID	Name	Abbreviation	RouteType	RouteDesignator
34	SR 470	SR470	State Road	State DOT
35	US 27	US27	US Route	AASHTO
36	SR 15	SR15	State Road	State DOT
37	East Main Street	E Main St	City Street	City of Auburn
38	Johns Ferry Road	Johns Ferry Rd	County Road	Stephens County

Table 15.1

The answer you get looks like a table and is called a result set. What you essentially told the RDBMS to do is show you everything in the Route table. A shorthand way to do the same thing is to use an asterisk character to replace the field names, which means "every field":

```
SELECT     *
FROM       Route
```

This query, or SQL statement, will return all field values for every row in the Route table. Usually, though, you do not want every row and every column. (Even if you do, explicitly listing the fields to include usually executes faster than using the * shorthand notation.) To restrict the response, you want to apply a condition to the query. Let's say you wanted to extract only the state roads from the Route table. Here is what you would say in SQL:

```
SELECT     *
FROM       Route
WHERE      RouteType = 'State Road'
```

The response would be:

RouteID	Name	Abbreviation	RouteType	RouteDesignator
34	SR 470	SR470	State Road	State DOT
36	SR 15	SR15	State Road	State DOT

Table 15.2

The WHERE clause provides the conditions and is referred to as a logical expression or a predicate. It may be easier to think of it as a filter through which table rows must pass. The result of applying the filter is True, False, or Unknown. Only rows that generate a True value are placed in the result set. An Unknown result happens when any field included in the filter contains a nullcontains a null, or no value for that field in the row.

Null is a very special thing. It is not a value; it is an absence of a value *right now*. If you consider a value contained in a field as being a color, null is not the absence of a color, it is the absence of knowledge of that color. It is more like what you cannot see outside your range of vision, like the thing behind your head. It has a color; you just don't know what it is. So, if someone asked you the color of something you could not see, the answer would be Unknown. That is what null represents.

The full list of comparator operators for a WHERE clause is:

+   Equal to
\>   Greater than
<   Less than
\>=   Greater than or equal to
<=   Less than or equal to
<>   Not equal to

OK, let's review:

- SELECT specifies the fields or columns to include in the result set;
- FROM specifies the tables to be examined (the source of the rows); and
- WHERE specifies which rows to include in the result set.

Enter multiple conditions by using a conjunction, like OR or AND. For example, U.S. Routes are also normally state roads, so you would want to include them in the SELECT clause to get all the state roads:

```
SELECT      *
  FROM      Route
 WHERE      RouteType = 'State Road'
    OR      RouteType = 'US Route'
```

RouteID	Name	Abbreviation	RouteType	RouteDesignator
34	SR 470	SR470	State Road	State DOT
35	US 27	US27	US Route	AASHTO
36	SR 15	SR15	State Road	State DOT

Table 15.3

Our SQL statement says to include any row of the Route table when the value of RouteType is 'State Road' or 'US Route'. You have to put single quotes on either side of the RouteType value because it is a string field. If it were numeric, you would just give the number without the quotation marks, as in this example:

Getting started with SQL

```
SELECT      *
  FROM      Route
 WHERE      RouteID = 36
```

RouteID	Name	Abbreviation	RouteType	RouteDesignator
36	SR 15	SR15	State Road	State DOT

Table 15.4

You should know that there is nothing special about the order of fields in the table and the result set. You can specify any order for the fields included in the SELECT statement; they do not need to be the same order as they appear in your source tables.

Of course, the record-level metadata fields are omitted from the example to save space. You can still use them, though, as this example query shows:

```
SELECT      RouteID, Name, RouteType
  FROM      Route
 WHERE      RecordDate BETWEEN '01-SEP-2008' AND '30-SEP-2008'
```

RouteID	Name	Abbreviation	RouteType	RouteDesignator
34	SR 470	SR470	State Road	State DOT
36	SR 15	SR15	State Road	State DOT
37	East Main Street	E Main St	City Street	City of Auburn

Table 15.5

The records for these three routes have been updated during September 2008. Notice that you do not need to include the field used in the WHERE condition clause within the SELECT statement as long as it is in the table listed in the FROM clause. BETWEEN can be used to define the limits of a range, which could be dates or numbers. You can also use NOT BETWEEN to find all the rows that do not meet the test condition.

Another SQL word that operates in a similar manner is IN, which allows you to specify a list of discrete values rather than a range. The same query with IN substituted for BETWEEN is:

```
SELECT      RouteID, Name, RouteType
  FROM      Route
 WHERE      RecordDate IN '01-SEP-2008', '30-SEP-2008'
```

The result set would include only rows with a RecordDate value that was either September 1, 2008, or September 30, 2008, and no values in between. As with most other operators, you can add NOT to reverse the conditional test.

Another typical query to pick filter values from a list by using the `IN` modifier is to select roadway segments based on a subset of federal functional class codes. Say you wanted to create a major road centerline feature class. To do so, you could list all the valid functional class codes for Urban Interstate, Urban Other Limited-access Highways, Urban Other Principal Arterials, Rural Interstate, etc. The result set would include only those functional class event records with the listed codes. You would then use dynseg to create the corresponding centerline segments for your new feature class.

Another comparator SQL operator is `LIKE`. A `LIKE` comparison enables you to specify part of a test and allow the rest to be open-ended. You use a wildcard character, the percent sign, to specify the part of the test that is undefined. (Note that not all RDBMS vendors use the same % wildcard character. An underscore is also commonly used as a single character placeholder, but it is not universal.) Consider the following example:

```
SELECT     RouteID, Name, RouteType
FROM       Route
WHERE      Name LIKE 'SR%'
```

RouteID	Name	RouteType
34	SR 470	State Road
36	SR 15	State Road

Table 15.6

You can also specify a range of values within brackets, as in:

```
WHERE      RCLink LIKE '%-[56-58]-%'
```

This clause will extract all the Segment table rows with an RCLink value that included county numbers 56, 57, and 58. `NOT LIKE` is also a valid operator.

Other useful conditions are `IS NULL`, `IS NOT NULL`, and `EXISTS`. A good application for the `IS NULL` test is in the Edit Geodatabase certification process, where you need to check for the presence of any null values in identifier fields. A simple "WHERE [ID field] IS NULL" test can list all the records you need to correct.

`EXISTS` allows the nesting of queries. In a nested query, the second part of the statement is used to identify the valid rows in the first part of the statement. Here is an example:

```
SELECT      RouteID, Name, RouteType
  FROM      Route
 WHERE      EXISTS
            (SELECT *
            FROM Segment
            WHERE Route.RouteID=Segment.RouteID)
```

The statement tells the RDBMS to go get a row in the Route table and extract the RouteID value. The EXISTS clause in parentheses, called a subquery, then tells the RDBMS to see if it can match that value to any of the values for RouteID in the Segment table. If it cannot find a match, then the EXISTS test is false and that row of the Route table is not included in the result set. (Indices on RouteID in both tables will help this query run faster.)

In addition to the EXISTS operator, the last SQL example introduces something new. The syntax 'Route.RouteID' refers to the RouteID column in the Route table. The table name is to the left of the period; the field name is to the right. Similarly, 'Segment.RouteID' means the RouteID field in the Segment table.

Now that you understand the query, can you guess what the result set will be? It will be all the Route table rows for which there is a record in the Segment table with the same RouteID value; i.e., the routes with segments. This is not really a meaningful query, but you can reverse the test to do something quite useful: find all the routes that do not have Segment table records. You can do this by using WHERE NOT EXISTS. The result set would look something like this, if you had not yet created any segment records for local roads:

RouteID	Name	Abbreviation	RouteType
37	East Main Street	E Main St	City Street
38	Johns Ferry Road	Johns Ferry Rd	County Road

Table 15.7

This quality-control test is useful for determining edit workload (how many routes still need segments), checking for broken foreign keys (there should be a match but is not), and similar processes. It should be part of the Edit Geodatabase certification process.

As you might imagine, result sets can get fairly long. Scanning screen after screen to find what you are looking for can be tedious. You can make the task easier by telling SQL to provide the results in a certain way using control commands like, ORDERED BY and GROUP BY. Here is an example for ORDERED BY:

```
SELECT      RouteID, Name, Abbreviation, RouteType
ORDERED BY  Name
```

RouteID	Name	Abbreviation	RouteType
37	East Main Street	E Main St	City Street
38	Johns Ferry Road	Johns Ferry Rd	County Road
36	SR 15	SR15	State Road
34	SR 470	SR470	State Road
35	US 27	US27	US Route

Table 15.8

You can additionally specify whether the order is ascending (ASC) or descending (DESC). Ascending order is the default when no preference is specified. It is generally a good idea when writing an SQL script to be explicit so you (or someone else) can readily follow the process at a later date. Always specify even the default parameter values.

Most filter conditions go in the WHERE clause, but some are applied to the SELECT clause. The DISTINCT command imposes a condition on the SELECT statement that tells it to select only those rows with unique values, as in this example, which selects all the route types that actually exist in the Route table.

```
SELECT      DISTINCT RouteType
FROM        Route
ORDERED BY  RouteType ASC
```

RouteType
City Street
County Road
State Road
US Route

Table 15.9

You can even do math by including such commands as COUNT, AVG, MAX, MIN, and SUM in the SELECT clause. The result set of all math operators is a single row with the numerical answer. The COUNT command tells you how many rows met the test condition:

```
SELECT  COUNT RouteID AS NumberOfRoutes
FROM    Route
```

NumberOfRoutes
5

Table 15.10

There's a new word in that SELECT statement. AS allows you to specify a name that may make more sense for the result. For instance, if you left the name of the column as RouteID, you might think there was a row with a RouteID value of 5 rather than understanding the answer as saying there were five rows in the Route table.

Another useful selection qualifier is HAVING. The following example SQL statement returns the rows in the Segment table where the length of the segment is more than 10 miles:

```
SELECT      CountyID, SegmentID, (ToMeasure - FromMeasure) Length
FROM        Segment
GROUP BY    CountyID
HAVING      (ToMeasure - FromMeasure) > 10
```

The result set will be a list of segments longer than 10 miles arranged in CountyID order. This query assumes that the length of a segment is equal to the mathematical difference between the ending and beginning measures, and that those measures are in miles. The necessary math is in the SELECT and HAVING clauses. There's also what appears to be an extra field (Length) in the SELECT clause. This is really an alias for the column that would have been called "(ToMeasure - FromMeasure)". SQL lets you give an alternative name for a result set column by listing that alias immediately after the real name in the table. This means the result set will have three column headings: CountyID, SegmentID, and Length.

To get the total length of state road segments in County 56, you would use something like this:

```
SELECT      SUM (ToMeasure - FromMeasure) Length
FROM        Segment
WHERE       CountyID = 56
            AND RouteType = 'State Road'
```

These various commands can be used in creative ways to test the quality of your data, not simply retrieve results for others to use. For example, if you have a candidate primary key field, such as RouteID, you could count the total number of rows using a SELECT COUNT statement, and compare that number to the number of rows returned when you did the same query with a `SELECT DISTINCT RouteID`. If the two numbers are different, then there is at least one duplicate value present in the RouteID column. Of course, you do not know which RouteID values may be duplicated. To get that answer, you have to do something like this:

```
SELECT      RouteID, COUNT(RouteID)
FROM        Route
GROUP BY    RouteID
HAVING      COUNT(*)>1
```

Can you see how this works? The `COUNT(RouteID)` addition to the `SELECT` clause works with the `GROUP BY RouteID` clause to tell the RDBMS to count the number of instances with each value of RouteID. The `HAVING COUNT (*) > 1` clause tells the RDBMS to include in the result set only those RouteID values where the number of instances—the value of `COUNT(RouteID)`—is greater than one. This list contains all the duplicated RouteID values.

Using relational joins

So far, you have dealt mainly with a single table, but sometimes you will want to work with multiple tables, such as for a table join. A table join connects two or more tables through a common field, called a foreign key. (See chapter 2 for more information on foreign keys.) As you may recall, denormalizing a geodatabase means that you take previously separate attributes and combine them in a single table, so it should be no surprise that this happens primarily through a relational join.

The primary type of join you will use is an equi-join using a parent-child (one-to-many) relationship. This means the foreign key will be present in both the parent (one) and child (many) tables and that the test will be one of equivalency. It is also referred to as an inner join. Every time you see a green connecting line between two classes, it indicates how to perform an inner join using the indicated field as a foreign key.

Say you wanted to add route names and types stored in the Route table to the records in the Segment table. You would do an inner join of the Route and Segment tables using RouteID as the foreign key:

```
SELECT    Name, RouteType, Segment.*
FROM      Route, Segment
WHERE     Route.RouteID = Segment.RouteID
```

Because the Name and RouteType fields only appear in the Route table, you do not need to specify the table in which those columns are found in the SELECT clause. Segment.* says to select all the fields in the Segment table, of course. The filter in our WHERE clause compares the values of RouteID in both tables included in the FROM clause.

You may be wondering where the table join occurs. It is the FROM clause, where you list two tables, and the SELECT clause, where you put fields from those two tables into the result set. Again, this is an example of how you use SQL to tell the RDBMS what you want in the output, not how to do the work. (The JOIN command in SQL is for special operations, not a typical equi-join.)

You can get a little fancier and specify only state roads. Figure 15.17 shows what happens.

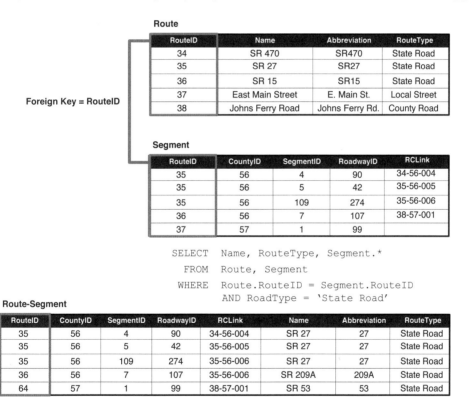

Figure 15.17 **SQL statement required to join the Route and Segment tables** What is treated in a flowchart as a two-step process really requires only one SQL statement. As discussed earlier, you can use **RouteID** as a foreign key to join the Route and Segment tables, and then select by **RouteType**. SQL will do both tasks at one time. The SELECT clause says which fields to include in the result set, while the FROM clause lists the tables that contain the fields and rows for the work. The WHERE clause provides the filtering test for selecting rows form those tables. Together, the SELECT and FROM clauses perform the join and the WHERE clause does the selection by road type.

Proceed with caution

A word of warning is in order: this chapter is NOT an SQL tutorial. Although SQL is a standard way of dealing with relational databases, each RDBMS vendor has varied a bit from the standard to differentiate its product from the crowd. This chapter has given you a sample of SQL to see what can be done as a general conceptual guide rather than a strict cookbook with explicit steps.

Here is a good example demonstrating why this part of the process is not a cookbook. Result sets have been shown as if they were output as tables. They are not. They exist only to the extent that you see them on the computer monitor or print them on a piece of paper. RDBMS platforms are masters of illusion, as discussed in chapter 2; nothing is as it appears. To actually put the result set into a table, you have to explicitly create the table and tell SQL to put the result set in it. There is no universal way to do this. With Oracle and MySQL, for instance, you would use something like this to store the results of our Route-Segment join:

```
CREATE    SRSegment AS
SELECT    Name, RouteType, Segment.*
  FROM    Route, Segment
 WHERE    Route.RouteID = Segment.RouteID
          AND RoadType = 'State road'
```

With MSSQL and Sybase, the statement looks more like this:

```
SELECT    Name, RouteType, Segment.* INTO SRSegment
  FROM    Route, Segment
 WHERE    Route.RouteID = Segment.RouteID
          AND RoadType = 'State road'
```

There is another reason this chapter cannot explicitly tell how you will compile the data for your agency's highway inventory. You will recall from chapter 3 that the geodatabase provides the appearance of a relational database but is not actually structured that way. What is presented as a single Route or Event table is really scattered across many data structures and object classes. You cannot communicate directly with the underlying data structure in a feature class or a table. You have to go through ArcGIS. This is part of the object-oriented paradigm.

So, why even talk about SQL? Because you will interact with many data sources at an enterprise level that will not be a geodatabase. For example, the Structures Inventory database that contains information about bridges, tunnels, and similar elements is likely to be stored in an external RDBMS database, such as that used by Pontis. This is also likely the case for traffic counts, HPMS section data, pavement management, work program projects, and many other parts of a complete highway inventory. Even if your agency has a strong plan to migrate away from legacy systems, it does not happen all at once.

Cross-platform interoperability is the basis for enterprise GIS at a state DOT. You will need to get data from non-ArcGIS users and supply data to users outside the ArcGIS community. There is no way for you to provide a complete highway inventory to the enterprise if you work solely in ArcGIS and the geodatabase. The Publish Geodatabase is only part of the complete inventory. You will need to use SQL to work with the rest of the inventory database. Whether you export all the table data in normalized form and then work in your RDBMS platform, or perform the publishing process through ArcGIS and export only the final result, you will eventually need to know how to work with SQL if you are doing enterprise GIS in a state DOT.

ArcGIS is exclusively positioned to be able to provide an integrating platform for this work through its numerous ways of interfacing with an RDBMS via both desktop and server platforms. Even if you go through just one ArcGIS mechanism for everything, it is likely you will have to deal with differences in the specific RDBMS and data structure used by a particular workgroup, including how various SQL query-processing approaches affect performance. ArcObjects supplies a whole list of functions, such as ISQLSyntaxProxy, that lets you examine the capabilities and syntax requirements of the underlying SQL so you can write scripts that provide the flexibility required to work with multiple RDBMS platforms. At the very least, this section should help you better use the Query Builder function in ArcGIS, which constructs SQL statements in the same form as shown here.

Data manipulation through SQL and the RDBMS cannot do the whole job. Steps in the transformation process, like applying calibration points, occur through an exclusively ArcGIS procedure and are not dependent on the underlying RDBMS. As a result of these vendor-specific issues, this chapter showed how to create the published geodatabase using flowcharts rather than SQL code.

Conclusion

Now you know how to convert the fully normalized editing geodatabase and combine it with external data to generate a complete highway inventory. Part of the published dataset should be a copy of the certified editing geodatabase that fed the publishing process. (See chapter 6 for additional guidance on the certification process.) Not only does this allow users to generate tables and feature classes that are not part of the published dataset, it also allows external users to extract updates made since their last update cycle by searching on RecordDate.

You also need to preserve all the retired records so the state of the database can be recovered at any historical point for which data exists within the geodatabase. You will certainly want to generate "now" versions of various published products, but many users will want to see how changes have occurred over time. These users will need all the retired records in addition to the active records that reflect the current state of the highway system. You could even use `RecordDate` in the publishing process to create special products that reflect only where things have changed.

A commonly requested map product is a straight-line diagram (SLD), where route geometry is a straight line and events are shown with tick marks and annotation along the length of the segment. (An example SLD is described in chapter 7.) You can automate the production of such products by creating a layout template with a straight line as the Centerline feature for applying event data from a denormalized table. Point symbols can denote the start and end of curves (PC and PT, respectively), bridges, and other elements along the segment. You can also employ the hatching capability of the Linear Referencing toolbox to generate tick marks at regular intervals along the segment.

Earlier chapters set the foundation for the concepts presented in chapters 14 and 15 on building, maintaining, and publishing a highway inventory for a state DOT. Foreign keys built into the editing geodatabase allow rapid reconstruction of a denormalized highway inventory that can be used for a number of legacy and new applications. Chapter 14 showed how to build the pieces. Chapter 15 showed how to put the pieces together in a variety of combinations. Along the way, you have seen how to use a little SQL to manipulate data.

Now that you have a published set of event-derived features created through dynseg, you can perform a number of spatial analyses that are not possible within a text-only legacy system. For example, you can do buffers to find events within a geographic distance from a central point, check to see that bridge features are properly oriented to cover the roadways to which they are related, create the appearance of divided roads by offsetting copies of a centerline feature, and generate event-laden features that can be fed into the network construction process of Network Analyst extension for ArcGIS. You can also use functions designed for one purpose to solve a different problem. For instance, using the Transform Route Events tool to transfer segment events to traversal centerlines. You can also use it to maintain event measures when routes are realigned.

Don't be bashful. Try things. You never know what great new data product you will discover. The full range of data products you can publish is limited only by your imagination.

chapter sixteen 16

A multipurpose transit geodatabase

- The central problem
- Expanding the network model
- The big picture
- Walk segments
- Fares
- Vehicles

Chapter 13 offered a transit network geodatabase structure example that can be implemented in ArcGIS Network Analyst extension. This chapter expands that example with a complete transit geodatabase design that can be part of a multi-modal database. The transit data model includes bus and rail conveyances with transit segments derived from highway and railroad facility inventories.

Chapter 16: A multipurpose transit geodatabase

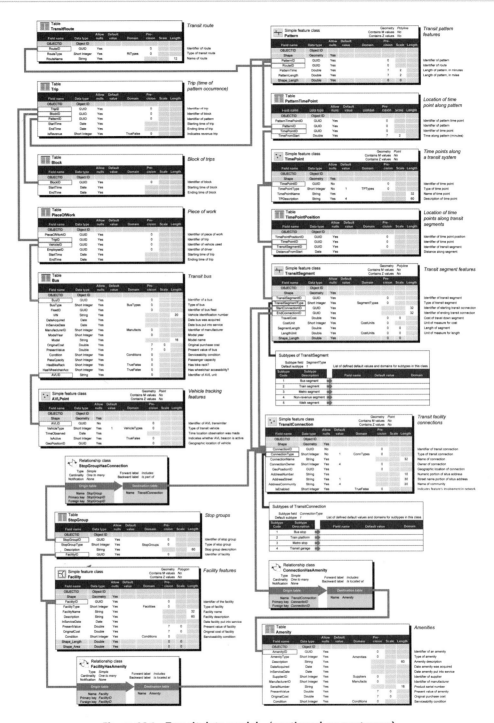

Figure 16.1 **Transit data model** (continued on next page)

The central problem

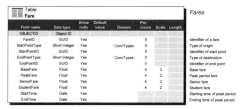

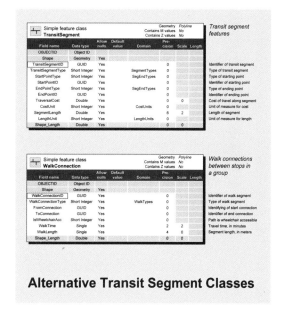

Figure 16.1 (continued) Transit data model
The complete transit geodatabase implementation presented in this chapter uses many ArcGIS class types and offers some alternatives to the basic structure that may be more in line with the business practices at some agencies. To view this figure in detail, go to http://www.esri.com/industries/transport/resources/data_model.html

The central problem

Transit segments may be composed of one or more roadway or railroad segments defined by terminal stops and/or time points. The main issue in designing a transit geodatabase is the organization of transit segments into higher orders of the structure (patterns, routes, trips, blocks, and pieces of work). These and other relevant terms need to be defined:

- Amenity—An element of a facility or transit stop, such as a bench, shelter, ticket kiosk, or message board.
- Block—One or more trip instances over a specified period of time.
- Facility—A building or other structure that usually contains one or more transit stops, perhaps arranged as a stop group. A bus station is an example of a facility that houses a stop group. A garage is a type of facility that does not include transit stops but may still be part of a pattern as conveyances depart to start a pattern or arrive to end one.

- Fare—The passenger's price to travel between an origin and a destination.
- Pattern—An ordered sequence of time points, transit segments, and/or transit stops (the path a vehicle follows).
- Piece of work—The assignment of a vehicle and an operator to a trip.
- Route—One or more pattern instances.
- Stop—A point where passengers may board or depart from a vehicle in scheduled revenue service. Stops are modeled as transit connections in a transit network.
- Stop group—A collection of closely spaced transit stops that work effectively as a single place for boarding and exiting conveyances.
- Time point—A location where a vehicle must be located at a specified time within a pattern. A vehicle that reaches a time point too early will normally be required to stay (dwell) at the time point until a specified departure time. A common use for time points is to maintain headways between vehicles along the service path.
- Transit segment—A directed uninterrupted path between two time points, two transit stops, or one time point and one transit stop. Each transit agency determines how to form transit segments.
- Trip—A pattern instance with a starting and ending time. Trips may be revenue or nonrevenue in nature. Unscheduled trips may be required to travel to or from a special destination, such as a charter operation, and may not be represented by a persistent pattern in the geodatabase.
- Vehicle—A motorized means of transporting people within the transit system. In this chapter, vehicles are buses that travel on roads and commuter trains, light-rail trains, and motorized vehicles that travel along an exclusive facility.
- Walk connection—A pedestrian path between two transit stops, usually located within a stop group.

Each of the terms defined above represents an entity for the transit data model. The basic elements from which other, more complex components are constructed are transit connections, transit segments, and time points. The most common arrangement is for transit segments to begin and end at transit connections, with time points located along transit segments.

Figure 16.2 illustrates that transit segments are directional. In this example, transit segments terminate only at transit stops, but including time points as segment termini would not alter the basic structure. The transit segment features have been offset from the street centerline to better illustrate the direction of travel. It would be fairly simple to derive the transit segments from the street segments, but the direction of vehicle movement along the segment within the pattern will need to be properly accommodated.

The sample portion of a transit system also includes a stop group at a street intersection and a walk connection within that stop group. Passengers can transfer directly from northbound Route 14 to westbound Route 3 at Stop 26, but eastbound Route 3 does not share a stop

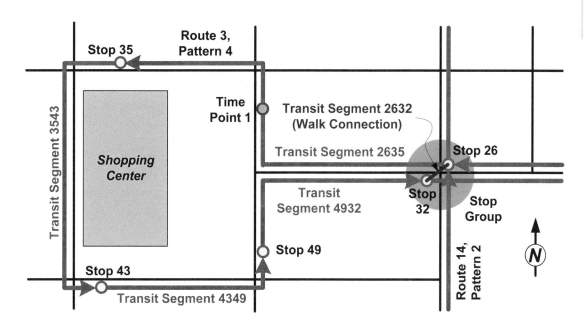

Figure 16.2 **Directional transit segments** The data model expects transit segments to have a direction implied by the ordering of connections. This figure expands the one in chapter 13. It shows that a given street segment may be traversed by multiple transit segments, which are defined by terminal stops (connections). A given transit segment may be "recycled" for use by multiple patterns only when the pattern has the same ordered pair of stops. This is because a pattern consists of an ordered aggregation of whole transit segments. A time point is located along Transit Segment 2635 to maintain headways.

with Route 14. The stop group includes the two bus stops located on opposite corners of the intersection: Stop 26 and Stop 32. The group supports transfers between these stops for all utilizing routes via a walk connection (Transit Segment 2632). Even though the two routes are just a few feet apart, the network cannot "see" the two routes as intersecting because they do not share a common stop. The stop group and walk connection provide the solution.

Chapter 13 presented example feature classes for transit segments and stops (connections) with a number of subtypes for each. Because the classes supported only the fundamental aspects of a traversable network, time points were not included. Stop groups are represented in this sample two-class model by including walk connections as a transit segment subtype. An alternative approach is to create additional connection subtypes and represent walk segments through connectivity entries. Other options are presented later in this chapter.

The transit network model indicates that a transit segment begins and ends at transit connections. A turn may serve as a connection between two transit segments, but most likely will be located along a segment. ArcGIS Network Analyst extension will subdivide the transit

Chapter 16: A multipurpose transit geodatabase

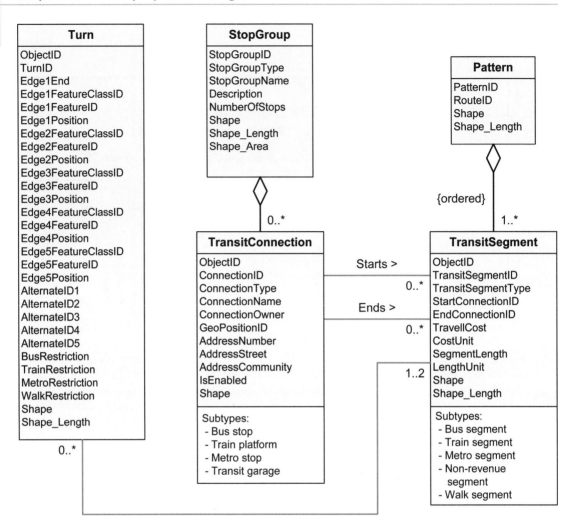

Figure 16.3 **Transit logical repeat**

segment into edges with turns and transit connections as end points, so transit segments are like complex edges in the geometric network model (see chapter 5).

As far as a transit network may be concerned, it is appropriate to terminate transit segments at connections in order to support a pathfinding application for a particular journey, as passengers may normally start or end such a journey only at stops. However, some agencies use time points to terminate transit segments with stops located along segments. Others may place time points along a transit segment, or treat one or more transit stops along a route as a coincident time point. To model such a network in the ArcGIS Network Analyst extension, you will need to treat time points as another connection subtype to provide the means to

The central problem

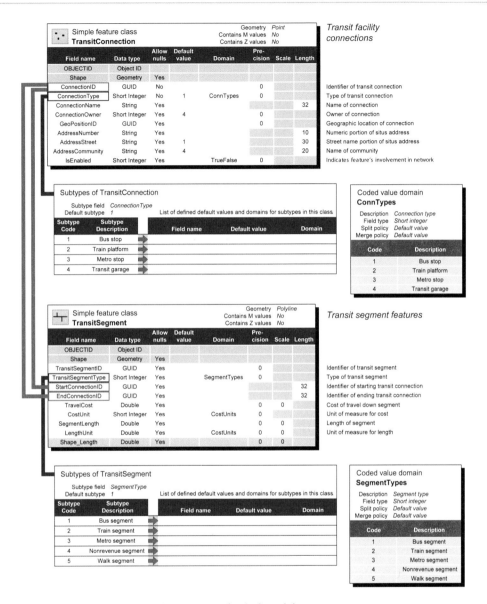

Figure 16.4 **Physical model repeat**

go from segment to segment. Under any of these various scenarios, time points can be an important element of a complete transit geodatabase. Nevertheless, implementation considerations require that transit segments in a network be terminated by connections.

Expanding the network model

The logical data model shown in figure 16.5 illustrates the many possible ways that transit segments, connections, and time points can be related. The original model has been modified to support both connections and time points as transit segment termini. Note that this is a logical model, unencumbered by implementation restrictions, such as the one noted above. The TransitSegment class has been expanded to allow starting and ending points to be defined by type and identifier. The TimePoint class does not include the many location description fields contained in TransitConnection, but this is merely to illustrate a minimal class specification. The additional fields may be added without changing the associative relationships illustrated in this model. To create a traversable network geodatabase, the TimePoint and TransitConnection entities may be implemented as a common TransitConnection point feature class, with Time Point being a subtype of that class. (This is why the TransitConnection class is not called "Stop": not everything you might represent in that class is a transit stop.)

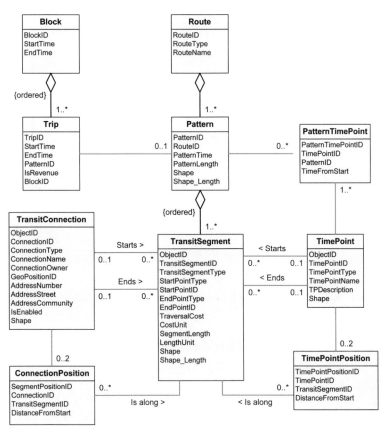

Figure 16.5 **Expanded transit logical data model** This logical data model expands the one presented in chapter 13 to include time points as transit segment termini or as something similar to point events along a transit segment. Not all possible associative relationships may be present in a single agency. The model also adds the additional classes required to provide structure to transit segments in a geodatabase design suitable for operational applications.

Expanding the network model

The ConnectionPosition and TimePointPosition classes are variations on the LRMPosition class, which could be used for the same purpose to provide a node-offset LRM as part of an enterprise geodatabase. The position classes indicate the minimal properties required to serve as an attributed relationship between TransitSegment and the two entities that may be located along segments. The DistanceFromStart field states the offset, in some unit of measure, for the time point relative to the start point given in the TransitSegment record.

As with LRMPosition, records in ConnectionPosition and TimePointPosition represent the spatial relationship between the located entity and the transit segments on which they exist. Other spatial relationships are also included in the logical data model, such as the placement of connections and time points at the start or end of a segment. It is likely that one or more of these relationships will need to be modified to reflect the practices of a particular agency during implementation. For example, an agency that has time points located only along transit segments will not instantiate the TimePointStartsTransitSegment and TimePointEndsTransitSegment relationships.

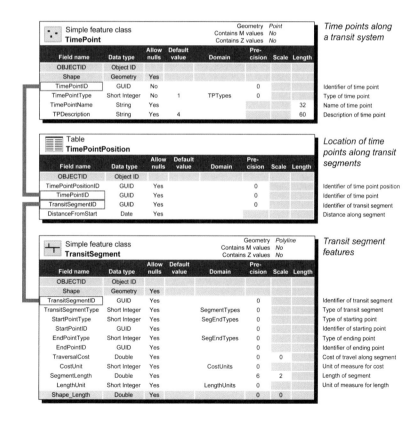

Figure 16.6

Time points and positions This figure shows the basic requirements for a time point feature class structure, with each time point located along a transit segment using a minimized version of LRMPosition. The TimePoint feature class may be expanded to include additional fields for such things as geographic location and street address. A time point may be related to more than one transit segment. This feature of the design may be useful when multiple patterns traverse the same facility.

Chapter 16: A multipurpose transit geodatabase

In the example classes shown in figure 16.6, TimePoint is a point feature class so that the location of time points can be readily illustrated. An alternative way to make a map of time points is to use dynamic segmentation and the TimePointPosition table as an event table to drive the process utilizing its node-offset LRM. Indeed, everything shown as a feature class could be implemented as an event table in a multimodal geodatabase. The only complication for such an approach is the need to construct transit segments as traversals. They will often need pieces of multiple roads to describe the path between transit stops along a pattern, which would also be a traversal.

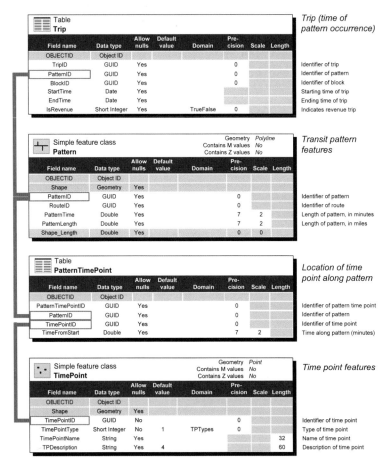

Figure 16.7 **Time points and patterns** TimePointPosition provides a physical location, describing the location of a time point along a transit segment using a distance measure. PatternTimePoint provides a logical location, locating a time point along the pattern based on time from the pattern origin. By adding the TimeFromStart value in the PatternTimePoint table to the StartTime found in the Trip table, you get the actual time when the conveyance should arrive or depart the time point.

ConnectionPosition and TimePointPosition are intended to supply physical location descriptions for transit connections and time points when these features are located along a transit segment. PatternTimePoint supplies the location of a time point relative to the start of the pattern in units of time. With the starting time for a particular trip provided in the Trip class (StartTime) and the time offset given in PatternTimePoint (TimeFromStart), it is easy to "do the math" and calculate the absolute time when the vehicle is to be at the time point.

The big picture

One of the primary differences between the geodatabase models presented in this chapter and those in chapter 13 is the expansion of classes to support the structure of inventory and operational databases used by transit agencies. In addition, the more complete design presented here must be compatible with the overall mission of this book to show how to create multimodal transportation geodatabase designs.

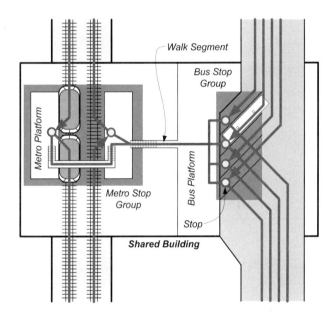

Figure 16.8 **Multimodal facility** Some transit facilities represent a point of interchange between different modes of travel. In this example, a bus stop group at ground level and a metro stop group located underground coexist within a single facility. Walk segments link stops within each group and the groups themselves. This example also illustrates why stops within a stop group must be explicitly modeled: the traveler must know which metro platform or which bus bay to use. An alternative to explicitly representing walk connections is implementing them as a matrix of stop pairs in a table. A network implementation would benefit from information to support the sign option in directions derived by ArcGIS Network Analyst extension.

One kind of multimodal facility is a transit station shared by bus and subway modes. The example shown in figure 16.8 has two stop groups, one for metro subway lines and another for buses. Walk segments link stops within each group and the two groups themselves. In an ArcGIS Network Analyst extension implementation, walk segments will provide connectivity at transit stops of the metro and bus subtypes. A walk segment must be constructed from each metro platform to each bus stop, and between bus stops. Walk segments must be directional in order to provide signpost support and other guidance mechanisms supported by ArcGIS Network Analyst extension.

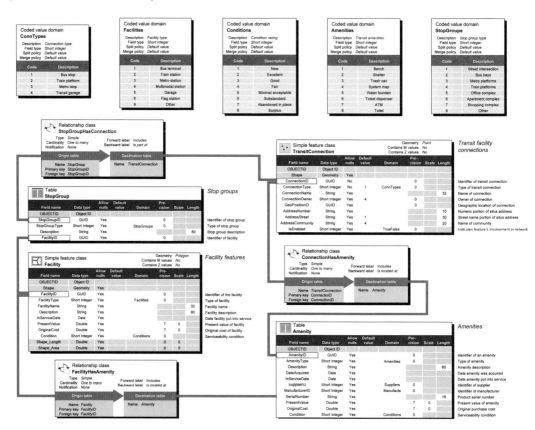

Figure 16.9 **Facilities, stop groups, and amenities** Facilities include bus and train stations, modeled here as polygon features. Most facilities will include one or more stop groups that will also be located at other sites, like street intersections. Some facilities may include multiple stop groups, depending on how the facility is arranged. Some larger metro stations, for example, could be accommodated by separate stop groups organizing adjacent train platforms, with movement between groups within the station defined by walk connections. Facilities and connections may both include amenities, such as shelters, ticket dispensers, and benches. You should assign an amenity to a facility when it is shared by multiple connections within the stop group. To view this figure in detail, go to http://www.esri.com/industries/transport/resources/data_model.html

Figure 16.9 illustrates a possible geodatabase design to store information about facilities and stops. It includes a Facility polygon feature class and the TransitConnection point feature class shown in earlier figures. It also includes two tables (Amenity and Stop Group), three relationship classes, and five domain classes. Both facilities and connections may have amenities. A feature class to illustrate amenities has not been included due to the expected scale of display for inventory data. Most amenities are not significant from a mapping perspective until the display scale becomes very large. However, it is a simple matter to assign a point symbol to an amenity type and use the FacilityHasAmenity or ConnectionHasAmenity relationship to place the appropriate point symbol on a map layer.

One relationship that is provided through foreign keys rather than a relationship class is that between StopGroup and Facility. This design decision is based on the relative rarity of stop groups being located in facilities, an outgrowth of there being few facilities in most transit systems. (Metro subway systems are an exception.) An alternative to implementing StopGroup as a table is to instantiate it as a polygon feature class, with some stop group polygons being located within the extent of facility polygons. You could then discover the association of stop groups and facilities through spatial operators.

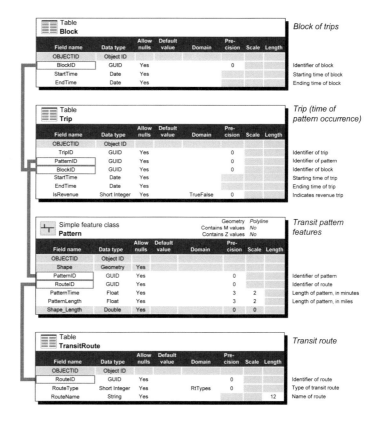

Figure 16.10 **Blocks and routes** At the top of the spatiotemporal structure of a transit geodatabase are the Block and Route tables. Although most transit users use the term "route" to refer to the path followed by a particular conveyance, they are really talking about a time-specific pattern within the route. For example, a route may go by a school for two hours in the early morning so children can get to school, and then switch to a pattern that takes the conveyance past a shopping center, bypassing the school. In the afternoon, the pattern might change again to go by the school and the shopping center. Blocks and routes may be illustrated by one or more Pattern features.

Chapter 16: A multipurpose transit geodatabase

Blocks and routes are at the top of the entity hierarchy displayed by the logical data model. Their implementation in a physical data model is shown in figure 16.10. Relationships between classes are provided through foreign keys. Explicit relationship classes could also be used, but foreign keys may be necessary to support applications outside the ArcGIS platform. All relationships are one to many. A block consists of one or more trips. Zero, one, or more trips can implement a pattern. A route consists of one or more patterns. Only Pattern is a feature class.

Walk segments

Walk segments are not part of any pattern. They are only a component of a path through the transit network followed by a passenger. So, if you use topology rules when editing the transit network, be sure to employ a rule that says a pattern cannot cover a walk segment. You could also use a rule that says transit segments must begin and end at transit connections (endpoint must be covered by). Of course, there are many other potentially useful topology rules, such as requiring transit segments to be a single part and not be self-intersecting, transit connections be covered by endpoint of a transit segment, and time points be covered by transit segment.

Simple feature class: WalkConnection

Geometry: Polyline
Contains M values: No
Contains Z values: No

Walk connections between stops in a group

Field name	Data type	Allow nulls	Default value	Domain	Precision	Scale	Length	
OBJECTID	Object ID							
Shape	Geometry	Yes						
WalkConnectionID	GUID	Yes			0			Identifier of walk segment
WalkConnectionType	Short Integer	Yes		WalkTypes	0			Type of walk segment
FromConnection	GUID	Yes			0			Identifying of start connection
ToConnection	GUID	Yes			0			Identifier of end connection
IsWheelchairAcc	Short Integer	Yes			0			Path is wheelchair accessible
WalkTime	Single	Yes			2	2		Travel time, in minutes
WalkLength	Single	Yes			4	0		Segment length, in meters
Shape_Length	Double	Yes			0	0		

Figure 16.11 **Walk connections** There may be advantages to treating walk connections as a separate feature class rather than a subtype of TransitSegment, such as when you want to include different attributes. In this example, a WalkConnection polyline feature class has been specified with a field that indicates whether the path is wheelchair accessible. The WalkTypes domain also defines a unique domain of segment description choices. WalkConnection could also be instantiated as a table to support the walk segment subtype of TransitSegment in a no-geometry approach.

Another approach to walk segments is to put them in their own polyline feature class rather than treat them as a subtype of transit segments. Part of the reason you will want to create a new class instead of a subtype is to provide different attributes. In the case of walk segments, you may want to know whether the travel path is, for instance, wheelchair accessible.

Fares

A prospective passenger is likely to want to know "How much?" when a pathfinding application shows the path through the system from origin to destination. The proposed Fare table in figure 16.12 is a list of all possible origin and destination pairs, each with its own set of costs to travel (fares). Although this sample table only includes four possible fare types, you can add any number of fare structures to the class specification.

Table: Fare

Field name	Data type	Allow nulls	Default value	Domain	Precision	Scale	Length	
OBJECTID	Object ID							
FareID	GUID	Yes			0			Identifier of a fare
StartPointType	Short Integer	Yes		ConnTypes	0			Type of origin
StartPointID	GUID	Yes			0			Identifier of start point
EndPointType	Short Integer	Yes		ConnTypes	0			Type of destination
EndPointID	GUID	Yes			0			Identifier of end point
BaseFare	Float	Yes			4	2		Base fare
PeakFare	Float	Yes			4	2		Peak period fare
SeniorFare	Float	Yes			4	2		Senior fare
StudentFare	Float	Yes			4	2		Student fare
StartTime	Date	Yes						Starting time of peak period
EndTime	Date	Yes						Ending time of peak period

Figure 16.12 **Fares** Many transit service agencies use a distance-based fare structure, where the longer trips cost more than shorter trips. This example Fare table illustrates how such an agency may include the fare structure in the geodatabase. Trips are assumed to begin and end at scheduled travel connections along normal revenue routes. Four different fare amounts can be accommodated, including a peak period fare that is higher than the base fare for nondiscounted travelers.

ArcGIS can help construct this table through a set of business rules. For example, if your system bases fares on total length, ArcGIS can create a set of traversals from each connection point to all possible destinations, add the total length traveled, and calculate the fare. If fares in your system are instead based on service zones, you could do polygon overlays to determine the number of zones crossed in each possible trip.

Vehicles

Chapter 13 described some vehicle classes that could be employed in a transit geodatabase. Chapter 3 illustrated how to create tracking objects for displaying temporal events. Figure 16.13 illustrates a portion of the transit geodatabase that implements both concepts.

The basic structure is that of a complex dynamic tracking object, where information about the tracked object is in one class and the location of the object is in another class. This split design precludes the need to repeat the inventory information with every location point

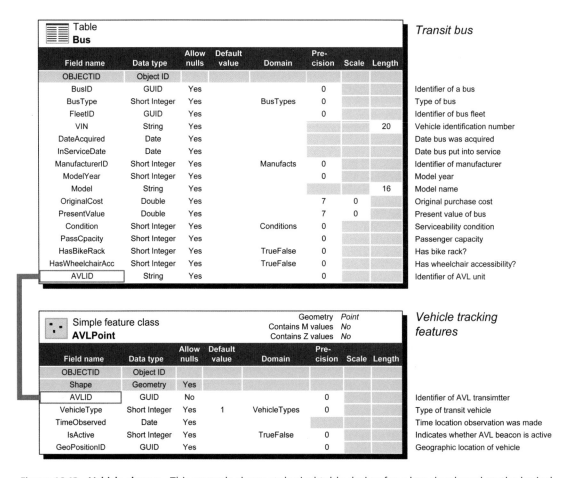

Figure 16.13 **Vehicle classes** This example shows a physical table design for a bus class based on the logical model described in chapter 13, except that the **GeoPositionID** field has been omitted because this design supports an inventory implementation, not a dynamic tracking application. However, the **AVLID** field offers a way to link the bus's automated vehicle location (AVL) signal to the inventory record to create a complex dynamic tracking event (described in chapter 3) as part of a bus tracking application.

380

feature. The design also anticipates that more than just buses may be tracked by including a VehicleType field. You could use a different symbol to display the location of various vehicle types. Through the use of transit connection buffers and facility polygon overlays, you could enhance the location information provided to users by indicating whether the vehicle was at a stop or facility. It would also be a fairly straightforward task to support an application that placed a vehicle along a pattern to supply an estimate of how long it would be before the vehicle reached a downstream connection point.

The PieceOfWork table completes the operational geodatabase design. A piece of work is the assignment of an operating employee and a vehicle to a trip. The starting and ending time fields in PieceOfWork are populated from the same fields in Trip. This redundancy is

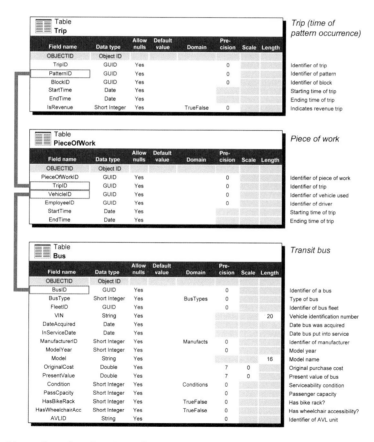

Figure 16.14 **Piece of work** The PieceOfWork table connects a trip to the driver and vehicle used to execute the trip. As such, it is functionally an attributed relationship between three classes. The **StartTime** and **EndTime** fields accommodate those relatively rare instances where the driver or conveyance used must change during the course of a trip due to such causes as mechanical failure of the vehicle or driver scheduling requirements.

intended to allow *Vehicle* and Employee classes to access this information without going all the way to Trip. This convenience allows a simple search and total of PieceOfWork records for a particular employee or vehicle identifier to see how much that vehicle was in service or which employee operated a vehicle.

A potentially less obvious reason for the duplicative starting and ending time fields is that they may not actually be duplicative. For various reasons, a single employee operator or vehicle may not complete a full trip. For instance, the vehicle could have a mechanical failure, requiring a replacement vehicle to complete the trip. Or an employee could become ill while operating a transit vehicle requiring a replacement driver to complete the trip. In both cases, the values of StartTime and EndTime will be different in the two classes.

chapter seventeen

Navigable waterways

- What's it for?
- Constructing waterway features
- River-reach data
- Where am I?
- Employing the new UNETRANS model
- Waterway events
- Channel geometry
- Gauging stations
- Bridges

Navigable waterways geodatabase design follows a model similar to that discussed in earlier chapters for highway-based transport systems. United States rivers are mapped using a linear referencing system based on statute miles, not nautical miles, in accordance with the rules of the U.S. Army Corps of Engineers. The full set of facility, event, and element options in the revised UNETRANS data model apply equally well to waterways as to highways.

A navigable waterways geodatabase suitable for pathfinding may extend the route event model to follow the geometric network structure of hydro features described in the ESRI Press publication *Arc Hydro* (David R. Maidment, Ed., 2002), or a version of the more flexible network model supported by ArcGIS Network Analyst extension, as described in chapter 13. The latter option would contain many of the elements of the transit data model of chapter 14.

Beyond using these other models as a general guide, you can also support multimodal and operational applications. Bridges across navigable waterways may impose restrictions on navigation, such as vessel height and width, just as they do on highway and railroad travel. Various modes of transport are affected by dynamic design limits imposed on marine traffic because of changing water conditions. Vertical clearance under a bridge depends on water level; a higher water level leaves less clearance for ships. Lift bridges may have limits on when they can be raised to allow taller ships to pass under them. Even relatively static elements, such as seaports, will likely appear as part of the inventory of railroad, transit, and highway systems that serve them, as well as the navigable waterways that provide access to waterborne vessels. The capabilities of ArcGIS Network Analyst extension can be fully employed in such multimodal environments by defining ports and bridges within multiple connectivity groups and calculating restriction factors using facility attributes.

Portions of the Arc Hydro geodatabase model are more directly applicable to the construction of a navigable-waterways geodatabase in a less rigorous environment. HydroEdge features in the Network feature dataset described in that publication can supply the geometry required to create waterway features for the navigable system, while the Bridge, Dam, and Structure point feature classes can offer representations of those entities. These class specifications are not repeated here.

This chapter will present a physical data model implementation using the UNETRANS concepts from chapter 12 and others, with elements of the Arc Hydro data model.

What's it for?

It is fine to offer a model for building a navigable-waterways geodatabase, but why would anyone do so? After all, there are very few governmental entities that manage the navigable-waterways of the United States or any other country. Accordingly, Arc Hydro was designed to manage water resources, which have a much broader application. Its geodatabase design is based on a directional network of edges and junctions oriented toward water flow analyses, although it also accommodates linear referencing. There are a few water bodies, monitoring stations, dams, locks, etc. scattered along the way, but it is really just like a utility system: water goes from an upstream source to a downstream sink.

Shipping—the principal waterway navigation activity—goes upstream and downstream. That is one major difference between a water resources design and a navigable-waterways structure. Indeed, it is the major attribute common to all transportation databases that differentiates them from utility networks and led to the development of ArcGIS Network Analyst extension.

The problem with using navigation, or pathfinding, as the focus application for a navigable-waterways geodatabase design is that people do not really use such an application. If you are going to St. Louis from New Orleans with a gravel-hauling barge, then you are traveling up the Mississippi River. Currents, water levels, and other concerns about the journey change continuously, but they are analogous to choosing which lane to be in rather than which road to take. There are lots of turns, but topologically, you are traveling in a straight line on a single facility. No, pathfinding is not a useful application around which to develop a navigable-waterways geodatabase model. Thus, a navigable network is not a component of the model presented in this chapter. If you need one, it can be constructed using the concepts offered in chapter 16, substituting ports of call for transit stops.

What this chapter will offer is a waterway inventory geodatabase design useful for local and state governments responsible for managing and signing rivers and lakes where mainly recreational activities occur. Local governments set waterway speed limits, establish no-wake zones, build and manage parks and boat launches, provide rescue services, and patrol the waterways for criminal activity. State governments build and maintain bridges across waterways, operate moveable bridges, regulate waterside development, and manage all the land they own to the high-water mark. There are only a handful of potential users of a navigable-waterways geodatabase that is, itself, navigable, but there are thousands of local and state governments that can use a design for a waterway facility inventory geodatabase.

Constructing waterway features

For most agencies, hydrographic features are a standard element of their spatial data resources. At the mapping scales that local and state governments typically use, navigable waterways are generally represented as polygons, what was traditionally called a double-line stream. There may also be floodplain polygons and a flowline that offers a linear abstraction (a single-line stream). In hydrographic analyses, the flowline is more correctly called the thalweg, which is a line that follows the lowest elevation of the streambed.

Chapter 17: Navigable waterways

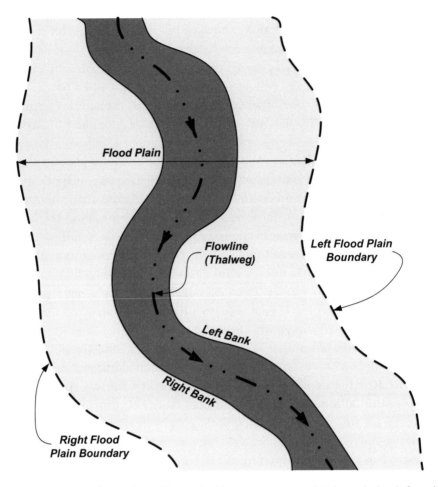

Figure 17.1 **Waterway abstraction** The navigable-waterways geodatabase design is founded on waterway abstraction rules. At the smallest scale of representation, a flowline or centerline shows where the waterway has its greatest depth. This line is more formally known as the thalweg. The direction of flow (illustrated by arrows on the flowline) determines left and right. The left and right banks indicate the normal limits of the waterway. Various high-water lines may exist beyond the banks, such as indicating the 100-year flood area (flood plain).

One major difference between navigable waterways and land-based modes of transport is the need to know the facility's depth. This difference can be characterized as a continuously variable load-bearing surface that changes in both width and depth along its length. Not only that, it varies over time at the same location because of changing water depth and flow speed. A complete geodatabase design for navigable waterways must be capable of storing information that can be used in a dynamic reporting of waterway capabilities. This means minimally being able to convert water surface elevation data into waterway width and depth.

Constructing waterway features

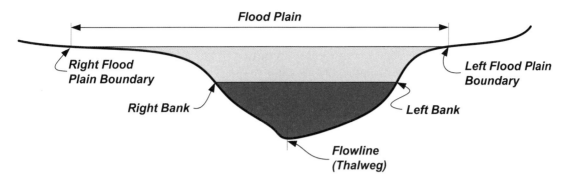

Figure 17.2 **Cross-section view** Although most geometry associated with transportation facilities is constructed from an aerial viewpoint in plan view, navigable waterways frequently need to be shown in cross-section to determine where a particular watercraft may travel. The navigable waterways geodatabase must accommodate cross-sections and flowline profiles.

Waterway width at a point along the feature is expressed as a cross-section that shows the profile of the streambed across the waterway. Waterway depth along a facility is described in a longitudinal flowline profile, which is defined by the thalweg. The capacity of a waterway is determined by its width at the shallowest point along the flowline.

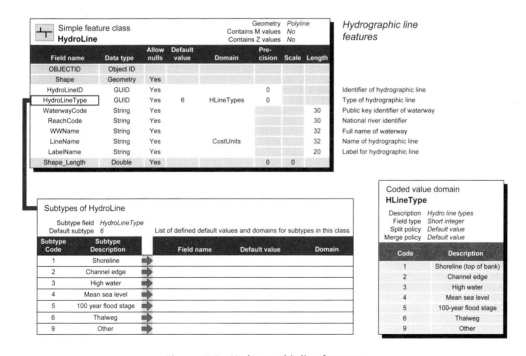

Figure 17.3 **Hydrographic line features**

Chapter 17: Navigable waterways

The Arc Hydro data model is primarily based on a geometric network foundation, but it also accommodates hydrographic line features that do not correspond to a navigable edge. The expanded form of that class is presented in figure 17.3 as a foundation for constructing all linear and polygonal representations of water features. One of the modifications to the original Arc Hydro version is to provide seven subtypes. Six of the subtypes represent a waterway boundary. The seventh is a waterway centerline, the thalweg.

The HydroLineID field offers a stable candidate primary key (foreign key) for linking members of this class to other classes, as is common practice throughout this book. In contrast, this is the first class to offer two other candidate primary keys that can serve as an explicit public key (WaterwayCode and ReachCode). This allows the expected common use of existing spatial data to construct HydroLine features and the need to maintain a foreign key link to that source. WaterCode serves to hold a locally generated key, while ReachCode accommodates the national identifier for navigable waterways. (See discussion below.) Both are string fields that provide the greatest flexibility.

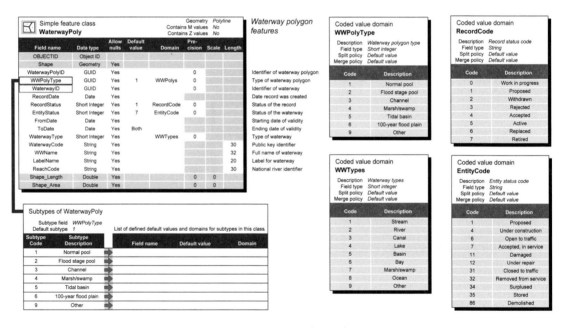

Figure 17.4 **Waterway polygon features**

Taking a page from the design book for cadastral editors, the model in figure 17.4 includes waterway and waterbody polygons constructed from HydroLine features or as original polygons without a HydroLine source. As with HydroLine, WaterwayPoly is structured again around seven subtypes. The example shown above also includes some of the editing support fields introduced in chapter 6, such as RecordStatus and the dates of validity for each record. These fields could be included in the HydroLine feature class, too.

HydroLine and WaterwayPoly subtypes only illustrate the range of values that could be accommodated. This is why there is not a one-to-one correlation between the two classes, which you would expect given that waterway polygons can be constructed from hydrographic line features. Some differences between the two classes have specific reasons. For example, the HydroLine feature class includes a LineName field that allows you to provide a full description for a given feature, such as "Right descending bank" or "Left edge of channel." Both feature classes include a WWName field for storing the full name of a waterway or waterbody ("Tennessee River") and a LabelName field for storing the label text you might want to put on a map ("Tenn. R.").

River-reach data

A reach is a continuous, unbroken expanse of surface water. There are approximately 3.2 million such segments delineated in the United States, each defining a fairly homogeneous set of attributes, such as width, depth, and water character. Each reach has a globally unique identifier. The U.S. Environmental Protection Agency (EPA) established river-reach numbers as public key identifiers for waterway segments. A river-reach number (RRN), also called a reach code, is composed of three parts: an eight-digit hydrologic unit code assigned by the U.S. Geological Survey (called "HUC"), which is itself composed of four two-digit parts; a unique four-digit sequence number (called "SEG"); and a river mile marker (called "RMI", in the form xxx.xx). The end result is a 17-character string defining a segment of a waterway or waterbody.[1] Although composed only of digits, the field must be defined as a string since some RRNs begin with a leading zero that would be dropped by a numeric field type.

The National Hydrography Dataset (NHD) is the source for river-reach data, which is known by the designation "RF3," for Reach File Version 3. Each version was based on map data of increasing scale; the RF3 dataset is correlated with 1:100,000-scale USGS maps in DLG format. Each increase in scale led to a corresponding increase in the number of segments defined. Some states have moved to a 1:24,000 dataset since RF3 was completed in 1993.

An NHD feature is the real-world entity and its geometric representation. Water features included in the NHD are naturally occurring waterways (stream or river), man-made paths along which water flows (canal or ditch), and standing bodies of water (lake or pond). Such features may be perennial—they contain water throughout the year except, perhaps, during infrequent periods of extreme drought—or intermittent. In both cases, water must be present at times other than rainfall or snowmelt, which means that dry ditches and many storm-water retention ponds are not included. Reaches are classified according to use: transport (stream, river, canal, or ditch); waterbody (complete boundary included); or coastline (part of boundary included).

Geometric representations may be points, lines, and polygons, although each feature can have only one kind of geometry; a change in geometric representation produces a new feature for the same waterway. Lines digitized in the direction of water flow, when it is known, represent transport reaches. The digitizing direction of lines may also be oriented in accordance with certain business rules. For instance, along U.S. coastal boundaries, the direction of flow is always oriented so as to place the waterbody to the right. Linear features must terminate at confluences, which are decision points along a waterway network. Bodies of water greater than 10 acres are represented by polygons. Line and polygon features may overlap.

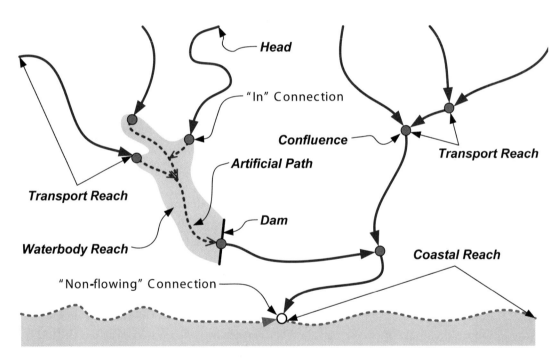

Figure 17.5 **Reach rules** This figure illustrates three basic rules for delineating reaches. The first is the confluence-to-confluence rule, which states that a transport reach cannot extend past the confluence of two reaches. Examples are shown on the right side of the figure. The second is the underlying feature rule, which breaks reaches when the manner of representations must change in conformance to geometry rules. Examples of this rule are included on the left side of the figure, where the transport reaches contact the waterbody reach. The third is the branched-path rule, which uses artificial paths through a waterbody feature to supply connectivity within the set of transport reaches. These paths are shown as dashed lines within the lake polygon.

Each feature is described by a set of characteristics, which are specific to each feature type. Hydrological connectivity is a part of the description for each transport reach, so it is possible to combine the segments into longer facilities. Artificial paths or connectors are transport reaches defined through intervening bodies of water, such as an impoundment along a river, in order to supply continuous connectivity along the segments. This allows flow relations to be encoded in the dataset. Such information can be used to construct a network for various applications.

There are six connection types based on flow direction expressed using the ordered statement of the two involved reaches' RRNs. The most common of these is for the end of one reach to flow into the beginning of the downstream reach. The next most common is the network start, which is at the head of the first reach in the sequence. (The upstream reach's RRN is null.) In addition to being stated in terms of the first reach and the second reach in the relationship, the nature of the flow relationship is described, such as through the terms "in" or "out." In figure 17.5, network start connections are defined for all water course heads and all but one of the remaining connections are of the "in" type. The exception is the nonflowing connection at the coastline. Nonflowing connections can be defined between coastline reaches and between a transport reach and a coastline reach.

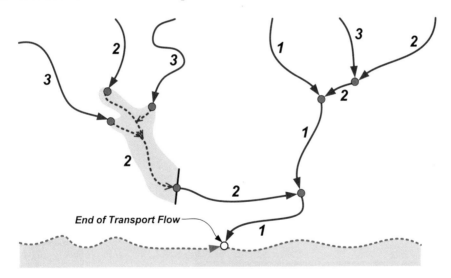

Figure 17.6 **Stream level** The stream-level characteristic supplied by the RF3 field offers a way to combine reaches into longer features. The process starts by assigning the number 1 to the outlet transport reach that is the one farthest downstream. Each successive upstream reach is assigned this same number based on water-flow volume. Whenever a confluence is reached, the number 1 is assigned to the upstream branch with the greatest water flow. Once this path has been traced, reaches intersecting the number 1 path are numbered as 2 and the process continues up each of these paths. Numbers keep increasing through the drainage network in the same manner until all reaches are assigned to a level.

One of the other useful characteristics of the network that is supplied by the RF3 dataset is stream level. The main transport feature is assigned to stream level 1 at its outlet or downstream terminus, directly intersecting transport features are placed in stream level 2, features that intersect level 2 features are assigned to level 3, and so on, until all reaches within the drainage basin have been assigned to a stream level. Where there are multiple choices as to which reach to assign to the lowest available level, such as at a confluence, the stream with the higher flow volume is chosen and the other reach is assigned to the next-lower level. You can use this characteristic to help identify the path of each major feature when constructing longer waterways from reaches.

Where am I?

The geographic location of features contained in the RF3 dataset is derived from the map upon which the dataset is based. You may similarly choose to compile features in this manner from USGS maps or your own aerial photography, but it is perhaps more likely that you will go out with a GPS receiver and locate features, like markers and other facilities to put in your inventory, using geographic coordinates. The GeoPosition table presented in earlier chapters remains a useful way to retain this information in a form that allows you to publish it in various datums, depending on the needs of the user communities you serve. Of course, you could do what the EPA did and rely on map coordinates and their projection to various datums. But that precludes direct analysis of locations outside ArcGIS.

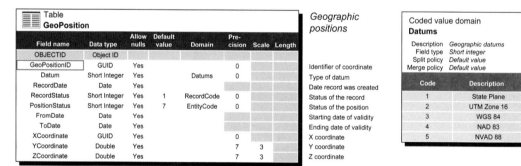

Figure 17.7 **Geographic position class** Perhaps the more common kind of position statement in waterway geodatabases is geographic, as it is difficult to clearly identify centerlines in the field. This is the same GeoPosition class that was presented in earlier chapters. It is proposed as the basic level of accommodation for geographic positions rather than the more common approach of adding coordinate fields to each feature class or relying on the projection of map coordinates. The reason is that it is likely that you will need to provide data in multiple datums for different user groups and are likely to receive data in multiple datums—or the same datum but with a slightly different set of coordinates for the same waterway element.

The alert reader will have noticed a reference a couple of pages back to a river-mile marker as part of the reach code. Based on statute miles and determined by the U.S. Army Corps of Engineers, river-mile markers are part of a linear referencing system for navigable waterways. The good news is that you can use these markers to extend the concept of routes and event tables to navigable waterways. The bad news is that this LRM has limitations to application, particularly in the field. The problem is that there is no painted centerline on a waterway, and with so many potential "lanes" from which to select, you will find it very difficult to actually travel a mile along the waterway between mile markers. And forget about any reasonable degree of repeatability. You also have to get used to the fact that, in most cases, river milelog measures increase numerically in the direction opposite the way water flows.[2]

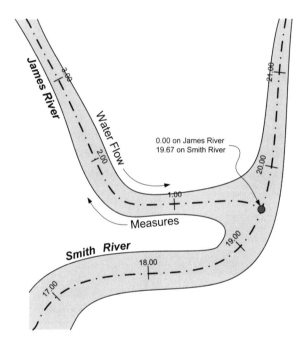

Figure 17.8 **Directions** Waterway centerlines are usually digitized in a downstream direction, which is suitable for most applications. But river measures almost always go in an upstream direction. The navigable waterways geodatabase design will need to keep track of the relationship between geometry direction and LRM measure direction.

There is one more distinguishing characteristic of the waterway LRM supplied as part of the RF3 dataset: it uses relative addressing. Unlike most transport LRMs, which use absolute addressing with measures stated in units of length, such as thousandths of a mile, relative addressing employs a percentage of travel of the total distance from the origin to the termini.

This practice resulted from having to deal with maps presented at different scales, which affected reach centerline length. As scale increases, so does the number of bends in the centerline. Each bend increases centerline length, which can "move" locations along that centerline if all you do is traverse the distance from the origin to the point of interest.

Dynamic segmentation (dynseg), as implemented in ArcGIS, avoids that issue by always implementing relative addressing between vertices where you have assigned a specific measure value, such as at calibration points. The first version of the river-reach file was created back in 1975—long before there was dynseg—so the design of RF3 does not have its roots in the GIS era.

As a result of all these limitations on the reach LRM, you can only calibrate the waterway centerline's LRM at the end of each reach segment using the last component of the reach code. All other LRM measures are relative to the origin measure for the reach. Fortunately, you can do the math yourself (or, better, let ArcGIS do it for you) to convert relative addressing to absolute addressing by multiplying the percent measure by the numerical difference between the reach's origin and terminus distance measures, and adding the result to the origin measure.

Employing the new UNETRANS model

The UNETRANS facility inventory data model is founded on route and event tables using an LRM, with additional support for geographic positions; any number of geometric representations can be explicitly stored or created on the fly through dynseg. The first task to adapt that general model to the specific needs of a navigable-waterways geodatabase is to create the route table. The Waterway table is the result.

Table Waterway

Field name	Data type	Allow nulls	Default value	Domain	Precision	Scale	Length	
OBJECTID	Object ID							
WaterwayID	GUID	Yes			0			Identifier of waterway
RecordDate	Date	Yes						Date record was created
RecordStatus	Short Integer	Yes	1	RecordCode	0			Status of the record
EntityStatus	Short Integer	Yes	7	EntityCode	0			Status of the waterway
FromDate	Date	Yes						Starting date of validity
ToDate	Date	Yes	Both					Ending date of validity
WaterwayType	Short Integer	Yes		WWTypes	0			Type of waterway
Length	Double	Yes			7	3		Official waterway length
FromMeasure	Double	Yes			7	3		Starting LRM measure
ToMeasure	Double	Yes			7	3		Ending LRM measure
WaterwayCode	String	Yes					30	Public key identifier
WWName	String	Yes					32	Full name of waterway
LabelName	String	Yes					20	Label for waterway
ReachCode	String	Yes					30	National river identifier

Waterways

Figure 17.9 **Waterways**

Employing the new UNETRANS model

It should be no surprise that the Waterway table looks much like the WaterwayPoly feature class. It also supplies the other end of the WaterwayID foreign key relationship implied by the WaterwayPoly feature class, which means there is a one-to-many relationship that says there may be zero, one, or more waterway polygons to supply the geometry for a waterway.

The notable difference between WaterwayPoly and Waterway is the addition of three fields required to support an LRM in the latter class: Length, FromMeasure, and ToMeasure. What may be a surprise is LRM fields are incorporated directly into the Waterway class rather than in a separate LRMPosition class, as has been done in previous example implementations of the UNETRANS model. This simplifying choice was made on the expectation that there would be only one longitudinal LRM, the one supplied by the RF3 dataset. (As will be shown later, there may also be any number of lateral LRMs.) The LRM extent of each waterway would be determined by the reach relationships of the RF3 file, or—if that source was not used—by the editor's own selection of the component reaches/segments.

The WWTypes domain class provides a general set of choices for the various waterway types that you may include in the navigable-waterways geodatabase. The set offered above includes feature types that would not be part of a transport system, such as marsh/swamp. Although it is possible to include every water feature within the Waterway table, only those with an LRM are contemplated for mandatory inclusion; others could be accommodated by the WaterwayPoly or HydroLine feature classes.

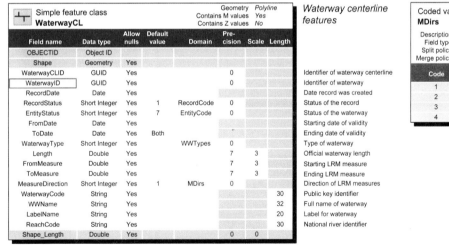

Figure 17.10 **Waterway centerlines** Waterway centerlines are geometric features that can be used like a route for locating events and employing dynamic segmentation. Existing waterway flowlines are typically adapted to create centerlines. If measure values are imputed from the geometry, then it will be necessary to indicate which direction the measures need to run. The MDirs domain class offers the range of values for that attribute, stored in the **MeasureDirection** field of the WaterwayCL polyline feature class.

Chapter 17: Navigable waterways

The next step in developing the navigable-waterways geodatabase is to describe the centerline feature class that supplies a geometric representation for each transport waterway included in the Waterway table. A waterway centerline can be constructed from the thalweg HydroLine subtype or one or more single-line stream features. You could add a MaxScale field to indicate the functional limit of displaying a class member on a map layer, thereby allowing you to have multiple, scale-dependent geometries for a given waterway. Specifying the appropriate combination of values for WaterwayID and MaxScale would select the geometry needed for a particular application.

In its basic form, WaterwayCL is essentially a polyline feature class with the attributes of the Waterway table. A notable difference is the inclusion of a MeasureDirection field that is used to indicate the relationship between the digitized direction of the line and the increasing measure direction for the LRM. A centerline feature that has not yet been calibrated with measure values would be uninitialized. A feature that cannot be calibrated is indeterminate.

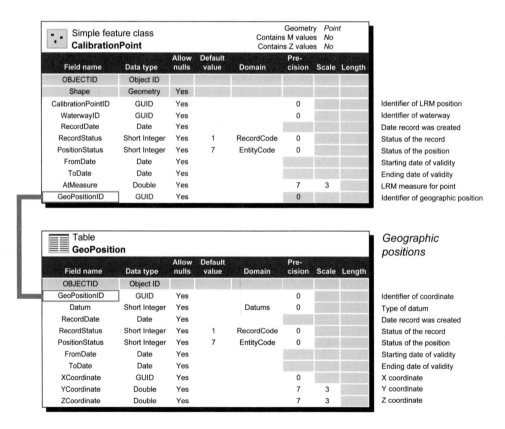

Figure 17.11 **Calibration point**

A CalibrationPoint feature class can supply the required point features that correlate a waterway LRM measure with a geographic position so as to apply measure values to the related WaterwayCL feature. An example of such a class is shown in figure 17.11 with its related GeoPosition class. An alternative design would put the three geographic coordinate fields into CalibrationPoint, but that design would forgo the benefits of the separate GeoPosition class, such as the ability to publish data in multiple datums more easily. Members of this class can be used during a manual or automated m-coordinate calibration session when WaterwayCL features are edited. (See chapter 14 for more information on this process.)

Waterway events

Now that the route equivalents in the Waterway class and their representative centerline features have been designed, the next step is to create event classes. The primary event class proposed here is a linear event representing a homogeneous channel segment, which has a fairly uniform width and depth. Although a linear event, the most likely representation of a channel segment is a polygon, given the display scales typically employed by local and state governments. The surface area of a channel segment may be described by a set of point features locating channel markers, so a class to store channel markers is also required to be able to construct the geometric representation of channel segments.

The U.S. Coast Guard designates channels for the safe movement of vessels along navigable waterways demarcated by lateral markers. The U.S. Aids to Navigation System (USANS) governs the placement of markers along navigable waterways to identify the limits of the route to be followed, essentially forming edge lines for the waterway route. USANS replaces prior systems, such as the Western River System previously used for the Mississippi River and its tributaries. There are also nonlateral aids to navigation that provide other information to boaters.

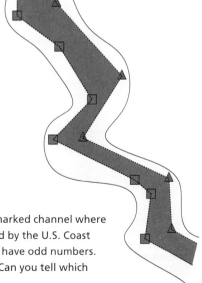

Figure 17.12 **Channel** Major navigable waterways have a marked channel where the deeper water is located. Channel markers are maintained by the U.S. Coast Guard. Starboard (right) descending buoys will be green and have odd numbers. Port (left) descending buoys will be red with even numbers. Can you tell which way the water flows here?

USANS is designed to work with navigational charts, and the meaning of some navigational aids on the waterway may not be discernable without reference to the appropriate chart. You will want to use the standard point symbols and accompanying notations on your maps. Each Coast Guard district additionally publishes a report called "Local Notice to Mariners" that includes reports on marker deficiencies (e.g., missing or moved by weather forces) and changes that may not appear on the navigational charts. These notices can help you keep your data current.

The lateral markers that define the navigable channels of U.S. waterways are buoys or daymarkers that indicate the edge or other important aspect of the shipping channel. Channel edge markers are green and square on the starboard (right) side of the channel and red and triangular on the port (left) side of the channel. Left and right are determined by the downstream (descending) direction of travel. It is critical that you know whether you are traveling upstream or downstream in order to properly respond to channel markers. A common saying, "Red, right, return," also known as the "3R-rule," is used by beginning mariners to recall that upstream travel keeps red triangular markers to the right, even though they are left side markers for the controlling direction of travel.

The ChannelSegment linear event table in figure 17.13 has the attributes of the Waterway class, but adds channel width and depth fields. It also has a ChannelName field, which accommodates those instances when a single waterway has multiple channels that allow

Table ChannelSegment

Field name	Data type	Allow nulls	Default value	Domain	Precision	Scale	Length	
OBJECTID	Object ID							
ChannelSegmentID	GUID	Yes			0			Identifier of channel segment
WaterwayID	GUID	Yes			0			Identifier of waterway
RecordDate	Date	Yes						Date record was created
RecordStatus	Short Integer	Yes	1	RecordCode	0			Status of the record
SegmentStatus	Short Integer	Yes	7	EntityCode	0			Status of the channel
ToDate	Date	Yes						Starting date of validity
FromDate	Date	Yes						Ending date of validity
FromMeasure	Double	Yes			7	3		Starting LRM measure
ToMeasure	Double	Yes			7	3		Ending LRM measure
ChannelWidth	Float	Yes			5	0		Channel width, in feet
ChannelDepth	Float	Yes			5	1		Channel depth, in feet
WaterwayCode	String	Yes					30	Public key identifier of waterway
ReachCode	String	Yes					30	National waterway identifier
WWName	String	Yes					32	Full name of waterway
ChannelName	String	Yes					32	Name of hydrographic line
LabelName	String	Yes					20	Label for hydrographic line

Channel segment elements

Figure 17.13 **Channel segment elements**

boaters to avoid an obstruction or to reach different internal destinations. For example, some harbor entrances have primary and secondary channels for reaching interior destinations.

The event table design philosophy embedded in the ChannelSegment class's LRM fields omits any reference to side of waterway or lateral offsets. This choice is based on the difficulty of reliably identifying reference objects. Even side-of-waterway presents a problem, since most waterways' direction of flow for determining side of channel (downstream) is the reverse of that presented by increasing measure values along the waterway centerline (upstream). Thus, any designation of left and right you might implement in your event tables will need to be carefully explained and well understood.

Channel marker point features

Simple feature class: **Marker**
Geometry: Point
Contains M values: No
Contains Z values: No

Field name	Data type	Allow nulls	Default value	Domain	Precision	Scale	Length	
OBJECTID	Object ID							
Shape	Geometry	Yes						
MarkerID	Long Integer	Yes			0			Identifier of marker
MarkerType	String	Yes		Markers			6	Type of marker
WaterwayID	GUID	Yes			0			Identifier of waterway
Measure	Double	Yes			8	2		LRM measure for location
GeoPositionID	GUID	Yes			0			Identifier of geographic position
Width	Short Integer	Yes			0			Width of marker face
Height	Short Integer	Yes			0			Height of marker
MarkerColor	String	Yes		Colors			10	Color of marker face
MountingType	String	Yes	POST	Mounts			10	Type of mounting
Legend	Float	Yes			3	2		Marker legend
BeaconType	String	Yes	NONE	Beacons			10	Type of beacon light
BeaconColor	String	Yes		Colors			10	Color of beacon light
ConditionRating	String	Yes	GOOD	Rating			10	Condition rating
DateChecked	Date	Yes						Date condition was determined
DateInstalled	Date	Yes						Date sign was installed

Figure 17.14   **Channel marker point features**

Figure 17.15 illustrates some of the common markers found along U.S. waterways. The green and red channel boundary markers are lateral aids to navigation. The other markers are nonlateral aids, which may be placed on buoys. Beacons presenting various light patterns can supplement most markers. Navigational aids without beacons are called daymarkers. You can extend the USANS domain of markers to include other locally used types.

Channel markers and many other navigational aids in USANS are labeled. Channel boundary markers have numbers, with odd numbers designating triangular red right-side markers and even numbers assigned to square green left-side markers. Nonlateral

Chapter 17: Navigable waterways

markers will have letter designations. Those labels are repeated on navigational charts so that a vessel's operator can more readily identify their present location as they travel along the waterway. Indeed, daymarkers with a checked pattern are put in place solely to tell the boater, "You are here."

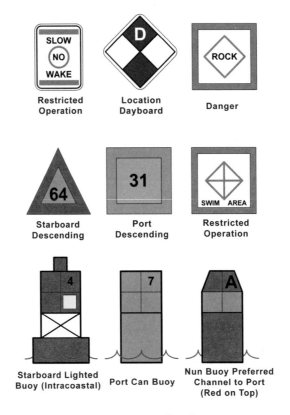

Figure 17.15 **Marker signs**

Channel markers along the Intracoastal waterways of the U.S. carry a yellow supplemental marker that indicates the channel from the perspective of the direction of travel for which the marker applies, not some seemingly arbitrary "downstream" direction. This can result in right-side channel markers, which are located on the right side of travel in the "descending" direction determined by the reach (ocean to the right) carrying a left side square Intracoastal supplemental marker for boaters traveling in the "ascending" direction (ocean to the left).

There are other colors used in the USANS marker system. Markers with a wide orange border denote hazards or restricted operations. In addition to being used for Intracoastal supplemental markers, yellow may also be used for markers alerting boaters to special features, such as an anchorage, jetty, or other important area to note.

A State Water Obstruction Marker using alternating black and white vertical stripes indicates an obstruction to navigation extends from the nearest shoreline to that point. Cardinal System Buoys were previously used for this purpose in the Western River System. In the opposite manner, red and white markers designate areas where water is unobstructed. Such safe-water markers may be passed on any side.

The last kind of marker likely to be found in your area is the range daymarker. These vertically striped markers are spaced one behind the other so that when a ship is properly aligned to enter the channel, the two markers will be aligned. Both range daymarkers use the same colors, but with alternating patterns. For example, the nearest marker may be green-white-green, while the more distant marker is white-green-white. Multiple color combinations distinguish between the various available channels to enter the harbor.

Markers typically have a single color and shape presentation specific to each type. There are variations, though, in both shape and color. For example, a can buoy, which has a square top, can be replaced by a red nun buoy with a conical top in order to reinforce its demarcation of the left side of the channel leaving a harbor. Another example is where there are multiple channels into a harbor. In such cases, a buoy with the lower portion being green and the upper portion having the appropriate color (red or green) for that side of the channel will note the preferred, or primary channel.

In addition to recording what marker is at a location, you may need to record how it is mounted. There are four basic ways to mount a marker: on a single daymarker post; on a system of three connected posts, called a dolphin; on a can or nun buoy, which does not have

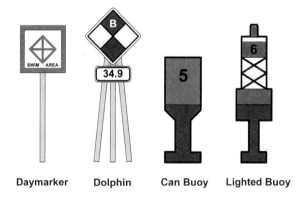

Daymarker   Dolphin   Can Buoy   Lighted Buoy

Figure 17.16   **Marker mounts**   The variety of water conditions and navigational needs requires different ways of mounting waterway markers. This figure illustrates the four basic types. A daymarker on a single post is typically used where water is shallow. A daymarker on three posts—an arrangement called a dolphin—is typically found in deeper water. Canned buoys and lighted buoys typically indicate channels. The weighted end that extends under the water surface helps keep the buoy upright. Buoys are anchored to the streambed using chains or synthetic ropes. Lights may be placed on several marker types.

Chapter 17: Navigable waterways

a beacon; and on a lighted buoy. Buoys are anchored to the streambed or seabed by a chain or synthetic rope. Weather and high water volumes may topple a post-mounted marker or move a buoy.

The foregoing discussion serves to explain the various domain values offered in the classes shown in figure 17.17 that support several Marker feature class fields. All use string values to make their native presentation more understandable within the Marker class.

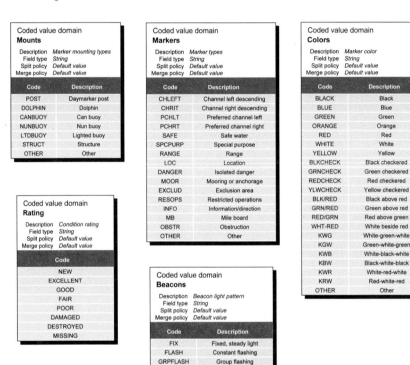

Figure 17.17 **Marker domain classes**

Channel geometry

It may seem illogical for a linear event not to have a linear geometry, but you will likely want to create WaterwayPoly features to illustrate channel segments. You can construct those polygons by connecting the "dots" (marker point features) along the waterway. Channel is a WaterwayPoly subtype.

A simple polygon for a 2D geometry may not be sufficient for some applications because depth is not continuous across a waterway, or even within a channel. Depth varies continuously across and along a waterway, and over time due to changing water levels. In a typical application, the width and depth values contained in the ChannelSegment table are minimums, and describe the most restrictive conditions that a boater might face during typical water conditions.

Arc Hydro included a 3D way to represent channel width and depth. That method is adapted for use here in a manner consistent with the other concepts of this book's geodatabase design philosophy. The resulting approach is based on waterway cross-sections and a thalweg profile section along the waterway.

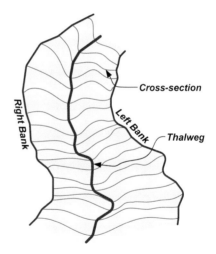

Figure 17.18 **Constructing the streambed** The thalweg is the flowline constructed by connecting the deepest point measured on each cross-section and can serve as a plan view centerline. Each cross-section is located by an event along the thalweg centerline, which forms an LRM route with an origin at the mouth (outlet) of the waterway. There is also a route across the waterway at each cross-section, with an origin at the bank or some other reference point. A 3D view of the waterway can be constructed from the sequence of cross-sections.

It was stated earlier that the thalweg is a centerline feature described by connecting the lowest streambed elevations. The only way to map the streambed for identifying the line of lowest elevation is to conduct depth soundings in a structured manner. Making depth measurements across the waterway along a lateral cross-section line allows a fuller description of channel configuration and a closer approximation of the thalweg than may be offered by other methods. It simultaneously provides information to marine interests that allow a fuller understanding of the dynamic characteristics of width and depth.

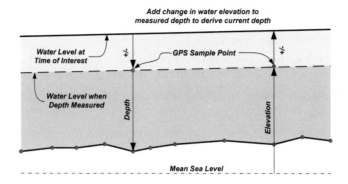

Figure 17.19 **Flowline profile** Profiles and cross-sections are derived from depth readings taken from the waterway's surface. By using GPS or some other mechanism to determine the 3D location of the depth sampling point, you can correlate the bottom to any water level that may exist later. The flowline (thalweg) is derived by connecting the deepest readings along the sequence of cross-sections.

The basic technique requires the simultaneous measurement and recording of the elevation of the water level relative to a standard, such as mean sea level, and the water depth. That way, any subsequent changes in water elevation can be detected and the depth reading changed accordingly.

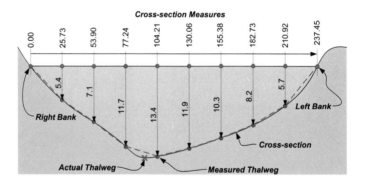

Figure 17.20 **Cross-section** Cut lines are constructed across the waterway to take depth soundings designed to produce a cross-section view. Each cross-section forms a route for measures across the waterway, which may extend beyond the banks to map the flood plain. Thus, a waterway inventory can include two LRMs: one along the waterway and one for a series of cross-sections. The thalweg is the deepest point along the cross-section cut line.

The construction of a cross-section profile of the streambed can be accommodated using a lateral LRM, with the origin (0.00) being on one side's limit of observation and measures increasing linearly across the waterway to the other side's limit of observation. Figure 17.20 illustrates evenly spaced depth soundings, but there is no requirement for that approach.

Channel geometry

Soundings can be spaced as warranted, with the dry portions of the survey area, such as the flood plain, derived from existing topographic sources.

Depth measure point features

Simple feature class: DepthPoint
Geometry: Point; Contains M values: No; Contains Z values: No

Field name	Data type	Allow nulls	Default value	Domain	Precision	Scale	Length	Description
OBJECTID	Object ID							
Shape	Geometry	Yes						
DepthPointID	GUID	Yes			0			Identifier of depth point
WaterwayID	GUID	Yes			0			Identifier of the waterway
CrossSectionID	GUID	Yes			0			Identifier of cross-section
CSMeasure	Double	Yes			7	2		Measure value on cross-section
Depth	Float	Yes			5	1		Water depth, in feet
Elevation	Double	Yes			7	2		Water surface elevation (MSL)
GeoPositionID	GUID	Yes			0			Identifier of geographic position
DPMethod	Short Integer	Yes		Methods	0			Method of measuring depth

Channel cross-section features

Simple feature class: CrossSection
Geometry: Polyline; Contains M values: No; Contains Z values: No

Field name	Data type	Allow nulls	Default value	Domain	Precision	Scale	Length	Description
OBJECTID	Object ID							
Shape	Geometry	Yes						
CrossSectionID	GUID	Yes			0			Identifier of cross-section
WaterwayID	Short Integer	Yes			0			Identifier of a waterway
WaterwayCode	String	Yes					30	Public key identifier of waterway
WWName	String	Yes					32	Identifier of a waterway
AtMeasure	Double	Yes			5	2		Reach LRM measure of cross-section
CSMethod	Short Integer	Yes		Methods	0			Method of deriving cross-section
Shape_Length	Double	Yes			0	0		

Channel profile features

Simple feature class: Profile
Geometry: Polyline; Contains M values: Yes; Contains Z values: No

Field name	Data type	Allow nulls	Default value	Domain	Precision	Scale	Length	Description
OBJECTID	Object ID							
Shape	Geometry	Yes						
ProfileID	GUID	Yes			0			Identifier of profile
WaterwayID	Short Integer	Yes			0			Identifier of a waterway
WaterwayCode	String	Yes			0			Public key identifier of waterway
WWName	String	Yes					32	Identifier of a waterway
FromMeasure	Double	Yes			5	2		From reach LRM measure of profile
ToMeasure	Double	Yes			5	2		To reach LRM measure of profile
PofileMethod	Short Integer	Yes		Methods	0			Method of deriving profile
Shape_Length	Double	Yes			0	0		

Figure 17.21 **Cross-sections and profile**

Three feature classes are described in figure 17.21 to store information about cross-sections and thalweg profiles: DepthPoint, Profile, and CrossSection. The source data consists of depth soundings made at points along a lateral line across the waterway. This information goes into the DepthPoint class.

The CrossSection class has a polyline geometry but is functionally a point event table with regard to the waterway in that it has one reach LRM measure location. Depth points along the cross-section are assigned a cross-section LRM measure, thus creating a second LRM type. Typically, the lateral (cross-section) LRM would be stated in units of feet or meters, while the waterway centerline LRM is stated in hundredths of a mile. Each cross-section creates a lateral route and corresponding LRM for locating the x-axis position of each depth sounding across the waterway. The measured depth is the y-axis coordinate. In addition to the LRM position, water depth, and a coded value indicating the manner in which the feature was derived, the DepthPoint feature class includes water elevation and a geographic location. The geographic location is on the surface of the water, but it can be projected to the streambed for 3D modeling of that surface.

The lateral line formed along the water's surface by depth points defines the cross-section, which forms a route for linear measures that represent the x-axis of the cross-section profile. Depth points are aligned along this section cutline, which is assigned a reach LRM measure location where it intersects the flowline (waterway centerline), to produce a plan view of the cutline; however, this horizontal line is not the geometry of the CrossSection class. The actual geometry stored in this class is the vertical streambed profile.

The process of creating a cross-section begins by selecting the location and angular orientation of the cutline, which need not be perpendicular to the waterway flowline. A geographic position and a reach LRM position define the location of the cutline; both values are stored in the CrossSection feature class. The angle across the waterway is not preserved in the CrossSection feature class because it can be derived from the geographic position of each included DepthPoint feature. The depth of each sounding provides the y-axis coordinate and forms a vertex along the cross-section's streambed geometry. Depth points arranged along the cutline provide the x- and y-axis information needed to construct the cross-section geometry, with each depth point forming a vertex on a cross-section polyline.

The Profile polyline feature class is functionally a linear event with a geometric representation so that it is an element of the waterway. It is a longitudinal cross-section with beginning and ending reach measures (linear event) and a geometry that describes not the planar view of the waterway's centerline, which could be derived using dynseg, but the streambed profile defined by the lowest elevation point of a series of cross-sections—the thalweg. The Profile class's geometry is constructed by connecting a straight line between the deepest point on each cross-section along the axis of the waterway's centerline. The planar view of each Profile feature can be constructed from the geographic coordinates of the selected DepthPoint features. The resulting geometry can be used to edit the waterway centerline.

Gauging stations

A common element of more complete hydrographic datasets is composed of gauging stations where water levels and other environmental conditions are monitored. Gauging stations produce a large quantity of hydrometric information. In the case of navigable waterways, they supply the critical piece of information for calculating channel depth: water elevation. Gauging stations directly measure the height of a water flow relative to a local origin, which is tied to a geographic elevation datum. RF3 uses the North American Vertical Datum of 1929.

Gauging stations operate continuously or intermittently. They may be read manually by an observer visiting the site or report their data through radio or landline communications on a regular interval, through polling by the remote site or when a trigger condition occurs. Stations come with a variety of additional instruments capable of measuring air and water temperatures, flow rates, and sedimentation load. Stations typically extend to the bottom of the waterway with the instruments housed at the top, well above the expected flood elevation. Stations are placed where they are somewhat protected from damage by debris flows, such as on the downstream side of a bridge pier, and where water flows are relatively protected from wave action so they better measure the mean water level.

Gauging Station (Temporal Object Table)

GSID	WaterwayID	Measure	GSType
1	21	23.84	Lake
2	21	15.86	Stream
3	19	5.03	Stream

Temporal Observation Table

GSID	Time	Level
1	00:15:00	16.25
1	12:15:00	17.19
2	04:20:15	5.02
2	12:07:18	5.15
2	18:28:03	6.52
2	23:45:04	7.19
3	12:10:05	12.02
3	23:41:27	11.34

Figure 17.22 **Complex stationary tracking event**
A water-level gauging station presents a classic example of a complex stationary tracking event. One table or feature class describes the gauging station, while another records the time-stamped observations of water level.

The best way to include gauging stations within the navigable-waterways geodatabase is with tracking events, as originally discussed in chapter 3. The type of tracking event you want to use is the complex stationary variety, where the locations of temporal observations are in one class and the observations are in a second class. In this application, the descriptive data regarding the gauging station is stored in the temporal object table and the water measurements made by that station are stored in the temporal observation table.

Chapter 17: Navigable waterways

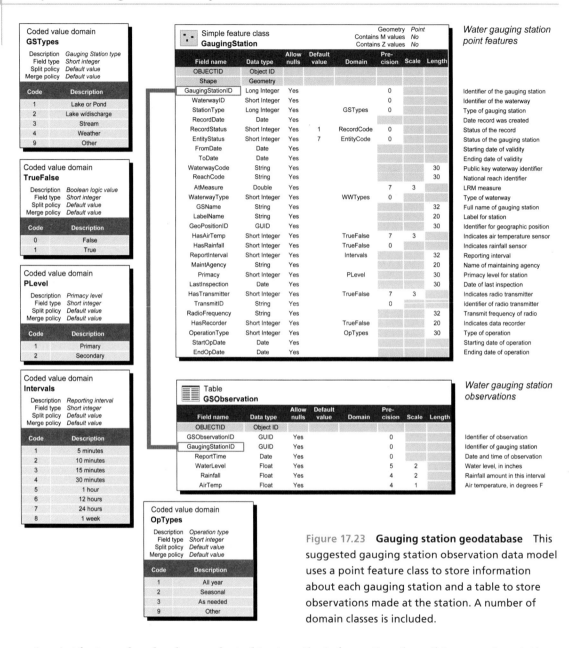

Figure 17.23 **Gauging station geodatabase** This suggested gauging station observation data model uses a point feature class to store information about each gauging station and a table to store observations made at the station. A number of domain classes is included.

A point feature class has been selected to store the information describing gauging stations and a table created to store all the observations made at those stations. Five domain classes are also included in the design. The GaugingStation feature class has all the fields found in other event tables, including an LRM position, plus fields storing other common attributes. Other fields may be needed for your design. For example, if gauging stations in your area use

408

landline communications, then you will want to add HasPhone and PhoneNo fields. You will also need to add fields to support additional sensors in both the GaugingStation feature class to describe that capability and in the GSObservation table to store the sensor readings.

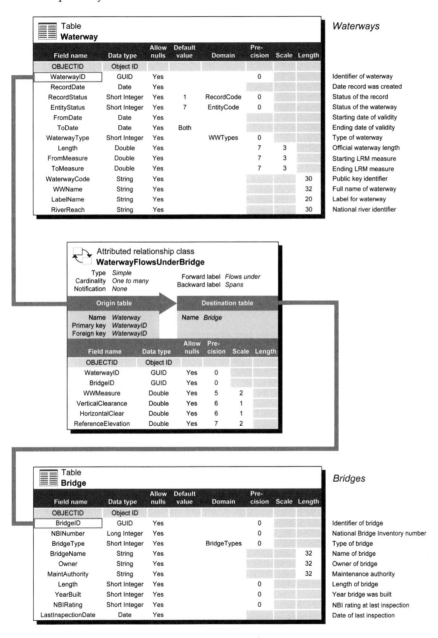

Figure 17.24 **Bridges**

Bridges

Since this is a book about multimodal geodatabase design, it is fitting to end the chapter on navigable waterways by discussing the primary point of interaction between waterways and land-based modes of transport: bridges. An attributed relationship class stores all the information required by the waterways geodatabase, assuming that the bridge's descriptive data and geographic position are stored in the highway or railroad portion of the geodatabase.

The first thing you need for bridges over waterways is a reach LRM position, which requires the waterway identifier and the measure value. You will also probably need to know horizontal and vertical clearances under the bridge, likely stated in feet or meters, as those parameters affect the ability of ships to pass under the bridge. Of course, vertical and horizontal clearances are dependent upon water elevation. The higher the water, the less room there will be under the bridge. To define the vertical and horizontal clearances, the relationship class also needs to know the water elevation at the time of measurement. As with the depth points, knowing the measurement basis for the dimensions allows them to be calculated for any water elevation.

In addition to bridges, your navigable-waterways geodatabase may need to accommodate docks, dams, locks, and similar facilities. Arc Hydro suggests a number of distinct feature classes and tables for use in storing information about such facilities. Alternatively, you could adopt a generic Facility feature class with subtypes for each distinct facility type.

Navigable waterways present unique requirements for designing a suitable geodatabase, such as storing polyline geometry for point events and representing waterway "routes" using 3D data structures. However, the basic UNETRANS data model can be readily adapted to meet these requirements.

Notes

[1] Originally, the RRN was a 14-character string at the 1:100,000 scale. That specification changed when the segment identifier component was increased to four digits from three and two numbers after the decimal were added to the reach mile marker component when 1:24,000-scale implementation began to occur. Data may be available for your jurisdiction in one or both forms.

[2] There is one notable exception: the Ohio River, which has origin for linear measures at the junction of the two rivers from which it is formed; i.e., at its upstream limit.

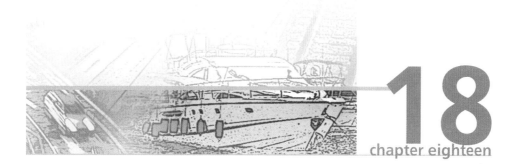

chapter eighteen

Railroads

- Tracks
- Railroad intersections
- Track geometry
- Railroad companies
- Yards and interchanges
- Trackside structures
- Railroad grade crossings
- Intermodal facilities

Railroads once supplied the only means of fast and reliable long-distance travel across land. They continue to be a low-cost way to ship bulk goods—and probably the most complex mode of transport for which to design a geodatabase. Railroads have all the requirements of the highway and transit modes, plus additional considerations for being both providers and users of transport facilities. A truck driver has a reasonable expectation of going wherever a highway can go, but railroads do not. They must balance the physical considerations of weight, motive power, availability of staff, and delivery requirements. Plus, there are contractual conditions imposed by a variety of competing railroads. This book presents useful principles for designing such a geodatabase, but a single chapter cannot begin to show how to support the complexities of building, maintaining, and operating railroads. It is a subject worthy of its own book.

This chapter, therefore, is not intended for railroad GIS staff, although it might be a useful introduction to railroad terminology for recent hires. Instead, the purpose of this chapter is to present the basics of what local, regional, or state agency employees need to know to incorporate railroad facilities into their multimodal geodatabase. It presents ways to identify and organize the rail-related facilities seen in aerial photographs or while managing facilities for other modes of travel.

As a GIS professional, your most frequent interaction with railroads may be at railroad grade crossings. But you may be called upon to quickly produce maps to deal with a natural or manmade disaster. For example, you may be asked to produce a map showing where emergency supplies may be offloaded from rail cars or evaluate the impact of a hazardous material spill based on a track and milelog LRM position description. If you work for a state DOT, you may be responsible for federal reporting related to rail-highway grade crossings, which requires a description of track visibility, a report on the number of crossing school buses, and the nature of present crossing protection.

Tracks

The most obvious railroad facility is the track. In North America, the standard gauge—the space between the rails—is 4 feet, 8½ inches. Rails are characterized as to weight based on a 3-foot length. Lighter duty tracks with lower speeds use lighter rail, typically in the range of 80 to 100 pounds. Main-line tracks use rail on the order of 116 to 132 pounds, with greater weights used for tracks supporting the highest volumes and speeds. Differences in weight reflect differences in cross-section dimensions of width, height, and thickness.

Rails are laid atop crossties (also called ties or sleepers), which were traditionally made of creosote-infused wood but are sometimes cast in concrete. Between the tie and the rail is a tie plate. In track made with wood crossties, four spikes are driven through holes in the tie plate into the crosstie, with the head of the spike extending over the rail flange. The purpose of the tie-plate connection is to spread the rail load over a larger area of the tie and to hold the rails in fixed lateral position but allow movement along the length of the rail. Rails are fastened to concrete ties with clips.

Rail comes in two basic forms: segmental and welded. Traditional rails are 39 feet long. End joints are staggered between the two parallel rails, which results in the rocking motion of railcars as they move down sectional track. To reduce this effect and for other reasons, main-line tracks have migrated to the use of welded rail, with segments commonly one-quarter-mile long.

Whether made of welded or segmental rail, or using wood or concrete crossties, all railroad tracks are laid on a bed of gravel, called ballast. The ballast has three primary purposes. First, it allows rainwater to drain away from the track and its supporting crossties very quickly. Placing the track above the level of the adjacent land helps in this effort. Second, it allows limited up and down movement of the track in response to the loads imposed by passing trains. Lastly, it keeps the track in place. Because the ties are set into the ballast, not simply laid on top of it, lateral movement is virtually eliminated.

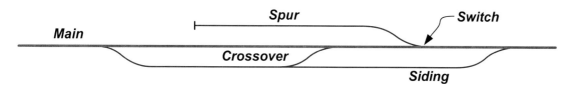

Figure 18.1 **Basic track types** Most of the railroad tracks you would include in your geodatabase may be labeled as mains, spurs, and sidings. Unless the railroad operates with a double-track main line, passing sidings are required to allow trains to meet or overtake each other. Train precedence rules typically say which train must wait in the siding. Other sidings may provide access to freight customers. Spur tracks also are installed in order to reach customers and to provide car storage. All connections between tracks occur at switches. A crossover is a pair of facing switches that allow movement between adjacent tracks.

There are three basic kinds of track, classified according to its function and design. The main is the track used for passage of trains over longer distances. A siding is a section of track connected to the main at both ends for inferior trains to await the passage of superior trains. Various rules determine who has to wait on the siding (inferior train, such as a coal drag) and who can keep going on the main (superior train, such as a hotshot merchant freight carrying perishable goods). A spur or stub track has only one connection and is used to access a customer's facilities. Some sidings may also provide access to customers, either directly or through branching spurs.

Both sidings and spurs are almost always set below the elevation of the main, in part to keep cars from accidentally rolling onto the main and causing a crash. If you see two parallel tracks over a distance of less than three miles, then one is the main and the other is a siding. (You can tell which is the siding by looking at the terminal switches. The track on the divergent side of the switch is the siding.) However, if the distance is longer and you can tell that both tracks are at the same elevation, then you are probably looking at a double-track main line.

A derail is sometimes used to kick a railcar that gets loose on a siding or spur from getting onto the main. The derail is hinged so that it can be flipped into position over the rail farthest from the main. When struck by a railcar, the shape of the derail picks up one wheel and diverts it off to the outside of the rail. The opposite wheel then drops down onto the crossties between the rails. The result is that the car is either stopped or overturned away from the main.

In addition to being generally classified according to function (main, siding, or spur), tracks are also classified according to the train speeds they can support. Maximum passenger train speeds are generally higher than freight train speeds within each class, due to difference in car design and load. Class 1 track has the lowest speed limit: 10 mph for freight trains and 15 mph for passenger trains. Most yard and spur tracks are assigned to this class, as is bad-order track normally assigned to other classes. Class 2 track has speed limits of 25 mph and 30 mph for freight and passenger trains, respectively. A lot of branch lines, tourist railroads, and sidings are assigned to this class. The secondary main lines of major railroads and many regional railroad mains are in class 3, which has maximum speed limits of 40 mph for freight trains and 60 mph for passenger trains. The main lines of major long-haul railroads are principally assigned to class 4, which allows speeds up to 60 mph for freight trains, 80 mph for passenger trains. Class 5 sets the limit at 80 mph for high-speed freight main lines in the western United States and Canada. Track in classes 6-8 are almost exclusively dedicated to passenger trains in the northeast United States and permit speeds up to 150 mph.

Railroad intersections

All connections between tracks occur through switches, also called turnouts, which are classified according to the frog number. The frog is the central meeting point where the rail of the diverging branch track crosses the rail of the main track. The two rails that intersect at the frog are called the meeting rails; the other rails are called the outside rail. Frogs are typically treated as a single casting. The frog number represents the length, in feet, required to separate the diverging rails by a distance of one foot. The higher the frog number, the more gradual the curve through the frog and the faster a train can travel through the switch

Adjacent switches may be in a facing or trailing configuration. A crossover composed of two facing switches allows the movement of trains between adjacent tracks. A scissor crossover, which looks like two simple crossovers on top of each other and arranged in opposite directions, may be used in congested areas, particularly approaching a passenger terminal, to allow trains approaching in any direction on either track to move to the other adjacent track. Slip switches and wye switches may also be found. A slip switch places the switch in the middle of a scissor crossover; there is no straight option: all traffic must go through the center of the switch. A wye switch has both output tracks diverging from the switch; there is no "main."

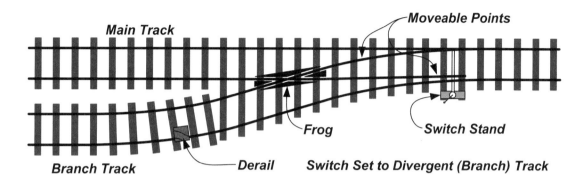

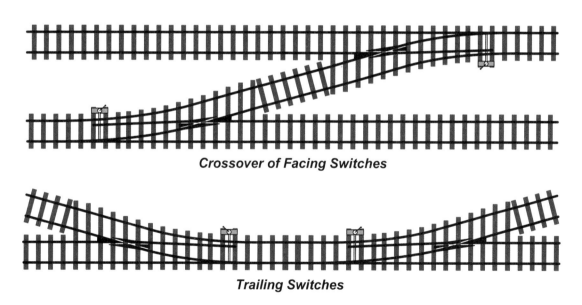

Figure 18.2 **The track switch: A railroad's intersection** Switches are characterized by the frog angle number. The frog number represents the distance, in feet, required to separate the divergent tracks by one foot. Movable points "bend the rail" to allow selection of the two outbound tracks. The switch stand includes the mechanism required to move the points and often have a mechanical connection to the switch position indicator. Main-line switches may be electrically connected to a lighted signal or semaphore that serves the same purpose. A derail may also be put in place to divert a runaway car from entering the main line.

Switches of all types operate through the action of two moveable switch points, which are rails that taper to a point at the end where they can abut the main track's rails. One switch point is straight and the other is curved. A switch stand is used to move the switch points at the input end of the switch. It can be set to one of two positions: either to keep the train on the main track or to divert it onto the branch track. In one position, the switch places the straight switch point against the main track's meeting rail so as to keep the train moving along the main track; the curved point has a gap between it and the main track's outside rail. In the other position, the switch stand places the curved switch point against the outside rail and places a gap between the curved switch point and the main's meeting rail.

Most switch stands have a target, which may be lighted, to indicate the switch's setting. A green aspect to the main track indicates that the switch is set to allow travel down the main. It is common for this aspect target to be shaped like a sideways figure eight. A switch stand set to divert traffic onto the branch will present a red aspect target, often shaped like an arrow pointing to the direction of divergence. Some switches are remotely operated by a motor, most use a mechanical drive that requires a person on the ground to "throw the switch" in order to set the direction of travel out of the switch. Some switches can be thrown in both ways.

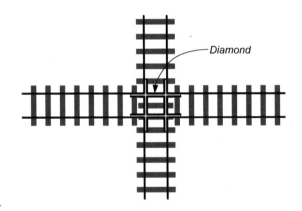

Figure 18.3 **The diamond: An intersection without turns** Railroads also have intersections where tracks cross without the local means to move between the tracks. (There may be nearby interchange tracks with switch connections that do allow movements between the main lines intersecting at the diamond.) Diamonds are typically protected by signals and require approval from a dispatcher at the controlling railroad's operations center to pass through the intersection. At one time, it was more common for a local crossing tower to manage the intersection, but centralization has occurred as a cost-saving measure.

There are some places where tracks intersect but travel between the tracks at the location cannot occur. These places, where two tracks cross each other, are called diamonds. Electrically operated switches and remote dispatchers using centralized train control (CTC) typically regulate travel through a diamond crossing today. Not too long ago, though, the work was done by a local operator housed in a nearby two-story building, called a tower, through an interlocking plant, which is a set of switches that require a specific sequence of actions to be taken in order to allow trains on a selected track to move through the crossing.

Track geometry

Creating polylines to represent track geometry by following the centerline of each track requires a steady hand but is conceptually simple. Railroad tracks are divided into segments that extend from switch to switch or from switch to the end of track. Although switches themselves would not be explicitly symbolized as a separate feature class, it may be useful to start the process by placing point features at the terminus of each railroad track segment. You can then use these features to snap the ends of all track segment features that end at that location.

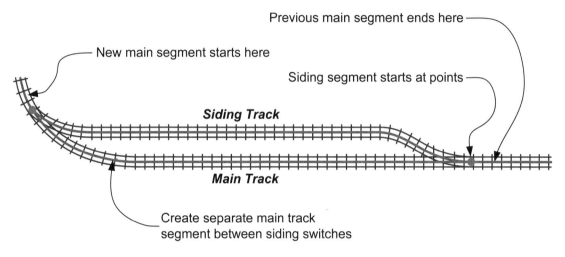

Figure 18.4 **Track geometry** Creating centerline geometry for railroad tracks requires following the midpoint along each track. The feature class will contain one member for each track segment. Each segment should extend from a starting switch to an ending switch, or from a starting switch to the end of the track. You may want to temporarily place point features at all switch points in order to support snapping of connecting feature endpoints to a common location.

Note that even though the railroad probably uses a linear LRM, there is no expectation that it be implemented on the track geometry you create. You will not be using dynseg on railroad features; that is a job for railroad company analysts. All you need to do with track geometry is show where the tracks are located, who owns them, and how they are used.

The RRTrack feature class consists of polylines representing track segments. The physical implementation example shown above includes the record metadata fields introduced in chapter 6. The TrackType field says how the track is used: main, spur, etc. The TrackNumber field allows you to distinguish between adjacent tracks serving the same function, such as two team tracks next to a freight depot, if you care to do so. RailWeight is just one of several descriptive attributes you could include, depending on your specific needs and interests.

Simple feature class **RRTrack**

Geometry *Polyline*
Contains M values *No*
Contains Z values *No*

Railroad track segment features

Field name	Data type	Allow nulls	Default value	Domain	Precision	Scale	Length	
OBJECTID	Object ID							
Shape	Geometry	Yes						
RRTrackID	GUID	Yes			0			Identifier of railroad track
RecordDate	Date	Yes						Date record was created
RecordStatus	Short Integer	Yes	1	RecordCode	0			Status of the record
PositionStatus	Short Integer	Yes	7	EntityCode	0			Status of the position
FromDate	Date	Yes						Starting date of validity
ToDate	Date	Yes						Ending date of validity
TrackType	String	Yes		TrackTypes			12	Type of track
TrackNumber	Short Integer	Yes			0			Sequential identifier of parallel tracks
RailWeight	Short Integer	Yes			0			Weight of rail, in pounds per yard
RRCompanyID	GUID	Yes						Identifier of railroad company
CompanyName	String	Yes					12	Name of railroad company
LabelName	String	Yes			0			Company name for map label
SubName	String	Yes		RRSubs			32	Name of railroad subdivision
FromMeasure	Double	Yes			7	2		Starting LRM measure of track segment
ToMeasure	Double	Yes			7	2		Ending LRM measure of track segment
Shape_Length	Double	Yes			0	0		

Coded value domain **TrackTypes**

Description *Track types*
Field type *String*
Split policy *Default value*
Merge policy *Default value*

Code	Description
MAIN	Main
SIDING	Siding
SPUR	Spur
INTERCHANGE	Interchange track
OTHER	Other

Figure 18.5 **Railroad tracks** The suggested feature class implementation for railroad tracks subdivides each track into segments that extend no further than from switch to switch, or from switch to terminus for a spur. The three major track types, plus interchange tracks and a catch-all "Other" type are included in the TrackTypes domain class. The from and to measures come from railroad sources and do not reflect an actual measure coordinate. (Note that the Contains M Values parameter says, "No.") Only a railroad GIS person would want to use dynseg, and this design does not support it.

The next three fields indicate the owning or maintaining railroad company's identifier, its full name, and the default label text to use for mapping. You could create a domain of values relevant in your area and use domain control to populate these fields from an RRCompany table (see figure 18.6). Railroads are organized into divisions and subdivisions.

The SubName field is equivalent to a route in a highway geodatabase, in that each subdivision has its own LRM. The FromMeasure and ToMeasure fields are based on that LRM, so you may be able to approximate the measures from known points along the track. For example, many railroads maintain explicit mileposts along their main lines. The proposed class design suggests the use of an RRSubs domain class to control the entries in the SubName field. An input dialog could use the RRCompanyID field as a foreign key to select those railroad subdivisions applicable to that company. However, you may need to override this function if the maintaining railroad is not the one that created the LRM.

Railroad companies

One common piece of information you will want to include in your railroad geodatabase—if, for no other reason than to label the tracks—is the name of the railroad company that uses the track. Once you know that name, you can refer to *The Official Railway Guide* (Commonwealth

Business Media) to get other information on the railroad and the track. You will need to know more than the company name for some railroads, though, such as the divisions, subdivisions, and branches in your area. Larger railroads divide their lines into divisions, and divisions into subdivisions and branches. The Guide will often provide you with information on railroad LRM measures for each station along an active line.

Finding the name of the present railroad is often complicated when the name of the owning railroad is some newly formed conglomerate or a historic remnant of a long-gone operating company unheard of outside the railroad industry. For example, BNSF Railway was formed from the merger of Burlington Northern and Santa Fe Southern Pacific, both of which were mergers of mergers dating back to 1928. BNSF has tracks in 27 states and Canada organized under 14 divisions with hundreds of subdivisions ranging in size from 10 to 300 miles of main line.

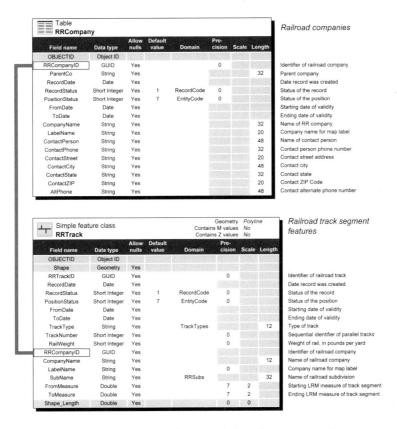

Figure 18.6 **Railroad company** The simple geodatabase design shown here seeks to answer two questions. First, what is the name of the railroad company you want to use to label the track? Second, whom do you contact when something bad happens? The relationship shown here will generally reflect who has maintenance responsibility for the track, which could be a lessor company.

Chapter 18: Railroads

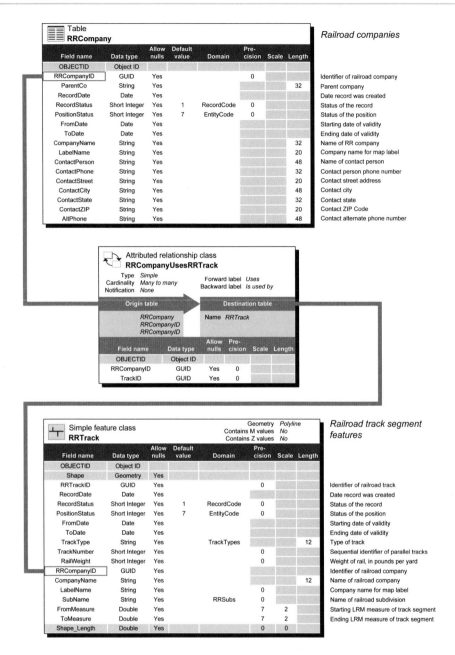

Figure 18.7 **Railroad company—part 2**
The simple foreign key relationship that establishes track ownership (or, at least, track maintenance responsibility) won't tell you which railroad companies use a track through trackage rights and other arrangements. That task requires an attributed relationship class in order to store the many-to-many relationship of tracks and the railroad companies that may use them.

While several large railroads have consolidated their ownership of component railroads in recent years, many relics of acquisition remain. Old records available to local governments may have the predecessor company's name, not that of the new, consolidated "megarailroad." Standing by the railroad tracks to read the name on the locomotives will not always resolve the matter. There may be more than one railroad operating over a given section of track. Tax records may show who owns the track, but there may be multiple tenant railroads with trackage rights over it. The track may also be leased to an operating company, which has happened with many low-volume branch lines that were leased by the big companies to regional operators.

Incidentally, railroad companies are classified based on their annual revenue, which affects their data gathering and reporting requirements and, as a result, the nature of the information you can find about railroads within your geographic area. The dividing line, as of this writing, between class 1 and class 2 is $319.3 million; it is at $20 million between classes 2 and 3. Nine North American railroads are currently in class 1.

The RRCompany class shown above provides information about each railroad company. In addition to the class identifier and the standard editing support fields, the class includes a field to store the name of the company, its parent company, and contact information. As noted earlier, the CompanyName and LabelName fields can be easily populated through a lookup process based on RRCompanyID.

The foreign key relationship between RRCompany and RRTrack expresses who owns or is responsible for maintaining the track segment. However, there may be several railroads that can operate over the track. If you need to store this information, an attributed relationship class like the one shown above is necessary due to the many-to-many cardinality of the association. This means you will find the owning or maintaining company for a track segment through the foreign key relationship supplied by the RRCompanyID field and all the companies with trackage rights through the RRCompanyUsesRRTrack attributed relationship class.

Yards and interchanges

A yard is a collection of generally parallel tracks designed to store and classify rail cars. In railroad parlance, "classification" means ordering cars in accordance with the destination as a way of creating trains. Cuts (groups) of cars, each going to a single destination, are arranged as they need to be set out along the train's route.

Large classification yards operate as hub airports do for the airlines. Long-distance trains move large groups of freight cars between these hubs, where they are broken down into smaller groups for delivery by local trains. Incoming trains enter on the yard lead, which is typically a siding off the main line, and park on a receiving track. The locomotives used by

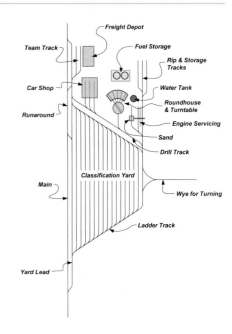

Figure 18.8 **Major classification yard** Besides serving as system hubs, classification yards are also often the location of major engine and freight car servicing facilities. Increasingly, large classification yards are being downsized or replaced by multimodal exchange facilities where shipping containers are transferred from trucks to rail cars.

the incoming train then are reassigned to pull an outbound train or are routed to adjacent engine-servicing facilities. Switcher locomotives that stay at the classification yard then start taking the train apart and placing the cars on different outbound tracks in the yard according to their destinations. Any bad-order car discovered in this process is sent to the rip track (repair, inspect, and paint) or the car shop for servicing before it can continue to its destination.

You may occasionally see a classification yard that actually looks like two V-shaped yards placed end to end, with one track connecting them. This is probably a hump yard, which uses a centrally located hill, or "hump," to classify cars. The inbound train is pushed up the hump to the crest, where each cut of cars is uncoupled and an operator in the yard tower identifies the car and the train it needs to go into to reach its destination. A set of electrically operated switches directs the free-rolling car to the right track. Electrically activated, remote-controlled wheel retarders are placed along the way to keep the car from going too hard or hitting the end of the outbound train too fast.

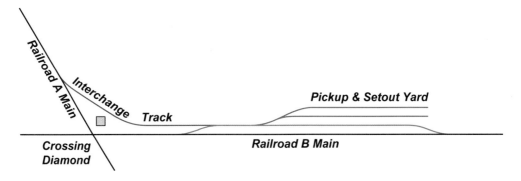

Figure 18.9 **Interchange point** Crossings of major railroads or connections to short lines will often include an interchange connection that allows movements between the two railroads. In this typical example, railroad A intersects railroad B at a diamond crossing. A siding on railroad B leads to an interchange track connecting the two railroads and serving as the lead to a small yard for pickups and setouts.

Trackside structures

Simple feature class RailYard				Geometry	Polygon				
				Contains M values	No				
				Contains Z values	No				
Field name	Data type	Allow nulls	Default value	Domain	Precision	Scale	Length		
OBJECTID	Object ID								
Shape	Geometry	Yes							
RailYardID	GUID	Yes			0				
RRCompanyID	GUID	Yes			0				
RailYardType	String	Yes		YardTypes			20		
RailYardName	String	Yes					32		
LabelName	String	Yes					20		
Shape_Length	Double	Yes			0	0			
Shape_Area	Double	Yes			0	0			

Railroad yard features

- Identifier of rail yard
- Identifier of railroad company
- Type of rail yard
- Name of rail yard
- Name for map label

Coded value domain YardTypes	
Description	Rail yard type
Field type	String
Split policy	Default value
Merge policy	Default value
Code	Description
CLASS	Classification
INTERCHANGE	Interchange setout/pickup
RRSERVICE	Railroad service facility
CUSTOMER	Customer owned
OTHER	Other rail yard type

Figure 18.10 **Railroad yards** You may include every track in a rail yard as a separate feature, but you still need a way to collectively describe the higher level facility they form. The RailYard polygon feature class is a way to do so. The polygon geometry should also include all the related buildings. One way to construct the RailYard polygon is to follow the outline of all the property parcels.

Like interstate highways, railroads have interchanges, with one or more interchange tracks serving the function of ramps to allow travel between intersecting main lines. A typical arrangement is shown in figure 18.10. Railroads A and B intersect at a crossing diamond. A siding on railroad B leads to an interchange track from railroad A. Adjacent to this track is a pickup and setout yard where each railroad can spot (place) rail cars for pickup by the other railroad. This arrangement allows local freight trains to service the pickup and setout yard without interfering with movements on the main lines.

A RailYard polygon feature class is proposed to store the area covered by each yard. It has few fields, as the standard editing support fields have been omitted. Although tracks may come and go, yards are rarely modified to the extent that the enclosing polygon would change. The few included fields indicate the owning railroad, the type of yard, and the name of the facility. The members of this class may be derived from the parcel polygons that embrace the property on which the yard is built.

Trackside structures

In addition to the passenger and freight depots that may be well known to you, there are smaller railroad structures scattered along the tracks. These structures, which should be visible on aerial photographs and easily identifiable on the ground, include equipment cabinets, warning detectors, signals, stockpiles, storage facilities, and minor buildings.

Major railroad tracks are subdivided into operating blocks to efficiently regulate train movement. Each block comprises an electrically isolated section of track. Detectors can sense the presence and movement of trains along the block. This information is used to set the aspect of trackside signals and, on modern equipment, inside the operating cab of locomotives. In the three-color signal system common in North America, a green aspect indicates that the next two blocks are unoccupied; a yellow aspect means the next block is unoccupied but a train is present in the second downstream block, so be prepared to stop; and a red aspect indicates that the next block is occupied and cannot be entered. The limits of a passing siding are almost always coterminous with the end of a block so that signals at each end can report the presence of a train on the siding and the setting of the track switches.

Railroads increasingly deploy a variety of sensors in order to detect dangerous conditions. The most common types on the track detect overheated axle journals (aka, "hot boxes"); dragging equipment detectors, which generally sense the impact of loose chains and other hold-downs; and excess dimension detectors for ensuring that all railcars in the train can pass through a downstream restriction (tunnel, bridge, etc.). Other sensors may be adjacent to the track to detect adverse environmental conditions. For example, mountainous areas may be served by landslide detectors, which look like big electric fences. A landslide will break one or more wires and send a signal to remote CTC operators alerting them to the condition. High-water detectors may also be used in flood-prone areas.

Equipment cabinets are typically associated with signals and trackside warning detectors. Many railroads have adopted the practice of stenciling the function of each cabinet on its side. You will most likely see these cabinets near railroad grade crossings.

In addition to structures to house equipment for signals and defect detectors, railroads will present other small trackside facilities. Such structures include rail flange lubers, which apply a lubricant to the side of wheel flanges to reduce rolling friction (saves energy) and rail wear, and visual readers that take a picture of the reporting marks (identifying label) on passing cars so that they may be tracked through the rail system.

No specific feature class is proposed to store these structures, as they are mainly of interest to the railroad. However, it would be a simple matter to trace the outline of each structure, determine its function, and assign a label for mapping purposes.

Railroad grade crossings

Switches and diamonds are located at rail-to-rail intersections. This section deals with rail-to-highway intersections, or railroad grade crossings. You will probably want to retain more information about these railroad features than any other type.

The above discussion of railroad structures is intended to help you understand what you see in aerial photographs so that you may properly identify, select, label, and arrange railroad components in your geodatabase. That is not the only reason information about railroad grade crossings may be included in your design. If you work for a transportation agency in an area with railroad facilities, then you will likely use information about grade crossings in several ways.

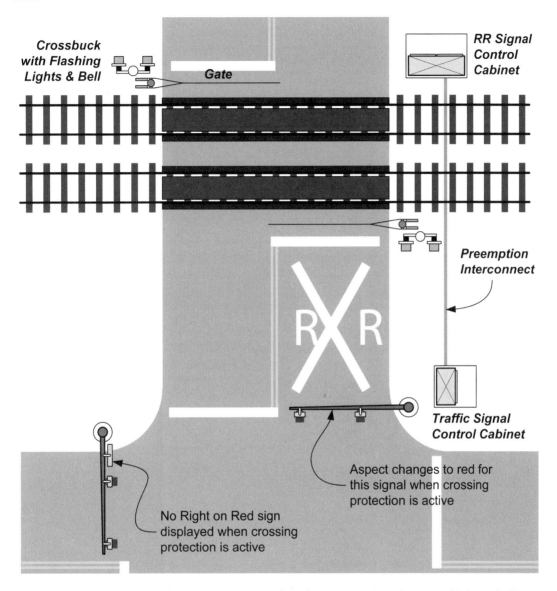

Figure 18.11 **Railroad grade crossings** A railroad grade crossing exists whenever a highway facility intersects a railroad. Each grade crossing is assigned a nationally unique identifier consisting of six digits plus one letter. One crossing may involve one or more parallel tracks. Crossing protection may be active or passive. Active protection includes flashing lights, bells, and gates that block traffic from crossing the tracks. Passive protection normally consists of signs, such as the classic crossbuck. Crossing materials are placed between the rails in order to provide a level surface for motorized traffic to travel. A flangeway must be retained in the crossing surface to allow trains to pass. Interconnection between the railroad grade crossing protection controller and any nearby traffic signal controller is required to provide coordinated preemption of the traffic signal so as to preclude conflicting indications to motorists.

Grade crossings are characterized by two major components besides tracks and roadway: the crossing material used by highway vehicles and the crossing protection used to signal the presence of a train. The crossing materials affect ride quality across the track. A rigid material, such as concrete, cannot be used because the track assembly must remain somewhat flexible. If asphalt—a flexible paving material—is the primary crossing material, the pavement edges adjacent to the track will need to be framed by a rigid material, such as timbers, metal edging, or hard rubber strips to prevent edge failures. A superior product for railroad grade crossings is recycled rubber panels that fit into a rigid frame built over the track.

Every U.S. railroad grade crossing is assigned a seven-character identifier as part of the National Crossing Inventory managed by the U.S. DOT and implemented by the state DOTs and the railroads. The identifier consists of six digits followed by one letter. Only one identifier is assigned regardless of the number of tracks involved. The identifier is supposed to be posted at the crossing. The Manual on Uniform Traffic Control Devices (MUTCD) lists all standard sign types. Ideally, an MUTCD-designated I-13 or I-13a sign will be posted with the emergency phone number and crossing identifier for notifying the railroad of a problem with the crossing protection or the presence of a stationary vehicle on the tracks. This information should also be in your geodatabase.

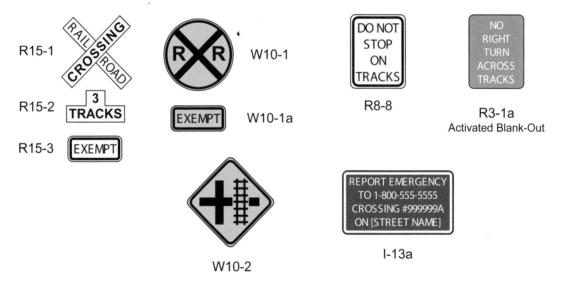

Figure 18.12 **Railroad grade crossing signs** The signs shown here appear at or near railroad grade crossings. The two yellow signs are warnings of a nearby grade crossing. The white and blue signs are located at the crossing. All signs are shown with their MUTCD designation, which starts with the sign type: R (regulatory), W (warning), and I (information). There are dozens of other, less frequently seen standard signs available. Many of the standard signs are contained in the three ESRI MUTCD fonts delivered with ArcGIS.

Highway vehicle warning and protection devices may be active or passive. The most common passive protection is the traditional "RAILROAD CROSSING" X-shaped crossbuck sign mounted on a post. Known in the MUTCD as Sign R15-1, the crossbuck should be supplemented by an R15-2 sign to indicate the number of tracks involved, if more than one is present. Tanker trucks, school buses, and several other highway vehicles are required to stop at every grade crossing unless the R15-3 "EXEMPT" sign is posted.

In many locations, a road and a railroad track are parallel to each other and in close proximity. A W10-1 sign, perhaps with the supplemental W10-1a sign, is located on the road that goes straight across the tracks. The W10-2, or its functional equivalent, can be located on parallel streets in advance of the intersection that leads to the crossing. The space between the road and the track is often so short that only one vehicle can be stored between the two facilities. In such cases, the local authority may also post an R8-8 "DO NOT STOP ON TRACKS" sign to remind motorists that they are prohibited by law from stopping on the railroad tracks.

If the adjacent intersection is controlled by traffic signals, it is important that the signal controller be interconnected with the railroad grade crossing to provide coordinated operation in a process called preemption. In a typical arrangement, a cable connects the grade crossing protection equipment to the traffic signal controller cabinet. Activation of the grade crossing protection system will send a signal to the traffic signal controller, putting it into preemption operation. The preemption signal phasing is intended to give immediate priority for a green signal aspect to any traffic on the affected approach. Once that traffic is cleared, the signal will likely rest in a phase that never gives that approach nor other conflicting movements a green aspect. Blank-out or changeable message signs, such as an R3-1a "NO RIGHT TURN ACROSS TRACKS," may also be displayed to traffic to preclude normally permitted movements that would lead to the grade crossing.

The GradeCrossing multipoint feature class has more fields than any other in this chapter. In addition to the standard fields, it includes a public key (NationalGCID), location data for both the road and the railroad, and several descriptive attributes.

Four fields support the road position description. If you use an LRM for roadway positions, then you can use the RoadID and AtMeasure fields to state the location of the crossing from the road's perspective. If you do not use an LRM for roads—or just not this road—you can use the StreetName field to identify the road and the StreetBlock field to say approximately the crossing's location along the street. (Railroad grade crossings are not given addresses, but stating the block in which the crossing falls is a reasonably close proxy.)

Four fields also support a position description from the railroad's perspective. RRCompanyID tells you who owns and/or maintains the crossing, while the SubName and RRAtMeasure supply the route and measure values, if they are known. The MainRRTrackID designates the track feature that determines the railroad measure value.

Chapter 18: Railroads

Simple feature class: GradeCrossing
Geometry: MultiPoint
Contains M values: No
Contains Z values: No

Railroad grade crossing features

Field name	Data type	Allow nulls	Default value	Domain	Precision	Scale	Length	
OBJECTID	Object ID							
Shape	Geometry	Yes						
GradeCrossingID	GUID	Yes			0			Identifier of the grade crossing
NationalGCID	String	Yes					7	National RR grade crossing number
RecordDate	Date	Yes						Date record was created
RecordStatus	Short Integer	Yes	1	RecordCode	0			Status of the record
CrossingStatus	Short Integer	Yes		EntityCode	0			Status of the crossing
FromDate	Date	Yes						Starting date of validity
ToDate	Date	Yes						Ending date of validity
RoadID	GUID	Yes			0			Identifier of roadway
AtMeasure	Double	Yes			7	2		Road LRM measure for crossing
StreetName	String	Yes					32	Name of crossing street
StreetBlock	Float	Yes			0	0		Block in which crossing is located
RRCompanyID	GUID	Yes			0			Identifier of railroad company
SubName	String	Yes						Name of RR subdivision
RRAtMeasure	Double	Yes			7	2		Railroad LRM measure for crossing
MainRRTrackID	GUID	Yes			0			Identifier of locating main track
NumberOfTracks	Short Integer	Yes			0			Number of tracks at crossing
ProtectionType	String	Yes		GCProtection			16	Type of crossing protection
HasMastArm	Short Integer	Yes		TrueFalse	0			Whether exempt from stops
CrossingMaterial	String	Yes		XingMats			16	Type of crossing material
NumberOfBuses	Short Integer	Yes			0			Number of school buses daily
NumberOfTrains	Short Integer	Yes			0			Number of trains daily
TrackSpeed	Short Integer	Yes			0			Maximum train speed
MinSightDistance	Double	Yes			0	0		Minimum road sight distance
IsExempt	Short Integer	Yes		TrueFalse	0			Whether exempt from stops

Coded value domain: XingMats
Description: RR crossing materials
Field type: String
Split policy: Default value
Merge policy: Default value

Code	Description
TIMBER	Wood timbers
ASPHALT	Asphalt pavement
GRAVEL	Gravel/unpaved
RUBBER	Rigid rubber mats
OTHER	Other material

Coded value domain: GCProtection
Description: Grade crossing protection type
Field type: String
Split policy: Default value
Merge policy: Default value

Code	Description
SIGN	Crossbuck sign
LIGHT&BELL	Lights & bell
2GATES	Lights, bell & 2-quad gates
4GATES	Lights, bell & 4-quad gates
FULLGATES	Lights, bell & 4-quad & ped gates

Figure 18.13 **Grade crossing feature class** Grade crossing is represented using multipoint geometry since there needs to be one point placed for each involved track, but the whole crossing gets just one identifier. Only the highest class track (**MainRRTrackID**) determines the location of the crossing on the railroad LRM. Other railroads using the crossing are discovered through the RRCompanyUsesRRTrack relationship class.

The next two fields describe the grade crossing's protection system. The ProtectionType field is supported by the GCProtection domain class. The Boolean HasMastArm field tells you whether an overhead mast arm is used to place flashing signals over the roadway for greater visibility, particularly on multilane roads. The following fields have fairly self-explanatory functions. The CrossingMaterial field uses a XingMats domain class. You could substitute a track class field for TrackSpeed in most cases. The final field, IsExempt, is a Boolean value that indicates whether buses, tankers, and other specific vehicles are exempt from regulations requiring them to stop before crossing.

Intermodal facilities

The National Highway System with its intermodal connectors under the Intermodal Surface Transportation Efficiency Act (ISTEA) of 1991 raised the level of interest among state DOTs and metropolitan area governments in the location of, and access to, intermodal facilities linking highway transportation to major railroads and navigable waterways to ship containerized cargo. Data reporting requirements for intermodal connectors substantially increased as a result of ISTEA's implementation and now include some level of analysis of each facility's capacity to move shipments, along with identification and treatment of deficiencies.

Much containerized cargo moves over railroads under the general label of trailer on flatcar (TOFC). At one time, the common method of loading trailers onto flatcars required the tractor to drive up onto the car itself and uncouple from the trailer. Today, most trailers and wheelless containers that ride on trailer frames are placed on flatcars using large mobile cranes that may look like houses in an orthophotograph. Be sure to check the shadows!

Final thoughts

As you've seen in this book, a central concept for traditional database design is the need to separate the editing database from the one used by analytical applications. This practice reduces risk in the IT world by eliminating data redundancy. Such risk is presented by having multiple versions of the same piece of information: one is likely to be different from the others. You may know all the details of the geodatabase you use for executive reporting and can "fix" its shortcomings when you generate a report, but someone else may not. As "your" geodatabase becomes the enterprise's geodatabase, those little details are rarely publicized. Your abstract understanding of the risks presented by spatial data errors will rapidly transform to a crystal-clear definition when the department head gets one mileage report from planning and another from finance. It is a definition you can apply in your next job.

Another advantage of separating the editing geodatabase from those used by other applications—editing is an application—is the ability to optimize the geodatabase design for each application. Leaving the applications to deal with the overhead of extracting data from a normalized editing geodatabase results in software redundancy and another source of risk: translation error. You increase risk to the organization when you leave it to the analytical application designers and other data users to handle data transformation in each application. Again, the primary risk arises from the potential for differences in understanding the data. A related form of risk is also present when you make data users create their own software for data transformation and extraction. Here the risk is duplicative cost and software errors.

These forms of risk have led to IT practices such as normalization, which helps editors by reducing their workload, and centralized data transformation and publication for use by others, which saves them all from doing the same work over and over—and sometimes getting it wrong. We have all read the standard estimate that GIS is 80 percent data. Yet, the structure and management of that data usually get little attention. We commonly use the data structure required by the analytical application. That approach works fine for single-purpose geodatabases, but we are trying to grow beyond workgroup GIS toward enterprise GIS. We do that by combining data resources to form enterprise geodatabases. The data structure that works well for one application may be very different from that optimized for another application, even when both use the same data.

Final thoughts

What must change for enterprise GIS to become possible is the way we structure our spatial data. *Designing Geodatabases for Transportation* has shown how to construct a spatial database optimized for the one application most of us ignore: data maintenance. Nowhere is this need greater than in transportation agencies, which are frequently collections of single-function divisions. While these divisions share an interest in a common transportation system, they generally operate in isolation. Data sharing—let alone data interoperability—is rare. Traditional data interoperability requires the data-using offices to change their behavior. However, if you separate the source geodatabase from the many versions of that geodatabase required by the disparate data-using workgroups, then you can effect enterprise GIS without requiring the data users to change their behavior.

The normalized data structures illustrated throughout this book have demonstrated the easiest way to maintain data. They strip the data down to its basics, free of all expectations beyond that of accuracy. Only then is it possible to construct views of the data that meet the needs of various user groups.

You are now armed with the knowledge required to construct an enterprise geodatabase in your agency in a manner that reduces the risk to you and your organization. *Designing Geodatabases for Transportation* is a cookbook. It is now time for you to pick a recipe, select the ingredients, and start cooking.

Index

0..* cardinality, explanation of, 25
0..1 cardinality, explanation of, 25
1:1 (one-to-one) relationship, explanation of, 20-21
1D location referencing systems, basis of, 83
1D versus 2D coordinate systems, 215
1:m (one-to-many) relationship
    cardinality of, 18, 20-21
    creating, 59
    example of, 23, 60
    foreign keys for, 22
1..* cardinality, explanation of, 25
1NF (First Normal Form), example of, 61
2NF (Second Normal Form), example of, 62
3NF (Third Normal Form), example of, 62-63
4NF (Fourth Normal Form), example of, 63-66
5NF (Fifth Normal Form), example of, 66-67
9-1-1 enhancement, support for, 92
20-27 data model. *See* NCHRP 20-27(2)

AADT (annual average daily traffic) statistic, use of, 184, 190, 203-204
AASHTO (American Association of State Highway and Transportation Officials), 54
abstract class names, representing, 25
abstract entities, representing, 30
abstractions
    centerlines, 73-79
    creating, 72-73
    determining in data modeling, 12-13
    example of, 72

Accepted (4) **RecordStatus** value, explanation of, 128
access point intersections, occurrence of, 79-80
accuracy, estimating, 209
Active (5) **RecordStatus** value, explanation of, 128
Activities package, use with UNETRANS, 234-235
address block linear events, 150
address geocoding. *See* linear referencing
address ranges, controlling for streets, 90
address segment linear events with street centerlines, 150
addressing design with node offsets, 83
Adds table with multiple editors, 119
aggregation relationship
    versus composition, 27
    explanation of, 26
    in pavement management system, 94
agile geodatabases, building, 7-9
agile methods, overview of, 4-7
agility, core concept of, 8
Ambler, Scott W., 4
amenities
    defined, 367
    modeling as polygon features, 376
American Association of State Highway and Transportation Officials (AASHTO), 54
anchor point
    accommodation by CalibrationPoint feature class, 321
    accuracy of, 212
    defining for linear datums, 214
    discovering LRM positions for, 253

Index

anchor point *(continued)*
    in NCHRP 20-27(2), 220
    in UNETRANS, 248
    use with linear datums, 211-212
anchor section
    in NCHRP 20-27(2), 220
    in UNETRANS, 248
    use with linear datums, 211
AnchorPoint, relationships for linear datums, 212, 214
AnchorSection table, with centerlines, 321-322
AncillaryRole field, in travel demand models, 107
AND, in SQL statements, 354
annexations, advisory relative to ESZs, 92
annual average daily traffic (AADT) statistic, use of, 184, 190, 203-204
application development, role of data modeling in, 4
application ontology, development of, 30
Arc Attribute Table, contents of, 206
Arc Data File, contents of, 206
Arc Hydro data model
    basis of, 388
    capabilities of, 384-385
    representing channel width and depth in, 403
ArcGIS
    developing data-maintenance routines in, 171
    maintaining data with, 169-171
    providing linear datums in, 212-214
    using as editing environment, 171
ArcGIS and ArcIMS architectures, diagram of, 140
ArcGIS geodatabases. *See* geodatabases
ArcGIS Network Analyst. *See* Network Analyst extension
ArcGIS Tracking Analyst, capabilities of, 67-69
ArcInfo route system, overview of, 206-209
ArcMap
    Calibrate Rotate Feature task in, 319
    meaning of route in, 150
ArcMap Validate Features command, with composite relationships, 55
ArcObjects classes, contents of, 24
area features, reflecting values of, 166
AS clause with SELECT statements, 359
Aspect table, inclusion in UNETRANS, 244

Aspects domain, inclusion in UNETRANS, 243
aspects versus elements, 244, 309
assets, role in UNETRANS model, 234
association relationships
    explanation of, 26
    in Inventory package, 251-253
    in pavement management system, 93
    of SimpleRelationship class, 55
    in UNETRANS, 231-232
association rules, implementation of, 39
associations
    defined, 21
    managing between classes, 40
    representing in physical data models, 34
associative tables, 22
asynchronous editing and publishing, explanation of, 140
at-grade intersection, defined, 79, 255
atomic objects, support in NCHRP 20-27(2), 217
atomicity
    challenge of, 74
    ensuring in segment attributes, 78
    preserving, 74-75
ATRs (automated traffic recorders)
    adjusting for traffic data, 197
    classification counter, 187
    description of, 185
    pavement design in, 188
    simple count site, 186
    in traffic monitoring systems, 188
    using in equipment inventory, 199
    vehicle classification site, 186
    WIM (weigh in motion), 186
Attribute class, to preserve history, 133-134
attribute data, storage of, 151
attribute domains, overview of, 49-52
attribute names, including in logical data models, 32-33
Attribute table fields, domain control for, 135
attributed relationship classes
    associative tables as, 22
    contents of, 54
    versus relationship classes, 56
    using, 56

AttributedRelationship class, specialization of, 54
attributes. *See also* network attributes
    accommodating multiple versions of, 178
    adding to geodatabases, 116-117
    components of, 12
    referring to, 24
    storage of, 12
    variation in transportation databases, 146-147
axle adjustment factors, providing, 196
axle detectors, use of, 186-187

backward versus forward notification, 57
Base Model, structure of, 225
base table as default version, 119
BETWEEN clause, in SQL statements, 355
bin counts, including in traffic monitoring systems, 191-192
binary large objects (BLOBs), 48
BinCount table, for traffic counts, 202
BLOBs (binary large objects), 48
blocks
    defined, 367
    in transit geodatabases, 377-378
Boolean indicators, 85, 174, 193, 194, 198, 201, 257, 269, 272, 306, 310, 311
boundary field in Network Analyst extension, 282
boxes, as used in data models, 30-31
bridge element, relationship with roadways, 312
bridge entities, migration to, 327
Bridge table as *Entity* stereotype, 130-131
BridgeMember associative class, contents of, 256-258
bridges
    avoiding turn features at, 279-280
    as facilities or elements, 247
    versus intersections, 256
    logical data models for, 258-259
    in navigable-waterways geodatabase, 410
    in state DOT example, 311
    in UNETRANS, 256-259
Brunnel subtypes, in geometric networks, 112-113
bus stop, including in stop group, 369
business rules, incorporating into migrations, 329-330

bytes
    for double-precision floating-point numbers, 47
    for floating-point numbers, 47
    for long integers, 47
    for short integers, 46

CabinetType field in equipment inventory, 199
Calibrate Route Feature task in ArcMap, 319
CalibratePoint
    lack of measures for, 323
    LRMPositionID value stored in, 323
calibration point feature class with routes, 159
calibration points
    applying, 338-339
    for navigable-waterways geodatabase, 396
calibration problem in state DOT example, 323-327
CalibrationPoint feature class
    joining LRMPosition table to, 323
    ReferencePoint as, 327
    using with Centerline feature class, 318-322
    using with navigable waterway geodatabases, 397
capacity, impact on traffic volumes, 184
capacity-constrained road, description of, 185
cardinality. *See also* relationships
    association relationship, 26
    for composite relationships, 56-57
    defined, 18
    determination of, 20
    expressing, 21
    versus multiplicity, 20
    of relationship classes, 55
    for simple relationships, 56
    types of, 20-21
carriageways
    example of, 73
    using, 74-75
cartographic representation in NCHRP 20-27(2), 218
census files, 86-87
census tracts
    combining, 106
    using in TAZs, 105

435

Index

Centerline feature class
  creating scale-specific versions of, 317-318
  joining SegmentLRM table with, 337
  making M-aware, 339
  for navigable-waterways geodatabase, 396
  preparing for highway inventory, 333
  publication process for, 335-336
Centerline feature geometry, in dynseg, 158
Centerline features, generating for highway inventory geodatabase, 332
centerlines. *See also* directional centerlines; logical centerlines
  choices for, 75
  creating for railroad tracks, 417
  creating for roadway segments, 294
  for divided roads, 78
  intersecting, 76
  logical versus carriageways, 73
  publication at record level, 340
  representing track segments with, 417
  in state DOT example, 317-323
  subtypes of, 73
  use of, 12
  using for linear facilities, 73-79
  using offset events with, 162
Centroid class, in travel demand model, 109
centroid features, defining for TAZs, 107
centroids, in TAZs, 105
changes
  managing for workgroups, 328
  managing in state DOT example, 300
channel, cross-section and profile features of, 405
channel geometry in navigable-waterways geodatabase, 403-406
channel markers
  point features for, 399
  use of, 397-401
channel width and depth, representing, 403
channelized turning movements, ignoring for centerlines, 76
ChannelSegment linear event table, 398
character limits
  setting for text fields, 14
  specifying, 46
character strings, defined, 14

check-in/checkout process, in disconnected editing, 122
child : parent relationship, in object-relational databases, 24
class descriptions, providing, 134
class interfaces, use of, 23-24
class model for geometric networks, 98
class names, including in logical data models, 32
class symbols, description of, 32
class tables, viewing fields in, 45
class templates
  for continuous versioning, 123
  stereotypes as, 25
class types
  support in ArcGIS geodatabases, 39
  for transportation data, 40
class-creation process, traditional approach toward, 118
classes. *See also* relationship classes
  establishing relationships between, 58
  managing associations between, 40
  mixing in UNETRANS, 233
  in object-relational databases, 23
  for Table classes, 44
  tables as, 40
  using subtypes for, 53-54
classification ATRs, use of, 188
classification counter, use with ATRs (automated traffic recorders), 187
Codd, Edgar, 18
CodedValue class, to preserve history, 133
coded-value domains, 50-51
codes with domains, 49
columns
  in fields, 15-16
  removing redundancies of, 63-66
  in tables, 17
comma-delimited text files, structure of, 17
Comment field, in continuous versioning, 126
comparator operators, in SQL statements, 354
complex candidate primary key, in state DOT example, 292
complex edges
  description of, 100
  using for road features, 109
  using for roadways, 101

Index

complex keys
    description of, 20
    lack of support for, 57
complex tracking events
    descriptions of, 67-68
    examples of, 68
composite classes, in NCHRP 20-27(3), 224
composite relationships
    cardinality for, 56-57
    defining, 55
    description of, 54
    using, 56
    using Validate Features command with, 55
composition relationship
    versus aggregation, 27
    explanation of, 26
compressing denormalized event-table rows, 167
concatenate option, in linear referencing, 167-168
conceptual data models. *See also* data models
    developing, 27
    extension for intersections, 290
    of geodatabase, 41
    for highway inventory, 289-290
    versus logical data models, 33
    for NCHRP 20-27(3), 224
    overview of, 29-31
    versus physical data models, 34
    for state DO, 292-293
    for TMS (traffic monitoring system), 194
    for traffic monitoring site maintenance, 200
    for traffic monitoring systems, 191-192
    for traffic signals, 314
conditions, entering in SQL statements, 354
ConnectionPosition class, in transit geodatabase, 373, 375
connectivity rules, specifying for UNETRANS, 267
"Contains M values," appearance in data dictionary views, 154
continuous versioning. *See also* versioning
    overview of, 123-132
    process of, 126
    RecordStatus values in, 127-129
    status codes in, 127
coordinates, relating to LRM positions, 214
count bins, repository for, 202

COUNT command, in SQL statements, 358
count sites
    assigning to traffic sections, 192
    using in traffic monitoring sites, 186
Count table, for traffic counts, 201
counting equipment inventory, managing, 198-199
counts. *See* traffic counts
coverage route system, shortcoming of, 207
coverages
    in ArcInfo route system, 206
    basis of, 17
    example of, 207-208
crash-reporting systems
    database design for, 263-264
    organizing, 82
CrossSection class, in navigable-waterways geodatabase, 406
cul-de-sac subtype, with turns, 278
cul-de-sacs, using centerlines with, 76-78
culvert, association example for, 26
Cursor class, instantiation of, 44
cursors, with Table class, 44

data
    checking based on location, 178
    concatenating, 167-168
    deriving from other data, 8
    maintaining, 169-171
data dependencies
    eliminating via normalization, 61
    removing, 59-67
data dictionary history, preserving, 132-136. *See also* history preservation design
data dictionary views
    "Contains M values" in, 154
    overview of, 45-46
data editing, viewing as separate application, 7-8
data entries by users, limiting, 49-52
data interoperability, concept of, 141
data maintenance, process of, 138
data migration, implementing, 328-330
data modeling. *See also* modeling
    deciding on abstractions for, 12
    defined, 12

data modeling *(continued)*
    process of, 13, 27-29
    role in application development, 4
data models. *See also* conceptual data models; logical data models; physical data models; transit data models; transportation data models; UNETRANS (Unified Network for Transportation)
    core classes in, 41-43
    databases as, 12
    overview of, 3-4
    primary relationships in, 25-26
    for route segmentation, 292
    testing, 28
    for traffic monitoring systems, 191-194
    using, 24
data objects, maximizing efficiency of, 38. *See also* objects
data redundancies
    advisory about, 60-61
    eliminating via normalization, 61
    removing, 59-67
    use of, 8
data replication
    example of, 122-123
    process of, 139
data scrubbing, preparing for, 168
data storage mechanism, Table class as, 42-43
data structures, translating user requirements into, 30
data types. *See also* field types
    character strings, 14
    date and time, 14
    numeric, 14
    specifying for domain control function, 50
database, defined, 12
Dataset abstract class
    in conceptual data model, 41
    description of, 42
datasets. *See also* network datasets; transport datasets
    defined, 12, 42
    storage of, 12
    in workspaces, 42
date and time information, storing, 47

Date data type, contents of, 14, 47
datums. *See* linear datums
daymarker, use of, 401
.dbf extension, explanation of, 17
default value policy, with domains, 51
default version
    base table as, 119
    editing directly, 120
Deletes table, with multiple editors, 119
DEM (digital elevation model), applying to linear facility, 13
denormalization. *See also* normalization
    of data structures, 60
    of geodatabases, 116-117
    of inventory segment class, 324
    of pavement layers, 95
    in published geodatabases, 138
    of StreetSegment feature class, 92
    of UNETRANS, 232-233
denormalized event-table rows, compressing, 167
denormalized/normalized design, example of, 147
Department table, normalizing, 63
dependencies
    eliminating via normalization, 61
    removing, 59-67
dependency relationship, defined, 25, 27
derail, use on railroads, 414
descriptors, with network attributes, 268
design alternatives, evaluation of, 121
destination tables. *See also* tables
    versus origin tables, 20, 58-59
    rows in, 21
d-factor (directional bias) statistic, use of, 184
diamond, use in railroad intersections, 416
digital elevation model (DEM), applying to linear facility, 13
Direction field, in geometric networks, 112
directional bias (d-factor) statistic, use of, 184
directional centerlines, defining, 298. *See also* centerlines
disconnected editing, example of, 122
dissolve option, in linear referencing, 167
DISTINCT command, in SQL statements, 358
divided roads, handling in state DOT example, 298-300. *See also* roads

Index

DMIs (distance measuring instruments), use by state DOT, 325-326
Domain class, with attribute domains, 50-51
domain classes
    adding for route events, 172-173
    for bridge, 257
    creation at workspace level, 42
    including in TMS, 193-194
    for linear datums, 214
    for Street edge, 272
    using in physical data models, 34
    using with valid value tables, 53
    for waterway markers, 402
domain control classes, examples of, 50
domain control function, 50-51
domain controls
    methods of, 50
    for multiple LRMs, 178
    for preserving history, 135
    using in traffic monitoring site maintenance, 201
domains
    defined, 49
    for street intersections, 278
    using codes with, 49
DOTs. *See* state DOT example
double-precision floating-point numbers, description of, 47
driving directions, generating in Network Analyst extension, 281
duplicate value policy, with domains, 51
dynamic route, defined, 159-160
dynamic traversal, 348
dynseg (dynamic segmentation)
    defined, 150
    dependence on LRM relationships, 157
    overview of, 153-158
    publishing centerline features for, 335-340
dynseg functions
    calculating angles with, 163
    line-on-line event overlays, 164-165
    line-on-point event overlays, 166
    offset events, 162-164
    rotating sign symbols, 163-164

dynseg operations
    conducting, 154
    driving, 177
    for pathfinding applications, 179-181
    providing measures for, 318-321
dynseg result features, offsetting, 162-163

edge elements, managing in geometric networks, 112
edge feature classes
    drawing polylines in, 99
    in UNETRANS, 229
    using in travel demand modeling, 105
    using with network datasets, 271
edges
    defined, 97
    simple versus complex, 99-100
    using with turns, 274
edit actions, reflecting, 125
Edit Geodatabase
    certifying for highway inventory, 332-334
    versus Publish Geodatabase, 338, 352
edit records, removing from highway inventory, 333
editing, transactions involved in, 8
editing databases. *See also* versioning
    advisory about direct use of, 8-9
    moving data from, 8
    versus publishing geodatabase, 332
    separating from published geodatabases, 136-139
    separation from published database, xi
editing environment
    designing and building, 8
    migrating to ArcGIS as, 171
editing history, preserving, 123-132
editing process
    maintenance of, 125-126
    traditional approach toward, 116-118
editing support in UNETRANS, 239-241
editors, versioning for, 118-119
editors and users, managing, 119
EditReason field, in continuous versioning, 126
edits, managing via versioning, 120

Index

element tables, adding, 173-175
elements
    versus aspects, 244, 309
    complex, 312-316
    defining and identifying, 173
    versus facilities, 244, 246
    storage of, 175
    traffic sections as, 310
elevation (Z) values, viewing, 45
elevation fields, explanation of, 267
emergency service zones (ESZs), tying street segments to, 91-92
Enabled field, in geometric networks, 112
EnabledDomain domain class, in mapping applications, 85-86
encapsulation, defined, 23
end of file characters, examples of, 16
end of record characters, identifying in files, 16
Enhanced 9-1-1, support for, 91-92
EnteredBy field, in continuous versioning, 125-126
enterprise, applying agile method to, 6
enterprise data model, development of, 4
enterprise databases, separating editing and usage portions of, 8
enterprise geodatabases, transportation challenges solved by, 1-2
enterprise information resources, governance of, 141
enterprise level, agile design process at, 6
enterprise multimodal geodatabase, objective of, 141
entities
    clarifying in conceptual data models, 29-30
    converting relational tables into, 43
    defined, 12
    identifying, 27
    representing core aspects of, 72-73
Entity class template, in continuous versioning, 123-124
Entity classes, use in UNETRANS, 239
entity status, in continuous versioning, 127
Entity stereotypes, 130-131
Entity template
    maintaining editing process with, 125-126
    selecting portions of, 126-127

EntityID value, using in continuous versioning, 124-126
EntityStatus field, using in continuous versioning, 125
equi-joins, using in SQL statements, 360-361
equipment inventory, managing, 198-199
ESAL (equivalent single-axis load) values, basis of, 187-188
ESRI transportation industries Web site, accessing, xi
ESZs (emergency service zones), tying street segments to, 91-92
evaluators, using with network attributes, 269-270
event points
    defined, 150
    use of, 151
event records, using with intersections, 307
event table publication
    foreign keys for, 342
    at record level, 343
event tables. *See also* multievent tables
    creating, 167
    creating for published geodatabases, 340-347
    creating for routes, 158
    creation of, 117
    editing, 171
    generating for highway inventory, 332-333
    joining, 166
    position descriptions in, 172
    prototype for published geodatabase, 344
    publishing, 345
    separating position descriptions from, 177-179
    using in TMS, 194-195
event types
    publishing as part of multievent tables, 347
    separating elements from, 173-175
events
    mapping, 347
    in navigable-waterways geodatabase, 396-402
    relationship to roadway elements, 308
    in state DOT example, 305-311
    in UNETRANS, 259-260
Events package
    inclusion in UNETRANS, 239
    using with Inventory Package, 260-261

Index

EventType and **EventValue**, with elements, 174-175
EventTypes table, description of, 173
exception events, creating for SRs, 303
EXISTS condition, in SQL statements, 356-357
explicit relationships
    using in physical data models, 34
    using in TMS (traffic monitoring system), 193
explicit state versioning, implementing, 121

facilities
    aggregation of, 251
    defined, 367
    versus elements, 244, 246
    examples of, 245
    including in routes, 158-159
    modeling as polygon features, 376
    segmenting at intersections, 82
    storing information about, 377
    support for travel modes, 2
Facility, relationship to StopGroup, 377
facility attributes, characteristics of, 151
facility classes, use in UNETRANS, 245
facility elements, inclusion in UNETRANS, 244
facility identifier, including in UNETRANS, 239
Facility polygon feature class, 377
facility realignments, strategy for, 80-81
facility routes
    identifying points of interest on, 153-154
    identifying segments on, 153-154
facility status field, adding, 80
factor groups, purpose of, 196
t-factor statistic for trucks, use of, 184
fares
    defined, 368
    in transit geodatabase, 379
feature classes
    creating, 43
    defined, 25
    in geodatabases, 39
    in geometric networks, 98
    impact on performance, 69
    for linear-datum physical data model, 212
    in physical data model, 35
    for railroad grade crossings, 428
    subtypes for, 53-54
    as table classes, 40
    in UNETRANS, 233, 241
feature datasets
    characteristics of, 40
    maintaining, 145
FeatureClass object
    in conceptual data model, 41
    using with FeatureCursor class, 44
feature-creation process, traditional approach toward, 117
Feature-Cursor, with Table class, 44
features
    defined, 25
    merging based on identifiers, 158
features classes
    for dynseg, 154
    using, 40
FGDC (Federal Geographic Data Committee), overview of, 220-221
Field classes, displaying in data dictionary view, 46
field expressions, with network attributes, 269
field measures
    migrating to, 327
    resolving reference points related to, 327
field parameters, modification of, 45
field types. *See also* data types
    BLOBs (binary large objects), 48
    date and time, 47
    double-precision floating-point numbers, 47
    floating-point numbers, 47
    GUID (globally unique identifier), 48
    long integers, 47
    short integers, 46-47
    single-precision numbers, 47
    text, 46
field-measured distances, handling in state DOT, 325-326
fields
    defining in variable-length files, 17
    displaying in class tables, 45
    in files, 15-16
    requirements for creation of, 45

Fields class, with Table class, 43
Fifth Normal Form (5NF), example of, 66-67
file structure for coverages, 207
files
    appearance of, 15-16
    components of, 15
    records in, 15-16
    versus tables, 17
filter values, picking from lists, 356
First Normal Form (1NF), example of, 61
fixed-length records, example of, 16
Float data type, 47
floating-point numbers
    description of, 47
    storage of, 15
Florida Standard Urban Traffic Modeling System (FSUTMS), 108
flowline
    defined, 385, 403
    profile of, 404
foreign keys
    as candidate primary keys, 60
    for event table publication, 342
    maintaining values for, 124
    for multiple LRMs, 177
    in one-to-one relationships, 21
    in origin tables, 20
    versus primary keys, 19
    providing for one-to-many relationships, 22
    in relational databases, 19
    in tables, 18
    using, 60
forward versus backward notification, 57
Fourth Normal Form (4NF), example of, 63-66
fractions, with Float data type, 47
Framework Transportation Segment (FTSeg), 220-221
FreeSpeed field, in travel demand model, 108
frog, use in railroad switches, 414
FROM clause, in SQL statements, 352-360
FromDate
    using in continuous versioning, 124-126
    using with coordinates and LRM positions, 214

FromML value, matching to **ToML** for concatenation, 168
FSUTMS (Florida Standard Urban Traffic Modeling System), 108
FTRP (Framework Transportation Reference Point), 220-221
FTSeg (Framework Transportation Segment), 220-221
functional classes, managing, 53
functional hierarchies, with network attributes, 268
funding sources for public works projects, 261

gauging stations in navigable-waterways geodatabase, 407-409
GDF (Geographic Data File) format, overview of, 221-222
generalization arrow, depicting inheritance with, 25
geodatabase data models. See data models
geodatabase design
    preparing for, 24
    process of, 27-29
Geodatabase Toolset, functions in, 69
geodatabases. *See also* transportation databases
    benefits for transportation datasets, 38
    capabilities of, 13
    components of, 40
    conceptual data model of, 41
    continuous datasets in, 38
    design process of, 5
    determining performance of, 69-70
    feature classes in, 39
    feature dataset example, 39
    features of, 12
    framework for, 41-44
    geometric network in, 39
    modeling, 3
    modifying, 121
    overview of, 38-40
    quality assurance for data entry in, 38
    relationship classes in, 39
    simultaneous editing of data in, 38
    tables in, 39
    temporal, 130

geodatabases *(continued)*
    uniform management of data in, 38
    viewing, 45
Geographic Data File (GDF) format, overview of, 221-222
geometric attributes, considering, 76
geometric elements, creating via dynseg function, 156
geometric features, segmenting, 78
geometric networks
    class model for, 98
    components of, 97
    feature classes in, 98
    in geodatabases, 39
    intersection subtypes in, 113
    intersections in, 112
    junction classes in, 112
    limited-access highway edges in, 111
    local road edges in, 110
    managing edge elements in, 112
    map geodatabase for pathfinding, 109-113
    nodes in, 112
    state road edges in, 110
    travel demand model, 101-104
    in UNETRANS, 228-229
geometry, associating in NCHRP 20-27(2), 219
geometry datasets, examples of, 41, 43
geometry object, explanation of, 25
geometry ratio policy, with domains, 51
geometry weighted policy, with domains, 52
GeometryDef class, with Table class, 44
GeoPositionID, 242, 246, 248-252, 256, 262, 284, 286, 289, 290, 305, 308-310, 320, 321, 326, 341-344, 366, 370-372, 376, 380, 392, 396, 399, 405, 408
GeoPosition, relationships with, 253
GeoPosition table, with navigable waterway geodatabases, 392
GIS applications, purpose of, 7
GIS-T (geographic information systems for transportation), capabilities of, 2
GIS-T field, former limitations on, xi
grade crossings for railroads. *See* railroad grade crossings
GradeCrossing multipoint feature class, 427

GROUP BY control command, using in SQL statements, 357-358
GUID (globally unique identifier), using, 48

HAVING clause, using in SQL statements, 359
header in files, 15
highway inventory, conceptual data model for, 289-290
highway inventory geodatabase
    certifying Edit Geodatabase for, 334
    publishing, 332-333
    publishing centerline features for dynseg, 335-340
    removing edit records from, 333
Highway route system, implementation of, 208-209
highway segment features, example of, 130
highway segments, creating, 291
highway system inventory, views of, 327
highway systems, paths through, 348
highway vehicles, warning and protection devices for, 427. *See also* vehicles
Highway.rat file, contents of, 207-208
history, preservation by state DOT, 325-326
history of data meanings, preserving, 133
history preservation design, update process for, 125. *See also* data dictionary history
HPMS (Highway Performance Monitoring System), 53
hydrographic datasets, gauging stations in, 407
hydrographic line features, 387
HydroLineID field, in navigable-waterways geodatabase, 388

I prefix, meaning of, 24
icons, building block and rocket, 10
implementation environment, creating physical data model for, 35
implicit relationships, in physical data models, 34-35
IN modifier, in SQL statements, 355-356
Incident package, use with UNETRANS, 234-235
increasing measures, direction of, 298. *See also* measures

443

incremental publication process, 351
independent segments versus link-node, 82
indexes, creation of, 43-44
inheritance in object-relational databases, 25
inheritance relationship, UML diagram of, 33
inspections, conducting for traffic monitoring sites, 200
instance, example of, 12
instantiable class, defined, 25, 31
instantiation relationship, explanation of, 26
intelligent key, defined, 19
interchanges
    components of, 318
    for railroads, 422-423
interfaces in object-relational databases, 23-24
intermodal facilities, overview of, 429
interoperability, concept of, 141
intersection attributes, storage of, 82
Intersection class, 175-177
Intersection entity, in highway inventory, 291
intersection features, support in UNETRANS, 229-230
intersection subtypes
    example of, 79
    using, 80
    using in geometric networks, 112-113
    using with turns, 277
"intersection" terminology, application of, 2
intersections. *See also* complex edges
    as complex elements, 313
    attributes, 82
    conceptual model extension for, 290
    defined, 74, 175
    domains for streets, 278
    entities, 327
    example of, 79
    point features, 314
inventory maintenance, sources of, 324
Inventory package. *See also* UNETRANS
    association relationships in, 251-253
    inclusion in UNETRANS, 239
    intersections in, 247
    position descriptions in, 242
    subtypes in, 247
    using with Events Package, 260-261

Inventory package classes, mode-specific, 244-248
inventory segment class, denormalization of, 324
inventory segments
    creation of, 324
    problem with, 325
IS NULL test, in SQL statements, 356. *See also* null in SQL statements
IsAnchorPoint attribute, with centerlines, 322
IsElement field, 174

JOIN command, in SQL statements, 361
join relationships, in physical data models, 34-35
joins, in SQL statements, 360-361
junction classes, in geometric networks, 112-113
junction element, forming, 99
junction of edges, movements possible at, 273
junction types, template in UNETRANS, 229-230
junctions
    correcting, 275
    defined, 107
    overlapping for bridges, 280
    perceiving line intersections as, 267
    requirement for, 97
    and simple edges, 99
    simple versus complex, 99

keys. *See* complex keys; foreign keys; primary keys; table keys

LabelName field, with street names, 88
LastInspectionDate field, including in TMSEquipment table, 199-200
legacy database design, example of, 171-172
legacy databases, dealing with, 328-330
legacy structure, duplication of, 171-172
LIKE comparator operator, in SQL statements, 356
LimitedAccessHwy network edge, 111
line features, presentation controls for, 347
line intersections, perceiving as junctions, 267
linear datums. *See also* LRMs (linear referencing methods)
    in NCHRP 20-27(2), 218-219

linear datums *(continued)*
    overview of, 210-212
    physical data model for, 212
    providing in ArcGIS, 212-214
    using with association relationships, 253
linear event tables
    checking for gaps and overlaps, 169
    example of, 345
linear events
    creating from polygon overlays, 167
    handling by dynseg operations, 154
    use of, 166
    using offsets with, 162
linear facilities
    defining extents of, 145
    overlapping of, 146
    representing, 13
    using abstractions with, 73-79
linear features, with dynseg, 154
linear location referencing, use by DOTs, 78
linear measurement systems, purpose of, 65-66
linear referencing
    using concatenate option in, 167
    using dissolve option in, 167
linear referencing functions
    compressing denormalized event-table rows, 167
    concatenation, 167-168
    data scrubbing, 168
    quality assurance, 168
linear referencing measure, updating, 145
linear referencing methods (LRMs). *See also* linear datums; multiple-LRMs
    accommodating, 177-179
    in coverage file structure, 207
    imposing on traversal, 160
    maintaining for reference points and field measures, 327
    "multidimensional" model for, 223-224
    NCHRP 20-27(2) requirements for, 215, 217
    pavement management system for, 260-261
    using with centerlines, 75
linear referencing, purpose of, 150
linear referencing systems (LRS), overview of, 148-153

linear transportation facilities, representing centers of, 75
LINEAR_EVENT table, normalization of, 65-66
line-on-line transfers, 164-166
line-on-point transfers, 164-166
link-nodes versus independent segments, 82
links, locating in NCHRP 20-27(2), 219
lists, picking filter values from, 356
LocalRoad network edge, 111
locate, requirement of NCHRP 20-27(2), 215
location data, keeping in one table, 178
location referencing, support in UNETRANS, 230-233
logical centerlines. *See also* centerlines
    best practices for, 75-76
    creating for directions of travel, 76
    examples of, 78
    versus facility location, 78
    using, 73-74
logical data models. *See also* data models
    attribute names in, 32-33
    boxes used in, 32
    for bridges, 258-259
    class names in, 32
    versus conceptual data models, 33
    developing, 28
    for events in state DOT example, 305-306
    NCHRP 20-27(2) as, 220
    overview of, 31-34
    for PMS, 94-95
    relationship classes in, 96
    relationships in, 33
    simplification of, 25
    for traffic monitoring systems, 193
    for transit geodatabase, 370, 372
    for transit networks, 283-284
    for UNETRANS revision, 266-267
long integers
    description of, 47
    storage of, 15
long transactions, editing, 121
lookup tables, with domains, 49
LRM (linear referencing methods) datums, specifying in Inventory package, 248
LRM. *See also* linear referencing methods

Index

LRM positions
    describing in UNETRANS, 241-242, 248
    discovering anchor points for, 253
    relating coordinates to, 214
LRM relationships, dependence of dynseg on, 157
LRM values, storing for published traversals, 350
LRMDatum class, in association relationships, 253
LRMPosition class
    in state DOT example, 307
    using in UNETRANS, 241
    using with linear datums, 214
LRMPosition rows, selecting, 177
LRMPosition table
    joining Segment table with, 336
    joining to CalibrationPoint feature class, 323
    using in state DOT example, 308
LRMPositionID, 177, 178, 214, 242, 243, 246-252, 254, 256-258, 260, 261, 264, 289, 290, 293, 302, 305, 307-309, 311-313, 317, 320, 322, 323, 335
LRMs (linear referencing methods). *See also* linear datums; multiple-LRMs
    accommodating, 177-179
    in coverage file structure, 207
    imposing on traversal, 160
    maintaining for reference points and field measures, 327
    "multidimensional" model for, 223-224
    NCHRP 20-27(2) requirements for, 215, 217
    pavement management system for, 260-261
    using with centerlines, 75
LRMType, 65, 177, 178, 213, 214, 242, 243, 248-253, 256, 259, 260, 289, 290, 305-308, 311, 320, 321, 323, 336, 337, 340, 342, 343, 351
LRS (linear referencing systems), overview of, 148-153
LRS components, use in NCHRP 20-27(2), 216-217
LRS concept, measurement rules added to, 210-212

M (measure) values, explanation of, 154
m coordinate, explanation of, 154
mainline and ramp centerlines, intersecting, 79
maintenance inventory, for traffic monitoring sites, 199-201

Make Route command, with traversal features, 350
MakeRouteEventLayer tool, with dynseg, 154
many-to-many (*m:n*) relationships. *See also* relationships
    explanation of, 21-22
    resolving, 56
map locations, computing for route events, 153-158
map metaphor, inapplicability of, 144
mapping applications
    classes for maps in, 84
    domain control in, 85
    in mapping application, 85
    polyline feature classes in, 85
    requirements for, 12
    RoadsCartoDissolve feature class in, 84
    support for editing process in, 84-86
mapping events, 347
marker mounts, 401
marker signs, examples of, 400
markers, placement along navigable waterways, 397-401
M-aware, making Centerline feature, 339
measure (M) values, explanation of 154
measure fields, location of, 208
"measure once, cut twice" philosophy, xi, 141
measure values, managing in dynseg, 156
measurement errors, correction of, 211
measurement rules, addition to LRS concept, 210-212
measurements, repeatability via precision, 210
measures. *See also* increasing measures
    assigning via calibration points, 318-319
    providing for dynseg operations, 318-321
    statement of, 150
    using in LRS (linear referencing systems), 149
merge policies, with domain control function, 51-52
methods
    adding with ObjectClass, 43
    using in logical data models, 32
Microsoft Visio, 35
migrating data, 328-330
milelog. *See* route-milelog referencing

Index

milelog measures
    calibration point feature class for, 322
    use of, 150

m:n (many-to-many) relationships
    explanation of, 21-22
    resolving, 56

Mobile Objects model, Pedestrian class in, 263-264

Mobile Objects package in UNETRANS, 239, 262-263

mobile objects, support in UNETRANS, 236.

modal connection, points of, 2

modeling, defined, 3. *See also* data modeling

MPOs (metropolitan planning organizations), needs of, 116

multiedge turns, types of, 273-274

multieditor environment. *See* editors

multieditor operations, ArcGIS support for, 171

multievent tables. *See also* event tables
    building, 346-347
    publishing event types as part of, 347

multimodal facilities
    example of, 375
    types of, 376

multimodal network, example of, 268, 284-285

multiple-LRMs, accommodating, 158-159. *See also* LRMs (linear referencing methods)

multiplicity of tables
    classification of, 20
    defined, 21
    displaying in object-relational data models, 25
    explanation of, 18

multiplicity versus cardinality, 20

multipoint geometry, for railroad grade crossings, 76

multitier version trees
    requirement for, 122
    using with long transactions, 121

MUTCD (Manual on Uniform Traffic Control Devices), 426

NameStatus field, with street names, 89

NameStatusCodes domain, with street names, 89

National Emergency Number Association (NENA), address standards of, 225-226

National Hydrography Dataset (NHD), water features in, 389

National Spatial Data Infrastructure (NSDI), use of, 225

navigability, defined, 25, 27

navigable waterways
    channels for, 397
    cross-section view of, 386
    markers along, 397

navigable-waterways geodatabase. *See also* waterways
    basis on waterway abstraction, 386
    bridges in, 410
    calibration point for, 396
    Centerline feature class for, 396
    channel geometry in, 403-406
    gauging stations in, 407-409
    overview of, 384-385
    position statements in, 392
    river-reach data in, 389-392
    waterway events in, 396-402

NCHRP 20-27(2)
    absence of measures in, 220
    anchor sections in, 220
    geometry component of, 218
    geometry of, 219
    linear datum in, 218-219
    as logical data model, 220
    LRMs in, 217
    LRS components of, 216-217
    overview of, 215-216
    placement of Traversal Measure attribute in, 219
    traversals in, 217-218

NCHRP 20-27(3), overview of, 222-224

NENA (National Emergency Number Association), address standards of, 225-226

Network Analyst extension
    adding vertices in, 272
    boundary field in, 282
    generating driving directions in, 281
    route numbers in, 281
    signpost point feature class in, 281
    signpost streets table in, 281-282
    simplified view of, 266
    using with network attributes, 269-270

447

Index

network attributes in UNETRANS, 268-270.
network connectivity in UNETRANS, 267-268
network datasets.
    building for UNETRANS, 267
    building from street feature sources, 275, 279-280
    building in UNETRANS, 270-272
network edge features, examples of, 110-111
network edges, including for TAZ, 108
network element weights, calculation of, 108
network elements, characteristics of, 98-99
network features
    relationships of, 99
    using, 99
network junctions, in travel demand model, 107
network links, in travel demand model, 108
network model, expanding for transit geodatabase, 372-375
Network package, inclusion in UNETRANS, 239
network pathfinding application, requirements for, 12
NetworkLink features, creating for TAZ, 107
networks, components of, 98. *See* geometric networks
NextID Table, with unique identifiers, 48
NHD (National Hydrography Dataset), water features in, 389
node offsets, 83
nodes, in geometric networks, 112
nongeodatabase tables, examples of, 151. *See also* geodatabases
normalization. *See also* denormalization
    benefits of, 61
    defined, 59
    versus denormalized data structures, 60
    forms of, 61
NOT BETWEEN clause, in SQL statements, 355
NOT LIKE comparator operator, in SQL statements, 356
notifications, with relationship classes, 56-57
NSDI (National Spatial Data Infrastructure), use of, 220-221
null in SQL statements, 354. *See also* IS NULL test
number fields, characteristics of, 15
number type, interaction with precision and scale, 15

numbers. *See* field types
numeric data types, defining, 14-15
numeric fields
    changing parameters of, 47
    parameters for, 45
numeric range domains, 50-51
numeric ratio, expressing via association, 26

object behavior, controlling with subtypes, 53-54
Object class instance, use with SimpleRelationship class, 55
object classes
    communication between, 24
    in physical data model, 35
    in UNETRANS, 231, 248
    visibility of, 31
ObjectClass
    capabilities of, 43
    in conceptual data model, 41
    interfaces in, 23
OBJECTID primary key
    creation of indexes on, 43-44
    inclusion of, 39-40
    using, 60
object-relational databases
    inheritance in, 25
    interfaces in, 23
    versus relational databases, 37-38
    relationships in, 25-27
objects, features of, 25. *See also* data objects; mobile objects
ObsSpeed field, in travel demand model, 108
The Official Railway Guide, consulting, 418-419
offset distance, determining, 172
offset events
    right versus left, 163
    using with dynseg, 162-164
offset locations, indicating, 172
offset referent, defined, 172
one-to-many (1:m) relationships. *See also* relationships
    cardinality of, 18, 20-21
    creating, 59

one-to-many (1:m) relationships *(continued)*
    example of, 23, 60
    foreign keys for, 22
one-to-one (1:1) relationships, explanation of, 20-21
Oneway field, with network attributes, 269-270
ontology, development of, 30
OR, in SQL statements, 354
ORDERED BY control command, in SQL statements, 357-358
origin versus destination tables, 20, 58-59. *See also* tables
origins of sections, values assigned to, 151-152
overpasses, including in geometric networks, 113
OverpassLevel attribute, in mapping applications, 86

parental relationship, expressing in object-relational databases, 24
parent-child relationship, inheritance as, 25
pathfinding
    constructing network for, 82
    creating traversals for, 179-181
    map geodatabase for, 109-113
    navigable-waterways geodatabase for, 384-385
patterns
    defined, 368
    in transit geodatabase, 374
pavement management example in UNETRANS, 260-261
pavement management system (PMS)
    denormalization of, 95
    designing, 93-96
    physical data model for, 95-96
pavement segment, defined, 261
PAVEMENT_CONDITION table, normalizing, 65
Pedestrian class in Mobile Objects model, 263-264
pedestrian signals and indicators, managing, 314
performance
    determining for geodatabases, 69-70
    improving relative to centerlines, 317
    improving with subtypes, 53
    in travel demand model, 109

person-trip travel demand, generating, 102
phases in traffic signals, 314, 316
physical data models
    for bridge, 257
    versus conceptual data models, 34
    creating, 130
    developing, 28
    for equipment inventory, 198
    for geometric networks, 110-112
    for linear datums, 212, 214
    overview of, 34-35
    for pavement management system, 95-96
    for TAZ (traffic analysis zone), 107
    for TMS (traffic monitoring system), 193-194
    for traffic monitoring site maintenance, 200-201
    for transit geodatabase, 371
piece of work
    defined, 368
    in transit geodatabases, 381
piezoelectric cable, use of, 186-187
place, requirement of NCHRP 20-27(2), 215
PMS (pavement management system)
    denormalization of, 95
    designing, 93-96
    physical data model for, 95-96
    for LRM, 260-261
pneumatic sensors, use in volume counters, 187
point events
    explanation of, 150
    handling by dynseg operations, 154
    preparing for pathfinding traversals, 181
    selecting for pathfinding traversals, 181
    transferring line event values onto, 165
    using offsets with, 162
point feature class, with gauging station, 408-409
PointEventclass, with elements, 175
points of interest, identification of, 153-154
polygon features
    modeling amenities as, 376
    modeling facilities as, 376
    modeling stop groups as, 376
polygon-on-line transfers, 164-166
polygon overlays, creating linear events from, 167

Index

polygons, with roads, 311
polyline feature classes
    for pathfinding traversals, 180
    representing track geometry with, 417
    for transit segment, 284-285
    using in mapping applications, 85
polylines
    drawing in edge feature classes, 99
    operation of dynseg on, 154
position classes in UNETRANS, 242, 248
position data, separating in UNETRANS, 241-242
position descriptions, separating from event tables, 177-179
position numbers, identifying in files, 16
position statements in waterway geodatabases, 392
positions
    approximation of, 207
    differentiating via resolution, 210
    requirement of NCHRP 20-27(2), 215
    in transit geodatabase, 373
precision
    considering in number fields, 15
    default for short integers, 46-47
    explanation of, 209-210
precision parameters
    for double-precision floating-point numbers, 47
    entering, 45
    setting for floating-point number fields, 47
predicate, WHERE clause as, 353
primary keys
    versus foreign keys, 19
    foreign keys as candidates for, 60
    in relational databases, 19
    simplicity and complexity of, 60
    in tables, 18
    using route numbers as, 19
ProblemID foreign key, in traffic monitoring site, 201
Profile polyline feature class, with navigable-waterways geodatabase, 406
ProjectInvolvement relationship class, in pavement management system, 96

projects, relating in pavement management system, 94
properties
    adding with ObjectClass, 43
    using in logical data models, 32
PropertySet
    in conceptual data model, 41
    description of, 42
Proposed (1) **RecordStatus** value, explanation of, 127
protection and warning devices, active versus passive, 427
public keys, use of, 19-20
publication, process of, 332
Publish Geodatabase
    versus Edit Geodatabase, 352
    updating, 338
published geodatabases
    creating event tables for, 340-347
    versus editing database, 332
    event table prototype for, 344
    moving data to, 8
    separating from editing databases, 136-139
    separation from editing database, xi
publishing traversals, 348-351

quality assurance
    via linear referencing functions, 168
    via SQL statements, 169
queries, restricting, 137. *See also* SQL statements

R8-8 "DO NOT STOP ON TRACKS" sign, use of, 427
rail, forms of, 412-413
rail lines, identifying positions on, 148
railroad classes in UNETRANS, 250
railroad companies, naming in geodatabases, 418-421
railroad grade crossings
    defined, 79
    feature class for, 428
    as intersection subtype, 79
    overview of, 424-429

railroad grade crossings *(continued)*
    protection system for, 429
    road position description for, 427
    signs used for, 427
    using multipoint geometry for, 76
railroad intersections, incorporating, 414-416
railroad switch, example of, 255
railroad tracks
    connections between, 414
    division into operating blocks, 423-424
    feature class implementation for, 418
    identifying use of, 417
    incorporating, 412-414
railroads
    creating logical centerlines for, 75-76
    derail used on, 414
    getting information about, 418-419
    interchanges for, 422-423
    sensors used in, 424
    as suppliers, 2
    track geometry in, 417-418
    trackside structures in, 423-424
    as users of transport capacity, 2
    yards and interchanges in, 421-423
RailYard polygon feature class, use of, 423
rain gauge, as stationary complex tracking object, 68
ramp and mainline centerlines, intersecting, 79
range daymarker, use of, 401
.rat (Route Attribute Table) file extension, meaning of, 206
raw-count adjustment factors, considering for traffic data, 197
RCLink field, with route segmentation, 291
RDBMS (relational database management systems), language shared by, 12
reach code, river-mile marker in, 393
reach LRM, limitations on, 394
reaches
    classification of, 389-390
    defined, 389
    geometric representations of, 390
    rules for delineation of, 390
    using stream-level characteristic with, 391
realignments, strategy for, 80-81

record level
    event table publication process at, 343
    publishing centerlines at, 340
record status, in continuous versioning, 127
RecordCode attribute, in continuous versioning, 126
RecordCode domain, in continuous versioning, 127
RecordDate field, in continuous versioning, 124-125
records
    adding for continuous versioning, 125
    adding to highway inventory geodatabase, 333
    in files, 15-16
    fixed-length, 16
    removing from highway inventory geodatabase, 334
    retiring, 137
    selecting based on event and LRM types, 345
    updating to reflect edit actions, 125
    variable-length, 16
RecordStatus values
    assignment of, 137
    using in continuous versioning, 124-125, 127-129
recreation access road, description of, 185
recursive relationships, lack of support for, 57
redundancies, avoiding, 175-177. *See* data redundancies
reference points
    dealing with, 326
    problem with, 325
    resolving relative to field measures, 327
ReferencePoint feature class
    as CalibrationPoint feature class, 327
    relationship for linear datums, 212
    using, 326
referential integrity, avoiding loss of, 9
Rejected (3) **RecordStatus** value, explanation of, 128
relational algebra, functions supported by, 18
relational database management systems (RDBMSs), language shared by, 12

relational databases
    advance offered by, 19
    defined, 12
    invention by Dr. Edgar Codd, 18
    versus object-relational databases, 37-38
relational joins, in SQL statements, 360-361
relational tables, converting to entities, 43
Relationship class
    instance of, 55
    specializations of, 54
relationship classes. *See also* classes
    versus attributed relationship classes, 56
    cardinality of, 55
    establishing, 58
    examples, 55
    examples of, 56
    in geodatabases, 39
    limitations on, 57
    in logical data models, 96
    names of, 56
    parameters for, 56
    in physical data model, 35
    for street names, 89
    using, 40, 54-57
    using in travel demand modeling, 105
    using notifications with, 56-57
RelationshipClass, contents of, 54
relationships. *See also* cardinality; one-to-one (1:1) relationships; one-to-many (1:m) relationships; many-to-many (m:n) relationships
    aggregation, 26
    association, 26
    composition, 26
    instantiation, 26
    in logical data models, 33
    naming by roles, 23
    of network features, 99
    type inheritance, 26
relative addressing, with waterway LRMs, 393-394
remote users, permitting transaction updates by, 124. *See also* users
repair entities, including in traffic monitoring systems, 200
Replaced (6) **RecordStatus** value, explanation of, 128

replication process, support for, 139
resolution, explanation of, 209-210
restrictions
    for turns, 276
    using with network attributes, 268
result sets
    putting into tables, 362
    returning in SQL, 353
Retired (7) **RecordStatus** value, explanation of, 128
RF3 designation, meaning of, 389
RF3 field, stream-level characteristic supplied by, 391-392
river-mile marker in reach code, 393
river-reach number (RRN)
    connection types related to, 391
    parts of, 389
road classes in UNETRANS, 249
Road coverage, example of, 207
road edges in geometric networks, examples of, 110-111
road features, complex edges for, 109
road geometry, example of, 73
Road table, inclusion in UNETRANS, 243
RoadLevel domain class, in mapping applications, 85-86
roads. *See also* divided roads
    creating logical centerlines for, 75-76
    and intersections in state DOT example, 312
    in state DOT example, 311
    storing data beside, 75-76
    types of, 185
RoadsCartoDissolve feature class, contents of, 84
roadway elements, events related to, 308
roadway segments, creating centerlines for, 294
roadway system, graphical depiction of, 294
roadways
    attributes for, 110-112
    relationship of bridge element with, 312
    using complex edges for, 101
RoutableEdge class, use in UNETRANS, 228-229
Route Attribute Table (.rat) file extension, meaning of, 206
route event table, example of, 348

route events
- computing map locations of, 153-158
- defined, 150
- for facility inventories, 171-172
- separation of, 171

route length, calculation of, 325
route markers, use of, 149-150
route measures, storage of, 206
route name, creating static traversal for, 348
route numbers
- dealing with multiples, 303
- treating as linear events, 303
- using as primary keys, 19
- using in Network Analyst extension, 281

route overlaps, dealing with, 303-304
route realignment, managing in state DOT example, 294-298, 324-325
route segment table, with state DOT example, 293
route segmentation, annexation impacts on, 301
route segmentation in state DOT example, 291-294
route systems, implementing, 207-208
Route tables
- joining with Segment tables in SQL, 361
- using SQL statements with, 353
- using in state DOT example, 295-297

route types
- editing in state DOT example, 301-304
- overlaying on single facility, 177

route-based segmentation schemas
- avoiding, 303
- calibration problem with, 323-327

route-based versus segmented structures, 9-10
RouteDesignator, controlling default value for, 53-54
route-event tables, components of, 154
route-milelog datasets, illustrating data in, 152
route-milelog referencing
- use of, 78
- using in LRS (linear referencing systems), 148-149

routes. *See also* traversals
- creating event tables for, 158
- creating hierarchy of, 159-160
- defined, 283, 368
- eliminating as organizing element, 302

routes *(continued)*
- evolution of, 294-295
- including facilities in, 158-159
- making, 158-159
- meaning in ArcMap, 150
- sections of, 206
- subdividing, 153-158
- in transit geodatabases, 377-378
- in UNETRANS, 230-233

routes and segments, storing relationship of, 298-299
RouteTraversal table, 350-351
RouteType
- controlling with valid value table, 52
- treating as event type, 302

Row class
- GeometryDef class used with, 44
- using with Table class, 44
- working with instances of, 44

row identifiers, making unique, 48
rows. *See also* valid value tables
- in destination tables, 21
- identifying in tables, 19
- identifying uniquely as complex primary keys, 124
- in tables, 17
- unique identification of, 60

RRDiamond class, inclusion in Inventory package, 247
RRGradeCrossing class, inclusion in Inventory package, 247
RRN (river-reach number)
- connection types related to, 391
- parts of, 389

RRSwitch class, inclusion in Inventory package, 247
RRTrack feature class, with railroads, 417

sbx extension, explanation of, 17
scale, considering in number fields, 15
scale attribute, with centerlines, 318
scale of display, considering relative to abstraction, 12

scale parameters
    entering, 45
    setting for floating-point number fields, 47
SDTS (Spatial Data Transfer Standard), relationship to FGDC, 220
seasonal factor groups, overview of, 196
seasonal patterns, considering in traffic monitoring systems, 189
.sec file extension, meaning of, 206
Second Normal Form (2NF)
    example of, 62
    explanation of, 61
section length, quality-control effect of, 211
section table, example of, 208
sections of routes, contents of, 206
segment attributes, ensuring atomicity in, 78
segment centerlines, in state DOT example, 293
Segment class, in state DOT example, 292
segment sequence, preserving in state DOT example, 300
Segment tables
    joining with LRMPosition table, 336
    joining with Route tables in SQL, 361
    using alternatives, 302-303
    using in state DOT example, 295-297
segmentation methods, overview of, 81-83
segmentation schemas, problems with, 147-148
segmented versus route-based structures, 9-10
SegmentID column, observing, 65
SegmentLRM table, joining with Centerline feature class, 337
segments
    building for pathfinding traversals, 180
    correlating for street names, 89-91
    identification of, 153-154
Segments, summing, 297
segments and routes, storing relationship of, 298-299
SELECT clause, using with SQL statements, 352, 354-360
sensors, use in railroads, 424. *See also* vehicle sensors
service-oriented architecture (SOA), overview of, 139-141
Shape column, inclusion of, 39
Shape fields, viewing, 45

shapefiles
    characteristics of, 17
    copying, 17
short integers
    storage of, 14-15
    using, 46-47
short-segment design, use in UNETRANS, 233
short-term count data, adjusting in traffic monitoring systems, 189
.shp extension, explanation of, 17
side-of-road locations, indicating, 172
sign inventory, constructing, 174-175
sign symbols, rotation in dynseg, 162-164
Sign table, inclusion in UNETRANS, 243
sign types, use with railroad grade crossings, 426
signpost point feature class, in Network Analyst extension, 281
signpost streets table, in Network analyst extension, 281-282
simple edges
    and junctions, 99
    using, 100
simple junctions, description of, 99
simple key, description of, 20
simple relationships
    cardinality for, 56
    between same two classes, 57
    types of, 54
simple tracking event
    description of, 67
    example of, 68
SimpleRelationship class
    association relationship of, 55
    specialization of, 54
single-precision numbers
    description of, 47
    storage of, 15
sink, TAZ centroid as, 107
SiteType, in TMS (traffic monitoring system), 194
SLDs (straight-line diagrams)
    example of, 152
    use of, 153
SOA (service-oriented architecture), overview of, 139-141
source, TAZ centroid as, 107

spatial accuracy, increasing, 210
spatial data
    determination of, 12
    qualities of, 209
Spatial Data Transfer Standard (SDTS), relationship to FGDC, 220
spatial extents, problems with, 147-148
SPEED_LIMIT table, normalizing, 65
split policy, using with domain control function, 50-51
SQL (Structured Query Language), use in RDBMS, 12
SQL statements. *See also* queries
    advisories about, 362-363
    combining commands in, 359-360
    comparator operators for, 354
    conditions in, 356
    doing math in, 358
    entering conditions in, 354
    form of, 352
    null in, 354
    returning results sets in, 353
    AND and OR in, 354
    BETWEEN clause in, 355
    COUNT command in, 358
    EXISTS condition in, 356-357
    filter conditions in, 358
    for quality assurance, 169
    FROM clause in, 352-360
    GROUP BY control command in, 357-358
    HAVING clause in, 359
    IN clause in, 355
    IN modifier in, 356
    JOIN command in, 361
    LIKE comparator operator in, 356
    NOT BETWEEN clause in, 355
    NOT LIKE comparator operator in, 356
    ORDERED BY control command in, 357-358
    relational joins in, 360-361
    SELECT clause with, 354-360
    SELECT verb with, 352
    WHERE NOT EXISTS clause in, 357
    WHERE clause in, 353-360
SRName event, in traversal publication, 349-350

state DOT example
    alternative business rule of, 308
    bridges and roads in, 311
    calibration problem in, 323-327
    centerlines in, 317-323
    complex elements in, 312-316
    divided roads in, 298-300
    DMIs used in, 325-326
    editing needs of, 119
    editing route types in, 301-304
    events in, 305-311
    general business rule of, 298
    LRMPosition class in, 307
    LRMPosition table in, 308
    migrating to new structure in, 328-330
    preservation of history by, 325-326
    preserving segment sequence in, 300
    reference points versus field measures in, 327
    route realignment in, 294-298
    route segmentation in, 291-294
    segment centerlines in, 293
    segment direction and ordering in, 298-299
    traffic signals in, 313-316
    Tunnel feature class in, 308-309
state road segments, getting total lengths of, 359
StateRoad network edge, 111
states
    restoring databases to, 133
    tracking in continuous versioning, 126
StateWater Obstruction Marker, use of, 401
static route, defined, 159-160
static traversal, creating, 348
station features, support in UNETRANS, 229-230
status, differentiating in continuous versioning, 125
status codes, in continuous versioning, 127
stereotypes
    in logical data models, 33
    using as class templates, 25
stop groups
    defined, 283, 368
    example of, 368-369
    modeling as polygon features, 376
StopGroup, relationship to Facility, 377

Index

stops
  defined, 368
  storing information about, 377
  transit segments between, 283
stops in group, walk connections between, 378
straight-line diagrams (SLDs)
  example of, 152
  use of, 153
streambed, constructing cross-section profile of, 403-404
stream-level characteristic, using with reaches, 391-392
street addresses, management by NENA (National Emergency Number Association), 225-226
street centerlines, using address segment linear events with, 150
Street edge class, example of, 271
Street edge domain classes, 272
street feature classes, compiling, 117
street feature source, building network dataset from, 270, 275, 279-280
street intersections, domains for, 278
street names, accommodating, 87-89
street segments, tying to ESZs (emergency service zones), 91-92
Street table, example of, 58-59, 89
StreetID value, 90-91
StreetName table, example of, 58-59
StreetNameIsAssignedToStreet example, 58-59
Street.rat file, contents of, 207
streets
  controlling address ranges for, 90
  representing as collections of segments, 89
Street.sec file, contents of, 208
StreetSegment entity, in pavement management system, 93
StreetType domain, in geometric networks, 111
String data type, 46
Structured Query Language (SQL). *See* SQL statements
subtypes
  support in logical data models, 33
  using in geometric networks, 112
  using with classes, 53-54
sum values policy, with domains, 51-52
supertypes, exclusion from logical data models, 33

suppliers, railroads as, 2
switches, use in railroad intersections, 414-416

tab-delimited text files, structure of, 17
Table class
  in conceptual data model, 41
  Fields class used with, 43
  in geodatabase data model, 42-43
  GeometryDef class used with, 44
  using cursors with, 44
  using Feature-Cursor with, 44
table classes, feature classes as, 40
table joins, in SQL statements, 361
table keys, complex versus simple, 20
tables. *See also* destination tables
  associating, 19
  as classes, 40
  Employee example of, 18
  versus files, 17
  in geodatabases, 39
  identifying rows in, 19
  multiplicity of, 18
  nongeodatabase example, 151
  origin and destination, 20
  Position example of, 18
  rows and columns in, 17
TAZ (traffic analysis zone) network
  building, 104-109
  as centroids, 105
  creating NetworkLink features for, 107
  defining centroid features for, 107
  example of, 101-103
TAZ centroids as sources and sinks, 107
TAZ polygons, building, 106
TAZ trips, distribution of, 105
team-based approach, in agile method, 5
templates. *See* class templates
temporal aspects, expressing in NCHRP 20-27(3), 224
temporal data fields, including in UNETRANS, 239-240
temporal geodatabases, advantage of, 130
terminal measures, applying, 337, 339
terminal nodes, adding to segment links, 82

Index

testing data models, 28
text data, components of, 14
text fields
    controlling content of, 46
    number of characters allowed in, 14
thalweg, defined, 385, 403
theme design, implementation of, 144
Third Normal Form (3NF)
    example of, 62-63
    explanation of, 61
TIGER (Topologically Integrated Geographic Encoding and Referencing), 86-87
time. *See* Date data type
time and date information, storing, 47
time components, use of colon (:) with, 15
time points
    defined, 368
    location of, 283
    in transit geodatabase, 373-374
TimePoint point feature class, in transit geodatabase, 374
TimePointPosition class, in transit geodatabase, 373, 375
timers, availability in Toolset, 69
TIN (triangulated irregular network), applying to linear facility, 13
TMS (traffic monitoring system)
    conceptual data model for, 194
    conceptual database design for, 191-192
    data model for, 191-194
    description of, 184
    logical data models for, 193
    maintenance geodatabase for, 200
    overview of, 188-191
    physical data model for, 193-194
TMS event tables, 194-195
TMSEquipment table
    fields in, 199
    LastInspectionDate field in, 199-200
TMSite class, fields in, 194
TMSite point event table, description of, 195
ToDate value
    using in continuous versioning, 124-126
    using with coordinates and LRM positions, 214

ToML value, matching to **FromML** for concatenation, 168
Toolset, functions in, 69
Track feature class, as *Entity* stereotype, 130
track geometry in railroads, 417-418
track segments, representing with centerlines, 417
track switch, example of, 415
Tracking Analyst, capabilities of, 67-69
tracking events
    simple versus complex types of, 67-68
    using with gauging stations, 407
TrackMember associative class, use of, 255
tracks. *See* railroad tracks
trackside structures in railroads, 423-424
traffic analysis zone (TAZ) network
    building, 104-109
    as centroids, 105
    creating NetworkLink features for, 107
    defining centroid features for, 107
    example of, 101-103
traffic characteristics
    development of, 190
    generation of, 203
traffic control systems, overview of, 314
traffic counters, types of, 187
traffic counts, conducting, 186, 197, 201-204
traffic data, managing in geodatabases, 196-197
traffic factor groups, 197
traffic flow, describing via traffic pattern, 184
traffic monitoring sites
    assignment to traffic sections, 189
    inspecting, 199-201
    maintenance of, 199-201
    overview of, 185-188
traffic monitoring system (TMS)
    conceptual data model for, 194
    conceptual database design for, 191-192
    data model for, 191-194
    description of, 184
    logical data models for, 193
    maintenance geodatabase for, 200
    overview of, 188-191
    physical data model for, 193-194
traffic pattern information, developing, 188-191

traffic patterns, overview of, 184-185
traffic sections
    appearance as elements, 310
    assigning count sites to, 192
    describing in TMS, 194
    descriptive characteristics for, 310
traffic signal installations, using intersection point features with, 314
traffic signal inventory, organizing, 82
traffic signals in state DOT example, 313-316
traffic statistics, types of, 184
traffic volumes, cycle of, 184
TrafficSection class, in TMS, 194
TrafficSegment linear event table, description of, 195
TrafficSignal entity, in state DOT example, 314
transaction updates, allowing, 124
transactions, editing, 121
transform, requirement of NCHRP 20-27(2), 215
TransformEvents tool, with routes, 158
transit connection point feature class, 285
transit connections
    defined, 283
    using subtypes with, 286
transit data models. See also data models
    creating for UNETRANS, 282-286
    example of, 366-367
    subtypes for, 283-284
transit facilities, multimodal facilities in, 375
transit geodatabases
    blocks and routes in, 377-378
    fares in, 379
    logical data model for, 370, 372
    physical data models for, 371
    piece of work in, 381
    terminology of, 367-368
    time points and positions in, 373
    vehicles in, 380-382
    walk segments in, 378-379
transit networks, logical data model for, 283-284
transit segment polyline feature class, 284-285
transit segments
    beginning and ending of, 369-370
    constructing, 283
    defined, 368

transit segments *(continued)*
    directional, 368-369
    ordering of, 284
    termination of, 370
transit station as multimodal facility, 376
transit-supported trip, choosing, 102
transport databases
    overview of, 2-3
    railroad tracks in, 130
transport datasets, making, 81-82. *See also* datasets
transport facilities, problem in transportation datasets, 145-146
transport network classes in UNETRANS, 266
transport systems
    examples of, 2
    tracking evolution of, 130
transportation
    as feature dataset, 144
    as mixed dataset, 144
transportation challenges, solving via enterprise geodatabases, 1-2
transportation data
    class types for, 40
    modes of, 2
transportation data models. *See also* data models
    ArcInfo route system, 206-209
    Base Model, 225
    FGDC, 220-221
    multimodal design, 225
    NCHRP 20-27(2), 215-220
    NCHRP 20-27(3), 222-224
    NSDI (National Spatial Data Infrastructure), 220-221, 225
transportation database designs, implementing changes in, 132-133
transportation databases. *See also* geodatabases
    creating, 78
    design problem with, 146
    many-to-many relationships in, 22
transportation datasets
    migration of, 7
    transport-facility problem with, 145
transportation facilities, changing over time, 146
transportation geodatabases, capabilities of, 2-3

Index

transportation industries Web site, accessing, xi

transportation mapping database, example of, 84-86

TransportEdge class, use in UNETRANS, 228-229

TransportJunction class, use in UNETRANS, 229-230

travel demand model
    building, 104-109
    example of, 105
    network junctions in, 107
    network links in, 108
    overview of, 101-104

travel modes, support for, 2

travel time weight, calculation of, 108

traversal centerlines, creating, 348

traversal construction, saving without geometry, 351

traversal features, generating, 350

Traversal Measure attribute, placement in NCHRP 20-27(2), 219

traversal measures, storing, 350

traversal publication process, flowchart of, 349

traversals. *See also* routes
    creating and preserving, 350-351
    creating for pathfinding, 179-181
    creating for routes, 159-160
    defined, 179
    generating for highway inventory, 333
    getting total lengths of, 349
    in NCHRP 20-27(2), 217-218
    publishing, 348-351
    static versus dynamic, 179
    types of, 348

triangulated irregular network (TIN), applying to linear facility, 13

trips
    allocating on TAZ networks, 104
    assigning operating employees to, 381
    assigning vehicles to, 381
    defined, 368
    distributing, 103
    following lowest-cost path, 103
    producing in travel demand models, 102-103
    splitting, 103

trucks, use of *t*-factor statistic with, 184

TrueFalse domain, 174, 176, 193, 195, 197, 198, 200, 202, 271, 272, 276, 286, 305, 306, 309, 310, 320, 321, 322, 326, 338, 366, 371, 374, 376, 377, 380, 381, 408, 428. *See also* Boolean indicators

TS-1 standard, adoption by NEMA, 314

Tunnel feature class in state DOT example, 308-309

Turn class, example of, 276

turn feature class
    defining, 273-275
    including fields in, 277
    using, 275-276

turn features, avoiding at bridges, 279-280

turn restrictions
    examples of, 276
    providing information about, 278

turnouts, use in railroad intersections, 414-416

turns
    cul-de-sac subtype for, 278
    defining, 274
    intersection subtypes for, 277
    multiedge, 273-274
    types of, 273

type field, including in UNETRANS, 239

type inheritance relationship, explanation of, 26

UML (Unified Modeling Language) steady-state diagrams for logical data models, 31-33

UNETRANS (Unified Network for Transportation). *See also* data models
    Activities class in, 235
    association relationships in, 231
    bridges in, 256-259
    building network datasets in, 270-272
    changes in, 265
    Comment field in, 241
    creatable classes in, 229, 235
    denormalized structure of, 232-233
    describing LRM positions in, 241-242
    as design philosophy, 238
    development of, 3
    editing support in, 239-241
    EditReason field in, 241
    EnteredBy field in, 241

UNETRANS *(continued)*
    events in, 259-260
    Events package in, 239
    facilities, aspects, and elements in, 243-244
    facility classes in, 245
    feature classes in, 233
    geometric network in, 228-229
    Incident class in, 235
    innovations of, 235
    intersection classes in, 247
    intersections in, 254-256
    Inventory package in, 239
    mixing classes in, 233
    mobile objects in, 236
    Mobile Objects package in, 239, 262-263
    navigable waterway classes in, 251
    Network Analyst extension for, 266
    network attributes in, 268-270
    network connectivity in, 267-268
    Network package in, 239
    object classes in, 248
    object packages in, 228
    pavement management example, 260-261
    position classes in, 242, 248
    railroad classes in, 250
    revising class packages in, 238-239
    road classes in, 249
    RoutableEdge class in, 228-229
    routes and location referencing in, 230-233
    short-segment design of, 233
    specifying connectivity rules for, 267
    transit data model for, 282-286
    transport network classes in, 266
    TransportEdge class in, 228-229
UNETRANS classes
    grouping by mode, 249-251
    road classes in, 249
UNETRANS model
    assets in, 234
    implementation of, 233, 394-397
    improved, 265-286
    original, 227-236
    revised, 265-286
unique identifiers, creating, 48

upper bound in multiplicity, 20
urban center road, description of, 185
U.S. Census Bureau, registration of TIGER, 86-87
Usage attributed relationship, to preserve history, 133-134
USANS (U.S. Aids to Navigation System), overview of, 397-398
user requirements
    defining for data modeling, 27
    translating into data structures, 30
users. *See also* remote users
    identifying in continuous versioning, 125-126
    limiting data entries by, 49-52
users and editors, managing, 119
U-turn in cul-de-sac, cost for, 278

valid value tables, 52-53. *See also* rows
Validate Features command, with composite relationships, 55
variable-length records
    example of, 16
    field positions in, 17
V/C (volume-to-capacity) ratio, purpose of, 184
vehicle classification, basis of, 186
vehicle classification counts, subtotals supplied by, 187
vehicle sensors, types of, 187. *See also* sensors
vehicles. *See also* highway vehicles
    counting, 187
    defined, 368
    in transit geodatabases, 380-382
    weighing, 187
version tree, example of, 121
versioning. See also continuous versioning; editing databases
    disconnected editing through, 122
    explicit state, 121
    implementing for multiple editors, 119-120
vertices, adding in Network Analyst extension, 272
viewing tables and feature classes. *See* data dictionary views
visibility of object classes, explanation of, 31
Visio, 35

Index

Visual Basic script, with network attributes, 269
volume counters, use in traffic monitoring sites, 187
volume-to-capacity (V/C) ratio, purpose of, 184

walk connection, defined, 368
walk segments in transit geodatabases, 378-379
WalkingPath class, use in UNETRANS, 229
warning and protection devices, active versus passive, 427
water elevation, detecting changes in, 404
water features in NHD, 389
waterway abstraction, basing navigable-waterways geodatabase on, 386
waterway centerlines
    construction of, 396
    description of, 395
    digitizing of, 393
    editing, 406
waterway events in navigable-waterways geodatabase, 396-402
waterway features, constructing, 385-389
waterway LRM, using relative addressing with, 393
waterway markers
    domain classes for, 401-402
    mounting, 401-402
waterway polygon features, 388
Waterway table, 394-395
waterway types, choices for, 395
waterway width, expressing as cross-section, 387
WaterwayCL polyline feature class, 396
WaterWayPoly versus Waterway, 395
waterways. *See also* navigable-waterways geodatabase
    cross-section of, 404
    cross-sections and profile of, 405
    making depth measurements across, 403
weigh in motion (WIM) ATRs, use of, 186-188
WHERE clause, in SQL statements, 353-360
WHERE NOT EXISTS clause, in SQL statements, 357
WIM (weigh in motion) ATRs, use of, 186-188
WIM data, summarizing for traffic counts, 202

WIM sites, in traffic monitoring sites, 191
Withdrawn (2) **RecordStatus** value, explanation of, 127
Work in Progress (0) **RecordStatus** value, explanation of, 127
workgroups, managing change for, 328
workload, reducing with attribute domains, 50
WorkspaceFactory
    in conceptual data model, 42
    instantiating Workspace with, 42
workspaces
    datasets in, 42
    kinds of, 42
wye switch, use in railroad intersections, 415

yard for railroad, description of, 421-423

Z (elevation) values, viewing, 45

Related titles from ESRI Press

Designing Geodatabases
ISBN 978-1-58948-021-6

GIS, Spatial Analysis, and Modeling
ISBN 978-1-58948-130-5

Modeling Our World
ISBN 978-1-879102-62-0

Thinking About GIS, Third Edition
ISBN 978-1-58948-158-9

ESRI Press publishes books about the science, application, and technology of GIS. Ask for these titles at your local bookstore or order by calling 1-800-447-9778. You can also read book descriptions, read reviews, and shop online at www.esri.com/esripress. Outside the United States, contact your local ESRI distributor.